AF389468

Applications of SEM Automated Mineralogy

Applications of SEM Automated Mineralogy

From Ore Deposits over Processing to Secondary Resource Characterization

Editor

Bernhard Schulz

MDPI • Basel • Beijing • Wuhan • Barcelona • Belgrade • Manchester • Tokyo • Cluj • Tianjin

Editor
Bernhard Schulz
Department of Economic Geology and Petrology,
Technische Universitat Bergakademie Freiberg
Germany

Editorial Office
MDPI
St. Alban-Anlage 66
4052 Basel, Switzerland

This is a reprint of articles from the Special Issue published online in the open access journal *Minerals* (ISSN 2075-163X) (available at: https://www.mdpi.com/journal/minerals/special_issues/ Automated-Mineralogy).

For citation purposes, cite each article independently as indicated on the article page online and as indicated below:

LastName, A.A.; LastName, B.B.; LastName, C.C. Article Title. *Journal Name* **Year**, *Volume Number*, Page Range.

ISBN 978-3-0365-0622-7 (Hbk)
ISBN 978-3-0365-0623-4 (PDF)

Cover image courtesy of Bernhard Schulz.

Contents

About the Editor . . . . **vii**

Bernhard Schulz
Editorial for Special Issue "Applications of SEM Automated Mineralogy: From Ore Deposits over Processing to Secondary Resource Characterization"
Reprinted from: *Minerals* **2020**, *10*, 1103, doi:10.3390/min10121103 . . . . **1**

Lars Hans Gronen, Sven Sindern, Janet Lucja Katzmarzyk, Udo Bormann, André Hellmann, Hermann Wotruba and F. Michael Meyer
Mineralogical and Chemical Characterization of Zr-REE-Nb Ores from Khalzan Buregtei (Mongolia)—Approaches to More Efficient Extraction of Rare Metals from Alkaline Granitoids
Reprinted from: *Minerals* **2019**, *9*, 217, doi:10.3390/min9040217 . . . . **7**

Mathis Warlo, Christina Wanhainen, Glenn Bark, Alan R. Butcher, Iris McElroy, Dominique Brising and Gavyn K. Rollinson
Automated Quantitative Mineralogy Optimized for Simultaneous Detection of (Precious/ Critical) Rare Metals and Base Metals in A Production-Focused Environment
Reprinted from: *Minerals* **2019**, *9*, 440, doi:10.3390/min9070440 . . . . **31**

Bernhard Schulz, Gerhard Merker and Jens Gutzmer
Automated SEM Mineral Liberation Analysis (MLA) with Generically Labelled EDX Spectra in the Mineral Processing of Rare Earth Element Ores
Reprinted from: *Minerals* **2019**, *9*, 527, doi:10.3390/min9090527 . . . . **51**

Patrick Krolop, Anne Jantschke, Sabine Gilbricht, Kari Niiranen and Thomas Seifert
Mineralogical Imaging for Characterization of the Per Geijer Apatite Iron Ores in the Kiruna District, Northern Sweden: A Comparative Study of Mineral Liberation Analysis and Raman Imaging
Reprinted from: *Minerals* **2019**, *9*, 544, doi:10.3390/min9090544 . . . . **69**

H. Donald Lougheed, M. Beth McClenaghan, Dan Layton-Matthews and Matthew Leybourne
Exploration Potential of Fine-Fraction Heavy Mineral Concentrates from Till Using Automated Mineralogy: A Case Study from the Izok Lake Cu–Zn–Pb–Ag VMS Deposit, Nunavut, Canada
Reprinted from: *Minerals* **2020**, *10*, 310, doi:10.3390/min10040310 . . . . **83**

Nynke Keulen, Sebastian Næsby Malkki and Shaun Graham
Automated Quantitative Mineralogy Applied to Metamorphic Rocks
Reprinted from: *Minerals* **2020**, *10*, 47, doi:10.3390/min10010047 . . . . **117**

Shaun Graham and Nynke Keulen
Nanoscale Automated Quantitative Mineralogy: A 200-nm Quantitative Mineralogy Assessment of Fault Gouge Using Mineralogic
Reprinted from: *Minerals* **2019**, *9*, 665, doi:10.3390/min9110665 . . . . **147**

Markus Buchmann, Nikolaus Borowski, Thomas Leißner, Thomas Heinig, Markus A. Reuter, Bernd Friedrich and Urs A. Peuker
Evaluation of Recyclability of a WEEE Slag by Means of Integrative X-ray Computer Tomography and SEM-Based Image Analysis
Reprinted from: *Minerals* **2020**, *10*, 309, doi:10.3390/min10040309 . . . . **159**

Andrea C. Guhl, Valentin-G. Greb, Bernhard Schulz and Martin Bertau
An Improved Evaluation Strategy for Ash Analysis Using Scanning Electron Microscope
Automated Mineralogy
Reprinted from: *Minerals* **2020**, *10*, 484, doi:10.3390/min10050484 . . . . **179**

Bernhard Schulz, Dirk Sandmann and Sabine Gilbricht
SEM-Based Automated Mineralogy and Its Application in Geo- and Material Sciences
Reprinted from: *Minerals* **2020**, *10*, 1004, doi:10.3390/min10111004 . . . . **191**

About the Editor

Bernhard Schulz is Associate Professor at the Department of Economic Geology and Petrology at Technische Universität Bergakademie Freiberg, Saxony, in Germany. Since 2010, he has been acting head of a laboratory equipped with scanning electron microscopes and electron probe microanalyzers for automated mineralogy. His research interests concern the application of SEM-based automated mineralogical methods for geo- and raw materials. Further areas of interest include metamorphic geology, especially geothermobarometry, and monazite dating. Bernhard Schulz graduated in 1985 with a diploma in geology at TU Clausthal-Zellerfeld in Germany. After his promotion at the University of Erlangen-Nürnberg, Germany, and various appointments at Geosciences Rennes, France, and the former Institute of Mineralogy at Würzburg, Germany, he joined Bergakademie Freiberg in 2005.

Editorial

Editorial for Special Issue "Applications of SEM Automated Mineralogy: From Ore Deposits over Processing to Secondary Resource Characterization"

Bernhard Schulz

Institute of Mineralogy, Economic Geology and Petrology, TU Bergakademie Freiberg, Brennhausgasse 14, D-09599 Freiberg/Saxony, Germany; bernhard.schulz@mineral.tu-freiberg.de; Tel.: +49-3731-39-2668

Received: 29 November 2020; Accepted: 4 December 2020; Published: 9 December 2020

1. Introduction

Matter is particulate. Grains, particles and their compounds are fundamental constituents of many important materials in nature and technology. During the last decade, software developments in scanning electron microscopy (SEM) have provoked a notable increase in applications to the study of solid matter. The mineral liberation analysis (MLA) of processed metal ores was an important driver for innovations that led to several software platforms. These combine the assessment and analysis of the backscattered electron (BSE) image to the directed steering of the electron beam for energy dispersive spectroscopy (EDS) to automated mineralogy [1–6]. However, despite a wide distribution of SEM instruments in material research and industry, the potential of SEM automated mineralogy (SEM-AM) is still under-utilised. The MDPI with the journal *Minerals* generously provides a versatile platform to bring articles dealing with SEM-AM applications to a widespread audience in applied and basic research. The *Minerals* Special Issue *Applications of SEM Automated Mineralogy: From Ore Deposits over Processing to Secondary Resource Characterization* is one of the publications dedicated to this goal.

A bulk material (rock, ore) chemical analysis provides no specific information on the target element-bearing phases (ore minerals)—a crystallographic analysis by X-ray diffraction (XRD) will identify the target phase (ore mineral) if its mode is >1 wt%, but gives no target particle size information—sedimentological particle size analysis allows no distinction among the numerous phases (gangue and ore minerals) in compounds—and, studies by optical microscopy are time-intensive, hardly deliver statistically sound databases, and are affected by subjective bias. The demands of the mineral industry and of mineral processing were the main driving forces to overcome the methodological limits and gaps of these single, well-established and validated analytical methods. Specific analytical questions related to the identification of phases (ore minerals) and characterization of grain and particle properties are common to all disciplines and stages along the raw materials processing chain, from geoscientific exploration to mining, through flotation and concentration to metallurgy up to tailing deposition and potential environmental hazards [7–9]. An answer to the challenges is the development of SEM-AM, an analytical tool combining software for BSE image examination, with more important specific steering control of the electron beam for EDS spectral analyses. Different classification algorithms and four principal SEM-AM measurement routines for point counting modal analysis, particle analysis, sparse phase search and EDS spectral mapping are offered by the relevant software providers [1,10–12]. These measurement routines provide the potential for the SEM-AM applications to scientific topics outside of the raw materials processing chain [13–18].

2. The Special Issue

This Special Issue contains ten papers which cover a wide spectrum of SEM-AM case histories and innovative applications. The characterization of primary ores, and the optimisation of comminution, flotation, mineral concentration and metallurgical processes in the mining industry by generating quantified data, is still the major field of application for SEM-AM. An article by Gronen et al. [19] on Zr-REE-Nb ores from alkaline granitoids in Mongolia addresses the problem of post-magmatic alteration causing fine and intricate intergrowth of the ore minerals with associated gangue. This hampers the economic recovery of this deposit type, as intensive comminution is necessary to liberate the ore minerals. Automated mineralogy was applied to quantify the textural properties of the ore phases which occur in mineral clusters. The analytical results permit an efficient pre-concentration of rare metal ores at coarser particle size fractions, requiring less energy consumption during comminution.

An article by Warlo et al. [20] deals with a novel approach to overcome difficulties when trace minerals are detected within a single measurement when they are accompanied by the major ore minerals. Small grain sizes of the trace minerals complicate the task as they induce mixed X-ray signals.

The contribution of generically labelled EDS spectra in the mineral processing of rare earth element ores by Schulz et al. [21] provides a possibility for how to relate the EDS spectra from the complex and variable REE mineral phases in an ore to mineral names. The labelling of the spectra obtained from REE-bearing minerals is based on their contents of Si, Ca, F and P in a bulk normalised analysis. The labelled spectra are then combined into groups of REE-P (~monazite), REE-Ca-Si-P (~britholite), REE-Ca-F (~synchysite) and REE-F (~bastnaesite, parisite, fluocerite) and a mixed spectra group with low counts for REE. This classification scheme is applied in several case studies with SEM-AM data from beneficiation products by comminution and multistage flotation of REE carbonatite ores.

Iron ores require specific methodological solutions when analysed by SEM-AM as is outlined in the article by Krolop et al. [22]. Similar elemental compositions of hematite and magnetite and almost identical backscattered electron (BSE) intensities complicate their discrimination in compound ores. Raman imaging is applied and compared to the EDS spectral mapping data for control and the improvement of classification, iron ore discrimination and measurement time duration.

SEM-AM facilitates studies that are not feasible with other methods. This is exemplified in the article by Lougheed et al. [23] on the base metal exploration using heavy mineral concentrates (HMC). Traditional methods rely on visual examination of HMC fractions > 250 μm, but SEM-AM allows to investigate the finer fractions (<250 μm) as encountered in glacial till samples. Precise compositional and morphological data from a large number (10,000–100,000) of heavy mineral grains in a single sample can be collected, and rare specific indicator minerals can be reliably identified.

One motivation for the production of this Special Issue is to demonstrate the potential of SEM-AM beyond its classical field of applications in mining and mineral processing. The contribution by Keulen et al. [24] is entitled "Automated Quantitative Mineralogy Applied to Metamorphic Rocks" and deals with the application of the comparably new ZEISS Mineralogic software platform (Carl Zeiss Microscopy Ltd., Cambridge, UK) to petrological investigations of complete 25 × 45 mm sized petrographic thin sections. A direct quantification of the EDS spectra allows for the visualisation of element zonations in garnet and other minerals, and for detailed recognition of metamorphic textures. In this way, mineral maps, element concentration maps, element ratio maps and mineral association maps can be produced. Additionally, the conventional mineral modes, grain shapes, sizes, and orientations can be recognised at a quantitative level.

A paper by Graham and Keulen [25] amplifies the SEM-AM applications in petrology to the special rock type of a fault gouge. Such rocks are hard to investigate by conventional methods. Additionally, the EDS spectral analysis of the extreme fine grain sizes at the submicrometer scale in such rocks is hampered by the large interaction volume of the primary electron beam which leads to mixed signals. By applying a low primary beam acceleration voltage, combined with a large aperture, and a dedicated mineral classification, these mixed signals can be deconvoluted down to a grain size of 200 nm. This allowed the authors to obtain reliable quantitative mineralogy and grain size distribution data from the rocks by SEM-AM.

Recycling technology and secondary resources from waste treatment are emergent fields which require adopted analytical methods for phase identification, as well as grain and particle characterization.

This topic is addressed by a paper from Buchmann et al. [26] entitled "Evaluation of Recyclability of a WEEE Slag by Means of Integrative X-Ray Computer Tomography and SEM-Based Image Analysis". Waste of electrical and electronic equipment (WEEE) contains a lot of metals and rare earth elements. Smelting of WEEE leads to significant trapping of such elements in slags from which they are hard to extract again. To make recycling economically applicable it is crucial to bring these metal contents into deducible structures. The technical manipulation of the phase transfer processes necessitates detailed understanding and controls. Therefore, WEEE slag is investigated by X-ray computed tomography (XCT) and SEM-AM to understand the typical structures and their implications for further processing and extraction.

Ashes resulting from technical processes such as combustion in power plants are increasingly considered as secondary resources. This especially applies for sewage slush ashes which are a potential resource for phosphorus. Conventionally, a great deal of ash is analysed by X-ray diffraction (XRD) and Rietveld methods for phase identification and modal composition. However, as in slags, a considerable part of the ash is made up by amorphous or glassy particles, of which, sorts and sizes cannot be satisfactorily characterised by XRD and sedimentological grain size classification. A novel analytical approach to this topic is presented by Guhl et al. [27] in a contribution entitled "An Improved Evaluation Strategy for Ash Analysis Using SEM-AM". Sewage slush ashes are materials composed of polyphase particles. For the establishment of a recycling process, the characterization of the phosphorus-bearing particles with heterogeneous compositions is of major interest. This can be performed by EDS spectral mapping involving spectra classification and grouping related to the phosphorus-bearing target components.

The contributions to this Special Issue are restricted to a given time period. Additionally, manuscripts where SEM-AM is applied along with other analytical methods may be identified as contributions to more specialised topics. For the presentation of the potential of SEM-AM to a broader audience, a review contribution is needed. That is provided by Schulz et al. [12], entitled "SEM-Based Automated Mineralogy and its Application in Geo- and Material Sciences". It presents an overview of the SEM-AM measurement routines and their implementation in the available software platforms. The different EDS spectra classification algorithms, the pitfalls of spectra labelling and the often neglected challenges of sample preparation, are discussed. As demonstrated by case examples in this contribution, the EDS spectral mapping methods appear to have the most promising potential for novel applications in metamorphic, igneous and sedimentary petrology, ore fingerprinting, ash particle analysis, characterization of slags, forensic sciences, archaeometry and investigations of stoneware and ceramics.

3. Summary and Outlook

This Special Issue presents the view that SEM-AM allows for a wide spectrum of targeted studies on solid and particulate matter beyond its initial and traditional applications in the mining industry and process mineralogy. However, even though SEM instruments are widespread in various public, governmental and commercial research institutions, only a very limited number of working groups operate and have access to SEM-AM equipment. In Germany, this equates to only three working groups—two at universities and one at a federal geological survey institution. A similar situation can be found in other European countries, and indeed, worldwide. In Germany one university group is located at the RWTH Aachen. The other group is situated at Freiberg/Saxony, as an association between TU Bergakademie and the Helmholtz Institute Freiberg for Resource Technology. Members of these groups, among others, contributed various articles to this Special Issue. This is no coincidence, as both locations are well-known for base and applied research in all kinds of mineral resources and process mineralogy. Hopefully, the reports and case studies in this Special Issue will encourage further working groups in geo and material sciences to invest in corresponding SEM upgrades for automated mineralogy for the investigation of exciting new application fields.

Funding: This research was funded by the Helmholtz Institute Freiberg for Resource Technology, the Deutsche Forschungsgemeinschaft (DFG) Grant SCHU676/20 and through contract work for numerous enterprises. The Open Access Funding and APC was funded by the Publication Fund of the TU Bergakademie Freiberg.

Acknowledgments: The Guest Editor would like to sincerely thank all authors, reviewers, the editor Francis Wu and the editorial staff of *Minerals* for their timely efforts to successfully complete this Special Issue.

Conflicts of Interest: The author declares no conflict of interest.

References

1. Gottlieb, P.; Wilkie, G.; Sutherland, D.; Ho-Tun, E.; Suthers, S.; Perera, K.; Jenkins, B.; Spencer, S.; Butcher, A.; Rayner, J. Using quantitative electron microscopy for process mineralogy applications. *JOM* **2000**, *52*, 24–25. [CrossRef]

2. Gu, Y. Automated scanning electron microscope based mineral liberation analysis. An introduction to JKMRC/FEI Mineral Liberation Analyser. *J. Miner. Mater. Charact. Eng.* **2003**, *2*, 33–41. [CrossRef]

3. Lastra, R. Seven practical application cases of liberation analysis. *Int. J. Miner. Process.* **2007**, *84*, 337–347. [CrossRef]

4. Fandrich, R.; Gu, Y.; Burrows, D.; Moeller, K. Modern SEM-based mineral liberation analysis. *Int. J. Miner. Process.* **2007**, *84*, 310–320. [CrossRef]

5. Hoal, K.O.; Stammer, J.G.; Appleby, S.K.; Botha, J.; Ross, J.K.; Botha, P.W. Research in quantitative mineralogy: Examples from diverse applications. *Miner. Eng.* **2009**, *22*, 402–408. [CrossRef]

6. Sylvester, P. Use of the Mineral Liberation Analyzer (MLA) for Mineralogical Studies of Sediments and Sedimentary Rocks. In *Quantitative Mineralogy and Microanalysis of Sediments and Sedimentary Rocks*; Sylvester, P., Ed.; Mineralogical Association of Canada (MAC): St. John's, NL, Canada, 2012; Volume 42, pp. 1–16.

7. Gäbler, H.-E.; Melcher, F.; Graupner, T.; Bähr, A.; Sitnikova, M.A.; Henjes-Kunst, F.; Oberthür, T.; Brätz, H.; Gerdes, A. Speeding Up the Analytical Workflow for Coltan Fingerprinting by an Integrated Mineral Liberation Analysis/LA-ICP-MS Approach. *Geostand. Geoanal. Res.* **2011**, *35*, 431–448. [CrossRef]

8. Lund, C.; Lamberg, P.; Lindberg, T. Practical way to quantify minerals from chemical assays at Malmberget iron ore operations—An important tool for the geometallurgical program. *Miner. Eng.* **2013**, *49*, 7–16. [CrossRef]

9. Mariano, R.A.; Evans, C.L. Error analysis in ore particle composition distribution measurements. *Miner. Eng.* **2015**, *82*, 36–44. [CrossRef]

10. Graham, S.D.; Brough, C.; Cropp, A. An Introduction to ZEISS Mineralogic Mining and the Correlation of Light Microscopy with Automated Mineralogy: A Case Study using BMS and PGM Analysis of Samples from a PGE-bearing Chromite Prospect. In Proceedings of the Precious Metal Conference, Vienna, Austria, 18–20 October 2015; pp. 1–12.

11. Hrstka, T.; Gottlieb, P.; Skála, R.; Breiter, K.; Motl, D. Automated Mineralogy and Petrology-Applications of TESCAN Integrated Mineral Analyzer (TIMA). *J. Geosci.* **2018**, *63*, 47–63. [CrossRef]

12. Schulz, B.; Sandmann, D.; Gilbricht, S. SEM-Based Automated Mineralogy and its Application in Geo- and Material Sciences. *Minerals* **2020**, *10*, 1004. [CrossRef]

13. Pirrie, D.; Butcher, A.R.; Power, M.R.; Gottlieb, P.; Miller, G.L. Rapid quantitative mineral and phase analysis using automated scanning electron microscopy (QemSCAN); potential applications in forensic geoscience. *Geol. Soc. London Spec. Publ.* **2004**, *232*, 123–136. [CrossRef]

14. Knappett, C.; Pirrie, D.; Power, M.R.; Nikolakopoulou, I.; Hilditch, J.; Rollinson, G.K. Mineralogical analysis and provenancing of ancient ceramics using automated SEM-EDS analysis (QEMSCAN®): A pilot study on LB I pottery from Akrotiri, Thera. *J. Archaeol. Sci.* **2011**, *38*, 219–232. [CrossRef]

15. Von Eynatten, H.; Tolosana-Delgado, R.; Karius, V.; Bachmann, K.; Caracciolo, L. Sediment generation in humid Mediterranean setting: Grain-size and source-rock control on sediment geochemistry and mineralogy (Sila Massif, Calabria). *Sediment. Geol.* **2016**, *336*, 68–80. [CrossRef]

16. Pietranik, A.; Kierczak, J.; Tyszka, R.; Schulz, B. Understanding heterogeneity of a slag-derived weathered material: The role of automated SEM-EDS analyses. *Minerals* **2018**, *8*, 513. [CrossRef]

17. Pirrie, D.; Crean, D.E.; Pidduck, A.J.; Nicholls, T.M.; Awbery, R.P.; Shail, R.K. Automated mineralogical profiling of soils as an indicator of local bedrock lithology: A tool for predictive forensic geolocation. *Geol. Soc. London Spec. Pub.* **2019**, *492*. [CrossRef]

18. Minde, M.W.; Zimmermann, U.; Madland, M.V.; Korsnes, R.I.; Schulz, B.; Gilbricht, S. Mineral replacement in long-term flooded porous carbonate rocks. *Geochim. Cosmochim. Acta* **2020**, *268*, 485–508. [CrossRef]
19. Gronen, L.H.; Sindern, S.; Katzmarzyk, J.L.; Bormann, U.; Hellmann, A.; Wotruba, H.; Meyer, F.M. Mineralogical and Chemical Characterization of Zr-REE-Nb Ores from Khalzan Buregtei (Mongolia)—Approaches to More Effcient Extraction of Rare Metals from Alkaline Granitoids. *Minerals* **2019**, *9*, 217. [CrossRef]
20. Warlo, M.; Wanhainen, C.; Bark, G.; Butcher, A.R.; McElroy, I.; Brising, D.; Rollinson, G.K. Automated Quantitative Mineralogy Optimized for Simultaneous Detection of (Precious/Critical) Rare Metals and Base Metals in A Production-Focused Environment. *Minerals* **2019**, *9*, 440. [CrossRef]
21. Schulz, B.; Merker, G.; Gutzmer, J. Automated SEM Mineral Liberation Analysis (MLA) with Generically Labelled EDX Spectra in the Mineral Processing of Rare Earth Element Ores. *Minerals* **2019**, *9*, 527. [CrossRef]
22. Krolop, P.; Jantschke, A.; Gilbricht, S.; Niiranen, K.; Seifert, T. Mineralogical Imaging for Characterization of the Per Geijer Apatite Iron Ores in the Kiruna District, Northern Sweden: A Comparative Study of Mineral Liberation Analysis and Raman Imaging. *Minerals* **2019**, *9*, 544. [CrossRef]
23. Lougheed, H.D.; McClenaghan, M.B.; Layton-Matthews, D.; Leybourne, M. Exploration Potential of Fine-Fraction Heavy Mineral Concentrates from Till Using Automated Mineralogy: A Case Study from the Izok Lake Cu–Zn–Pb–Ag VMS Deposit, Nunavut, Canada. *Minerals* **2020**, *10*, 310. [CrossRef]
24. Keulen, N.; Malkki, S.N.; Graham, S. Automated Quantitative Mineralogy Applied to Metamorphic Rocks. *Minerals* **2020**, *10*, 47. [CrossRef]
25. Graham, S.; Keulen, N. Nanoscale Automated Quantitative Mineralogy: A 200-nm Quantitative Mineralogy Assessment of Fault Gouge Using Mineralogic. *Minerals* **2019**, *9*, 665. [CrossRef]
26. Buchmann, M.; Borowski, N.; Leißner, T.; Heinig, T.; Reuter, M.A.; Friedrich, B.; Peuker, U.A. Evaluation of Recyclability of a WEEE Slag by Means of Integrative X-Ray Computer Tomography and SEM-Based Image Analysis. *Minerals* **2020**, *10*, 309. [CrossRef]
27. Guhl, A.C.; Greb, V.-G.; Schulz, B.; Bertau, M. An improved evaluation strategy for ash analysis using scanning electron microscope automated mineralogy. *Minerals* **2020**, *10*, 484. [CrossRef]

Publisher's Note: MDPI stays neutral with regard to jurisdictional claims in published maps and institutional affiliations.

Article

Mineralogical and Chemical Characterization of Zr-REE-Nb Ores from Khalzan Buregtei (Mongolia)—Approaches to More Efficient Extraction of Rare Metals from Alkaline Granitoids

Lars Hans Gronen [1,*], Sven Sindern [1], Janet Lucja Katzmarzyk [2], Udo Bormann [1], André Hellmann [1], Hermann Wotruba [2] and F. Michael Meyer [1,3]

[1] Institute of Applied Mineralogy and Economic Geology, RWTH-Aachen University, Wüllnerstraße 2, 52062 Aachen, Germany; Sindern@emr.rwth-aachen.de (S.S.); udo.bor09@googlemail.com (U.B.); Hellmann@emr.rwth-aachen.de (A.H.); M.Meyer@rwth-aachen.de (F.M.M.)

[2] Unit of Mineral Processing, RWTH-Aachen University, Lochnerstraße 4-10 Haus C, 52064 Aachen, Germany; Katzmarzyk@amr.rwth-aachen.de (J.L.K.); Wotruba@amr.rwth-aachen.de (H.W.)

[3] German-Mongolian Institute of Resource Technology, GMIT Campus, 2nd khoroo, Nalaikh district, Ulaanbaatar, Mongolia

* Correspondence: Gronen@EMR.RWTH-Aachen.de; Tel.: +49-241-80-95775

Received: 11 February 2019; Accepted: 3 April 2019; Published: 5 April 2019

Abstract: Alkaline rocks are worldwide observed as hosts for rare metal (Zr-REE-Nb) minerals. The classification of the ore bearing rock type is challenging due to the fact that textures and mineral assemblage are obscured by post-magmatic alteration. In addition, the alteration causes fine and intricate intergrowth of the ore minerals with associated gangue. Hence, intensive comminution is necessary to liberate the ore minerals, which is one parameter hampering the economical use of this deposit type. This study provides a quantitative mineralogical investigation of the ore bearing rock suite at Khalzan Buregtei as an example of rare metal deposits. R1-R2 multication parameters are shown to be highly appropriate as quantitative mineralogical indicators based on readily available major element datasets to visualize and quantify alteration types of the ore bearing rock suite. The ore minerals were found to be associated with a cluster-forming assemblage of post-magmatic phases. Automated mineralogy was applied to quantify the textural properties of the ore mineral clusters. This finding permits efficient pre-concentration of rare metal ore at coarser particle size fraction, requiring less energy consuming comminution.

Keywords: Zr-REE-Nb deposits; alkaline rocks; automated mineralogy; Khalzan Buregtei

---

1. Introduction

Alkaline granite and syenite magmatic rocks are characterized by a remarkable enrichment of high field-strength elements (HFSEs), including Zr, Nb, Y, U as well as the rare earth elements (REEs). The unusual trace element composition of alkaline rocks is in line with elevated abundances of complex and rare HFSE minerals, which have attracted both petrologists and economic geologists for a long time [1–3]. Evolving technical capabilities (e.g., high performance magnets, super-alloys, phosphors, superconductors, ceramics) and a growing high-tech based economy have enhanced the global demand for HFSEs, such as Zr, Nb, Y, or, particularly, Nd, Eu, Tb and Dy among the REE. Notwithstanding, a substantial part of these commodities (i.e., REE, Y, Nb) has a critical character indicative of supply risk for political or environmental reasons [4]. As a response to this situation, alkaline granite and syenite complexes have also come into the focus of exploration campaigns and mining feasibility studies in the last decade (e.g., Khalzan Buregtei, Strange Lake, Motzfeld, Norra Kärr [1,5–7]).

Enrichment of HFSE to enhanced grades in alkaline intrusions is generally thought to be caused by magmatic fractionation, but mineralogical and petrologic evidence often points to post-magmatic pneumatolytic or hydrothermal processes. Post-magmatic mineralization events are commonly associated with intensive and highly variable alteration types of both intrusive and wall rocks, thereby superimposing primary petrographic features [1,2,6,8–10].

As a consequence, alkaline granites and syenites show a broad range of mineralogical and textural characteristics. Most apparent, these rocks display outstanding associations of rare minerals—among these are also ore phases incorporating the HFSEs as major elements (e.g., elpidite, eudialyte, gittinsite, zircon; Nb pyrochlore; REE + Y monazite, bastnaesite, synchysite, and several REE-silicates). Furthermore, formation in multistage and partially superimposing magmatic to hydrothermal processes frequently leads to narrow and complex intergrowths of fine grained and irregularly sized and shaped crystals. Thus, such rocks display a remarkable spatial variability and heterogeneity in terms of mineral composition, grain size, mineral association, and texture [1,9,10].

In general, these mineralogical and textural characteristics of alkaline rocks represent obstacles for efficient exploration and ore beneficiation, as will be outlined in this study. Owing to the marked mineralogical complexity, alkaline rock classification comprises a confusing multiplicity of partly unfamiliar names, many of these derived with reference to single locations only [11] (p. 558). The multitude of classifying terms is a source of incongruence in field data or core logs in exploration campaigns if work is carried out by different geologists or companies. Furthermore, classification of alkaline rocks implies magmatic genesis and does not reflect intensity of post-magmatic hydrothermal mineral formation. This leads to terminological indistinctness with potential negative effects for data handling, e.g., in 3D numeric deposit modelling.

As a consequence of small grain sizes and intricate intergrowth, liberation sizes of ore minerals can be low. Mining operators have to cope with intensive and energy consuming grinding as well as processing of fine fractions to produce marketable concentrates with acceptable recoveries. In addition to causing high energy costs, the processing of fine fractions of polymineralic ores, comprising phases with different physical properties (e.g., density, magnetic susceptibility), is likely to be inefficient. Such handicaps for efficient, economic, and sustained mining represent bars to the development of prospects associated with alkaline granites and syenites.

In this study the alkaline granite hosted Zr-REE-Nb deposit of Khalzan Buregtei (Mongolia), which was a target of exploration work by different companies in recent years, is used for a case study to demonstrate characteristic problems associated with these types of deposits. Based on results of published studies of the magmatic and post-magmatic (i.e., hydrothermal) evolution of the complex [5,12–15], detailed sampling work was carried out in the central part of the complex, which is most affected by Zr-REE-Nb mineralization. In contrast to a number of detailed petrological studies addressing genesis and potential processes of either magmatic or metasomatic enrichment of rare metals in the alkaline granite of Khalzan Buregtei [5,13–15], this study particularly focusses on the identification of ore properties, which are crucial for the economic exploitation of the mineralized rocks of this deposit.

Whole rock chemical major and trace element composition, chemical composition of HFSE minerals dominating the ore assemblage, and comprehensive textural datasets were determined in characteristic rock specimens. In particular, this study aims to investigate the influence of post-magmatic alteration on classification and to develop solutions to handle terminological indistinctness in such highly altered alkaline rocks. A second target of the study is to characterize mineralogical and textural effects of ore mineralization, to indicate handicaps for mineral processing, but also to discuss the potential of mineralogical and textural characteristics of Zr-REE-Nb mineralization in alkaline rocks to yield enhanced recoveries.

2. Geological Setting

The Khalzan Buregtei massif is situated in west Mongolia Altai mountains approximately 50 km north of the city of Hovd. It was discovered in 1984 by geologists of the former Soviet Union [13]. The complex covers an area of 30 × 8 km and is mainly built up by alkaline syenite to granitic rocks considered to be associated with early Devonian extensional tectonic activity [5]. The Khalzan Buregtei Zr-REE-Nb deposit itself is located in the centre of a roughly oval shaped intrusive body in the south of the massif (Figure 1). The part of the intrusion hosting the deposit is dominated by alkali granites, which are distinct to the surrounding syenitic rocks, classified as nordmarkites by the first investigators [13,14] and in the recent work of [15]. The latter authors assign the mineralized alkali granites to a fifth and seventh intrusive stage, whereas [5] classify the mineralized rocks as ore metasomatites, attesting to the strong imprint of post-magmatic hydrothermal processes. Syenites contacting the central granites were affected by hydrothermal activity as well. Major faulting of the working area postdates the ore formation indicated by the displacement of the rock suite as illustrated in Figure 1. According to [5] the major rare metal deposit of Khalzan Buregtei has an ore content of 2.4 × 10^6 t ZrO_2, 3.5 × 10^5 t Nb_2O_5, and 4.9 × 10^5 t REE_2O_3 + 1.3 × 10^5 t Y_2O_3. The mechanism of rare metal enrichment to final grades has been discussed controversially in recent years by several authors [5,14,15]. The work of [5] explains the ore formation by multistage post-magmatic alteration on the basis of a previous enrichment of alkaline magma with HFS and other rare metals. The post-magmatic alteration is triggered by fluids from further unspecified carbonatite plutons, which explains the abundance of Ca bearing Zr-silicates and carbonates. However, the recent work of [15] discussed the formation of the rare metal granite (V-Phase [14,15]) and the aforementioned ore minerals by differentiated crystallization of alkaline magma strongly enriched in HFSE and other rare metals. The formation of Ca bearing phases is assumed to be caused by enrichment of Ca following assimilation of adjacent limestones in the melt [15].

3. Materials and Methods

3.1. Sampling Method and Bulk Rock Chemical Analysis

The ore bearing rocks of the Khalzan Buregtei deposit were investigated in two particularly well exposed areas characteristic of the central domain (Figure 1A). The deposit covers an area of approximately 1 km^2 of mountainous arid landscape. The central part of the deposit is built of one central ridge with an elevation up to 1940 m bordering, to the west and east, two valleys filled with sediments and debris.

A sample suite taken from the central ridge was indicated by the addition of KB G to the sample numbers. Another sample suite was taken from one more shallow ridge beyond the western, valley having good accessibility and outcrop condition. Latter samples were named KB S. Samples were taken along outcrop sections, where a macroscopic change in mineralogy and/or texture was recognizable. Hand specimens were later investigated for homogeneity prior to further treatment. In total, 18 samples, each approximately 1 kg in weight, were taken in both outcrop areas. After splitting into two sub-samples one set of sub-samples was crushed by manual crushing and further comminuted using a rotating disc mill (Siebtechnik GmbH, Mühlheim an der Ruhr, Germany). The sample powder was analysed for bulk rock chemical composition at ALS Analytical Service, affiliation Loughrea, Ireland, applying a combined ICP-AES and ICP-MS method after Li-borate fusion (ALS Method Codes: ME-MS81d). Representative blocks were dissected from the other set of the sub-samples for the preparation of polished thin-sections for optical petrological microscopy, electron optical investigation by QEMSCAN© (FEI/Thermo Fisher, Hillsboro, OR, USA), and electron probe micro analyser (EPMA).

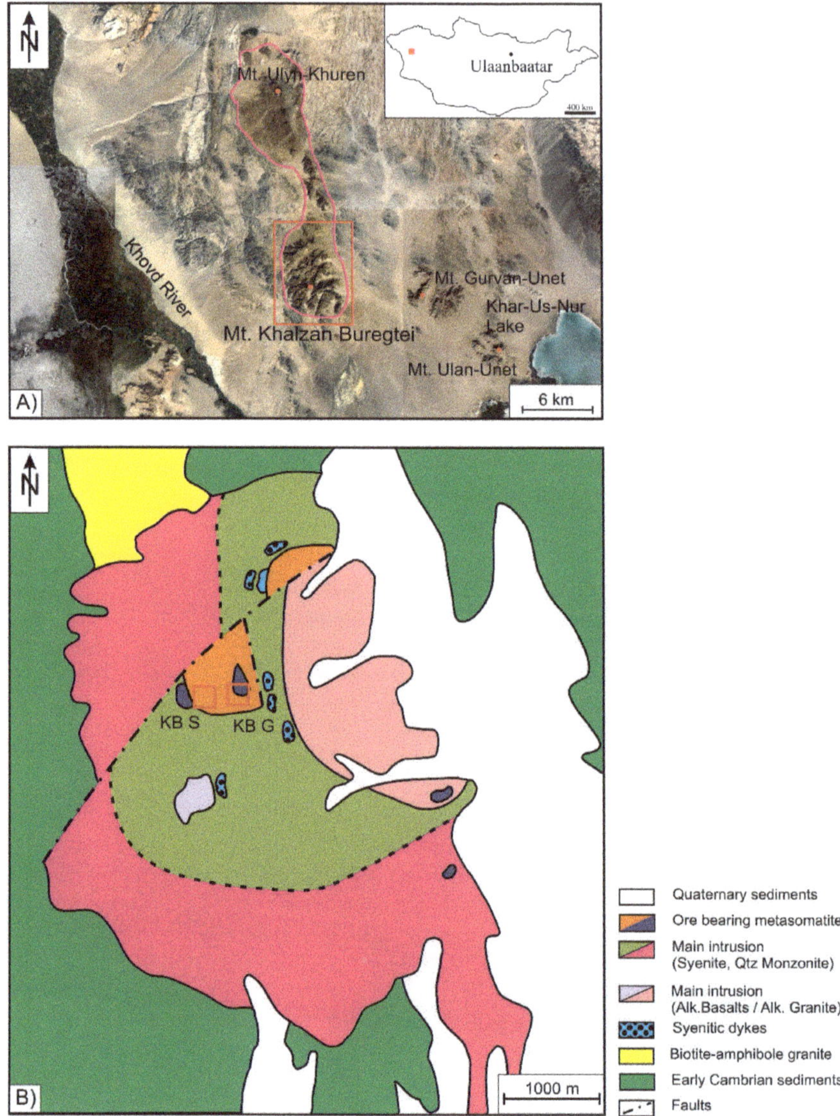

Figure 1. (**A**) Satellite image with the outline (magenta) of the entire alkaline massif of Khalzan Buregtei as proposed by [13,15]. The red box marks the area of the Khalzan Buregtei deposit shown as sketch map in B [16]. (**B**) Geological sketch map of the Khalzan Buregtei Zr-REE-Nb deposit with sampling areas KB S and KB G, see text for further details. Map basis according to [12,15] with nomenclature following [5].

Zirconium concentration data were obtained by X-ray fluorescence (XRF) measurement at the Unit of Mineral Processing, RWTH-Aachen University, Niton XL3t (Thermo Fisher, Waltham, MA, USA) with the handheld XRF system. The measurements were performed with three replicates on pressed powder pellets created with 8 g of air-dried sample powder. Data was validated by an external quality check using NIM-L lujavrite standard reference material [17] prepared and handled like the unknowns.

3.2. SEM Based Semi-Automated Mineralogy (QEMSCAN©)

Semi-automated mineralogical analyses were performed by applying a Quanta 650-F QEMSCAN© (FEI/Thermo Fischer) scanning electron microscope (SEM) at the Institute of Applied Mineralogy and Economic Geology, RWTH-Aachen University [18]. Polished thin sections and polished sections were analysed after carbon coating. The measurements were conducted with an acceleration voltage of 25 kV and a fixed sample current of 10 nA. The surface of each sample section was scanned with a spatial resolution of 5 μm. Back scatter (BSE) intensities and individual X-ray spectra were recorded for each pixel with a 4-quadrant BSE detector and two DualXFlash 5030 SDD (Bruker AXS, Karlsruhe, Germany) energy dispersive x-ray spectrometers (EDX). Phase assignment was carried out by comparison of spectral data obtained for each pixel with library information using the iDiscover (Version 5.3.2.501, FEI/Thermo Fisher, Hillsboro, OR, USA) software suite. Automated image analysis was applied to phase maps to compute quantitative mineralogical and textural parameters. The modal composition of each sample in volume (vol.) % was calculated from the volumetric abundance of each mineral phase. Particle size calculations were conducted by measurement of the diameter of the virtual sphere with equivalent perimeter length assigned to each individual target phase [19]. Particle populations were obtained by segmentation of the target minerals or mineral assemblage from QEMSCAN© phase maps.

3.3. Electron Probe Micro Analyzer (EPMA)

The JXA-8900R electron probe micro analyser (Jeol, Jeol Germany GmbH, Freising, Germany), of the Institute of Applied Mineralogy and Economic Geology, RWTH-Aachen University, was used for high-resolution element mapping and for quantitative chemical analyses of ore-forming minerals. For element mapping, the instrument was operated with a focused electron beam and an acceleration voltage of 20 kV. Each pixel of 1 μm size was measured with a dwell time of 50 ms. The setup for the five wavelength dispersive spectrometers of the EPMA is given in the Appendix A (Table A1).

Quantitative analyses of the mineral chemistry were conducted on different ore minerals. REE-carbonate minerals were analysed with an acceleration voltage of 15 kV and an electron beam of 10 μm diameter to avoid intensive damaging of the target minerals. Zircon, in contrast, was analysed with an increased acceleration voltage of 20 kV and a focused electron beam. The beam current was set in all analytical sessions to 24 nA. Detailed spectrometer setup parameters, like peak and background recording times, as well as standards, are given in the Appendix A (Tables A2 and A3).

4. Results

4.1. Petrography

Quartz, K-feldspar, and albite are the dominant rock-forming minerals and in total account for more than 70 vol. % in all samples. The mafic minerals arfvedsonite and aegirine are less abundant (<5 vol. %) and can only locally be observed (Figure 2). Despite the variable proportions of the rock forming minerals, mainly showing higher quartz abundances in the KB-S sampling area, all samples, excluding one, can be classified as alkali granites in terms of the terminology of [20] for magmatic rocks. Only the sample KB S10 was classified as quartz rich granitoid rock due to a significantly higher abundance of quartz. All samples show variable proportions of HFSE ore minerals, such as Zr silicates, REE-carbonates, pyrochlore, as well as of hematite, rutile, titanite, or fluorite (Figure 2).

Quartz is present in all samples and shows a variation in grain size. The coarse fraction is represented by quartz grains up to 2 mm in KB G and up to 5 mm in KB S. Some grains show snowball structures enclosing other mineral phases like K-feldspar and albite. Further, big crystals of quartz show evidence of ductile deformation, like bulging and the formation of sub grains (Figure 3A, yellowish). In addition to the coarse fraction, a fraction of fine quartz grains can be observed. The latter is strongly intergrown with HFSE minerals, as well as with hematite and fluorite (Figure 3A).

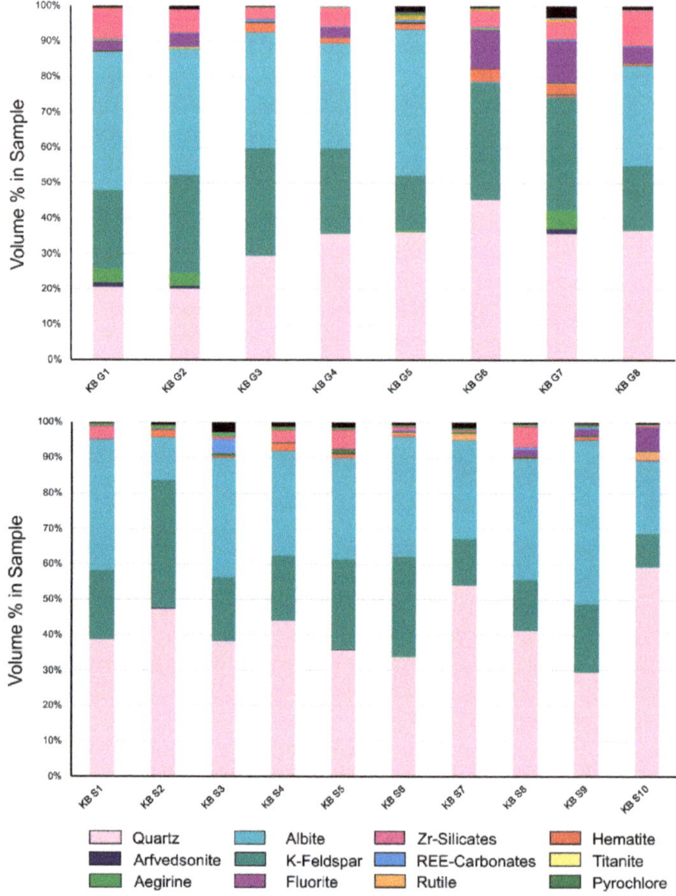

Figure 2. Modal composition (vol. %) of samples from ore bearing metasomatites in sampling areas KB G (**top**) and KB S (**bottom**), calculated from mineralogical phase maps.

K-feldspar shows a size distribution similar to the coarse grained quartz in both sample suites. Irregularly shaped crystal domains inside of K-feldspar grains may represent relict hatch-twin domains formed in earlier microclinization (Figure 3B). Euhedral K-feldspar is in close association to albite, which can be observed parallel to grain boundaries. K-feldspar often shares straight boundaries with coarse grained quartz (Figure 3B).

Albite forms lath shaped crystals varying in grain size from 50 μm to 400 μm (Figure 3C). Coarse laths of albite, however, are found at the edges of coarse grained snowball quartz and K-feldspar grains. Fine albite crystals can be observed in interstitial volumes of quartz and K-feldspar. Evidence of the replacement of quartz and K-feldspar by albite is dominant in all samples from the centre of the Khalzan Buregtei deposit. The close intergrowth of lath-shaped albite crystals with each other leads to the formation of a matrix-like texture observable in all samples containing high amounts of albite.

The mafic minerals arfvedsonite and aegirine occur in the rocks with a grain size up to 1 mm but with low modal abundance. As Figure 2 shows, significant amounts are only observed in the samples KB G1, G2, and G9. Arfvedsonite is euhedral and shows good cleavage. Aegirine forms halos surrounding arfvedsonite as illustrated in Figure 3D. Ongoing replacement reaction from arfvedsonite to aegirine is observed propagating along amphibole cleavage plains. Aegirine is further replaced by albite laths. Albite grains in contact with arfvedsonite and aegirine often show red staining (Figure 3D).

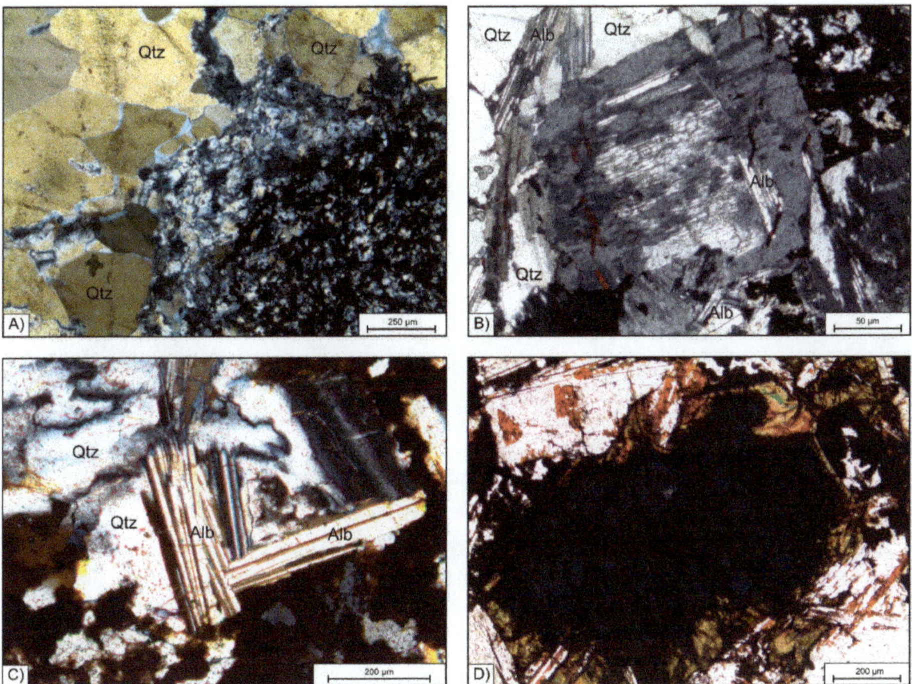

Figure 3. (**A**) Image of primary coarse quartz (Qtz) in sample KB G2 (yellowish interference color of quartz is due to section thickness > 30 μm) showing sub grain formation in adjacency to fine grained interstitial quartz intergrown with dark zircon-hematite aggregates. Optical microscopic image with cross-polarized light (xpl, transmitted light). (**B**) K-feldspar grain with irregular internal domains (xpl, transmitted light). (**C**) Lath shaped albite (Alb) crystals in interstitial volume of coarse grained quartz (xpl, transmitted light). (**D**) Arfvedsonite grain showing dark blue pleochroic color and typical cleavage, surrounded by aegirine (green pleochroic color). Lath-shaped albite (white to red) is colored red by fine grained hematite on lath planes. Microscopic image of KB G1 plain polarized light (ppl, transmitted light).

The group of HFSE-bearing minerals comprises Zr-silicates, REE-carbonates, and the Nb mineral pyrochlore. Zircon is the most abundant mineral with concentrations as high as 11 vol. %. The Zr-silicate gittinsite [$CaZrSi_2O_7$], in contrast, is only present in samples KB G1 and G2, in which it occurs in aggregates with zircon clustering around relicts of aegirine (green in Figure 4A). According to [14,16] and [6] more Zr-silicate minerals like elpidite [$Na_2ZrSi_6O_{15}·H_2O$] and armstrongite [$CaZrSi_6O_{15}·H_2O$] occur at the Khalzan Buregtei deposit; however, at the studied location, these minerals were found only on a very rare level. Due to the low abundance of these minerals at the sampled parts of the deposit, they are not considered as ore minerals in this study.

The REE-carbonates are represented by bastnaesite-(Ce) as the most abundant phase, as well as by parasite-(Ce) and synchysite-(Y). In total, their modal volume is <5 vol. %. Pyrochlore can be observed in nearly all samples, but with low abundance compared to zircon or REE-carbonates.

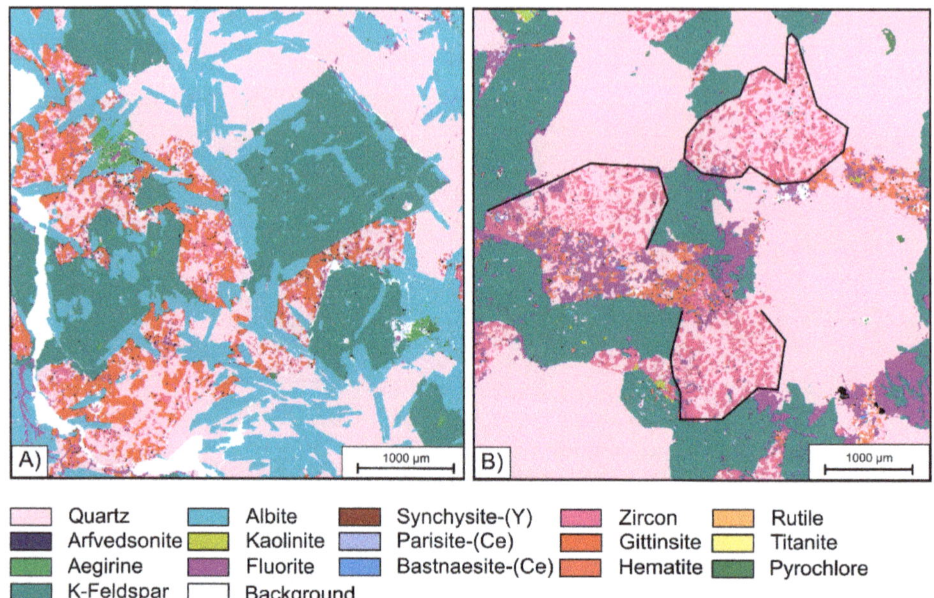

Figure 4. Representative QEMSCAN© phase maps of Zr-REE-Nb ore. (**A**) Zircon-gittinsite cluster in sample KB G1. The core of a residual aegirine (green) is recognizable within the cluster. (**B**) Zircon-quartz cluster containing fluorite (purple) and hematite (pale red) in sample KB G6. Lines indicate the preserved subhedral shape of precursor amphibole as straight grain boundaries to K-feldspar and coarse grained quartz.

Except for pyrochlore, the HFSE minerals occur concentrated within cluster-like aggregates together with fine grained quartz. It can be recognized in Figure 4B that such clusters form pseudmorphs after amphibole indicated by the subhedral shape of the aggregates, which have straight grain boundaries against coarse grained quartz and K-feldspar. These polymineralic ore clusters, which can be observed through the entire suite of samples from KB G and KB S, have variable sizes from 250 µm up to 2 mm, whereas the minerals within the clusters are mainly anhedral and show smaller diameters. While the REE-carbonates vary in grain-size from 90 to 200 µm, Zr-silicates and quartz grains are even smaller and rarely exceed 50 µm. In addition to Zr-silicate and the REE- and Y-carbonates, the ore clusters also contain abundant hematite and fluorite, as well as, in rare cases, the Ti-phases rutile and titanite. The latter two minerals are strongly intergrown with fine grained quartz. Rutile forms euhedral prismatic crystals characterized by oscillatory zoning, which occasionally reach a maximum diameter exceeding 100 µm (Figure 5A). Locally, hematite also forms individual clusters with fine grained quartz. Fibrous hematite aggregates can be observed exceeding diameters >50 µm of irregular fan or spherulite texture. However, it can be detected as very fine grained (<5 µm) pigments in interstitial volume of mineral phases like zircon, as illustrated by blue colors at the high-resolution Fe mapping in Figure 5C.

Purple colored fluorite can be observed in the cluster assemblage as well as associated with coarse grained quartz, K-feldspar, and albite (Figure 4B, Figure 6A). It is mostly of xenomorphic shape and can locally exceed a grain size of 500 µm. Besides the abundance in interstitial volume, purple fluorite is found in veins cross-cutting the central part of the Khalzan Buregtei deposit. Further, fluorite can be observed as a replacement of K-feldspar and coarse grained quartz.

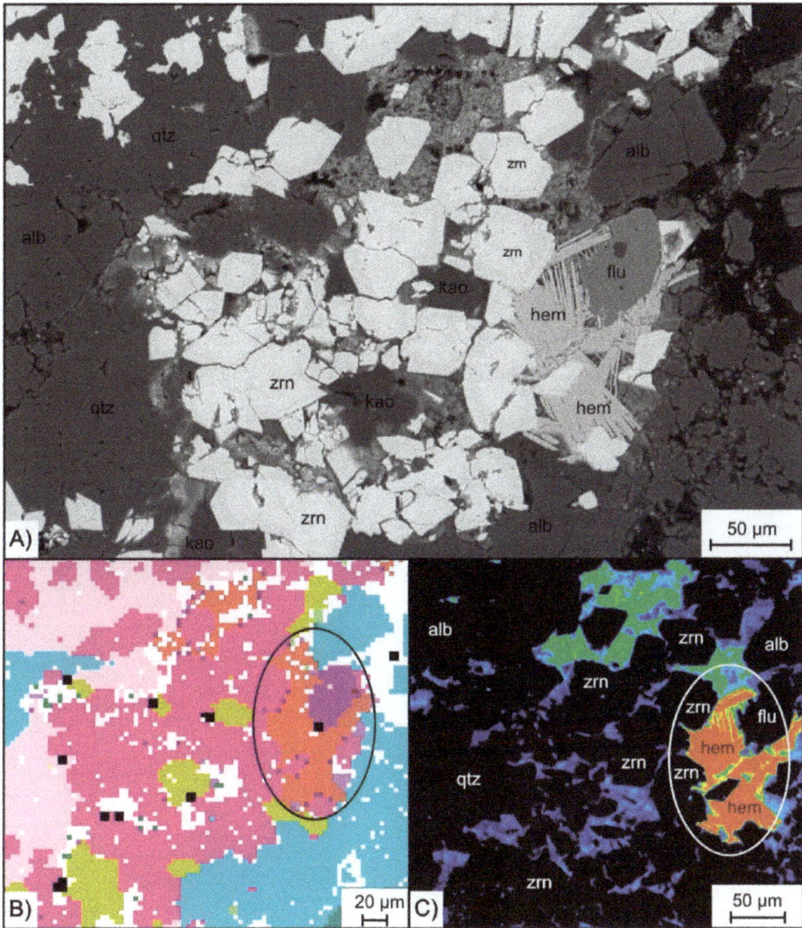

Figure 5. (**A**) Back scatter (BSE) image of a zircon cluster. Hematite (hem) as fibrous shape crystals in interstitial volume between zircons (zrn) and one fluorite (flu) crystal. (**B**) 5 QEMSCAN© phase map with 5 µm point spacing showing the hematite intergrowth in a close up of the cluster displayed in A. (**C**) Electron probe micro analyser (EPMA) high-resolution Fe map revealing Fe fine cementation as blue color surrounding individual zircon crystals. Other minerals: quartz (qtz), kaolinite (kao), albite (alb).

The Ti-phases rutile and titanite are strongly intergrown with fine grained quartz. Rutile forms euhedral prismatic crystals characterized by oscillatory zoning, which occasionally reach a maximum diameter exceeding 100 µm (Figure 6B). In contrast to the other ore minerals, pyrochlore is not accumulated in clusters, but is mostly associated with K-feldspar and coarse grained quartz. All grains of pyrochlore are euhedral (Figure 6C) and most of the crystals show evidence of brittle deformation. Furthermore, pyrochlore grains are observed being cross-cut by lath shaped albite. Like the other ore minerals, pyrochlore crystals are dominantly <50 µm but can exceed 140 µm in rare cases. As illustrated in Figure 6C, joints within pyrochlore are also healed by hematite precipitation.

Based in the observation of textural properties such as grain size, mineral assemblage, crystal shape, and grain boundaries indicative of equilibrium or replacement reactions, mineral phases can be differentiated into groups of primary or post-magmatic formation. Coarse grained minerals of subhedral shape sharing straight grain boundaries are considered as minerals formed under magmatic conditions. K-feldspar and arfvedsonite are two phases with coarse grain size as well as, in part, semihedral shape. Furthermore, the coarse grains of quartz are observed to share straight grain boundaries with K-feldspar. Thus, the precursor rocks of the metasomatites of the Khalzan Buregtei deposit can be identified as alkali syenite to alkali granites (QAPF, [20]). Owing to their coarse grain size, these rocks had a porphyritic appearance, which is in accordance with the observation of [5,13–15] regarding the unaltered rock suite of the main intrusive phase.

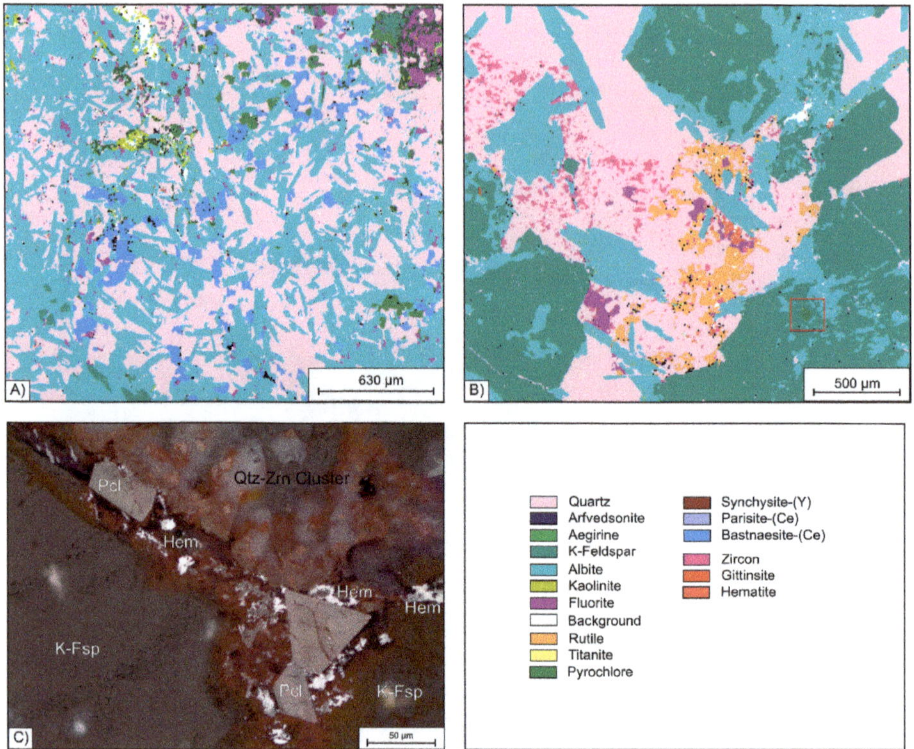

Figure 6. (**A**) Bastnaesite-quartz ore cluster in sample KB S9. K-feldspar is only found as relict. (**B**) Rutile-quartz cluster in granitic sample KB G6 containing fine grained quartz, zircon, fluorite and low amounts of hematite. The box indicates pyrochlore (Pcl) grain within the major association to K-feldspar (K-Fsp), being partly replaced by albite. (**C**) Euhedral pyrochlore on boundaries between K-feldsparand quartz-zircon (Qtz-Zrn) ore cluster. Joints inside the pyrochlore are healed by hematite (hem) precipitation, which also is found around the grains and within the ore cluster (ppl, reflected light, oil).

Post-magmatic phases are identified by significant change to finer grain size and irregular grain shape. Furthermore, the formation of polymineralic aggregates pseudomorphically replacing magmatic phases (e.g., amphibole) can also best be explained in a reaction of a solid mineral with a fluid, i.e., in a post-magmatic reaction. Thus, aegirine and albite, which grew at the expense of magmatic arfvedsonite and K-feldspar, respectively, are classified as post-magmatic minerals. Fine grained quartz, which shares straight grain boundaries with albite, is considered as cogenetic and thus also as post-magmatic. Post-magmatic formation is also indicated for ore and accessory minerals (e.g., hematite and fluorite) found in clusters intergrown with post-magmatic quartz. The semihedral shape of ore mineral clusters and their straight grain boundaries shared with magmatic phases are evidence that ore clusters mainly represent aggregates pseudomorphically replacing arfvedsonite and aegirine, which itself replaces arfvedsonite before (Figures 3D and 4A). The occurrence of clusters of post-magmatic minerals thus mirrors the texture of precursor porphyritic alkaline granite rock. Veins filled with carbonates were found cross-cutting all other mineralogical and textural features of the rocks and were interpreted to be late stage post-magmatic. It is important to note that these textural observations show that even with the enrichment of Zr, the REE or Nb were probably caused by magmatic fractionation [15]; the textural properties of the rare metal ores of Khalzan Buregtei are predominantly of post-magmatic metasomatic character, which is also in line with the characterization by [5].

4.2. Geochemistry

Whole rock major element composition as well as multication parameters (R1, R2 [21]) and rock classification based on major element signatures (TAS for plutonic rocks [22], R1-R2) are listed in Tables 1 and 2 for both sample series KB G and KB S.

Table 1. Major element composition of KB G samples from western part metasomatites. TAS classification for plutonic rocks according to [22]. QAPF classification according to [20] on the basis of volumetric mineral composition as shown in Figure 2. Multication parameters R1 and R2 were calculated according to [21] with: $R1 = 4Si - 11(Na + K) - 2(Fe + Ti)$, and $R2 = 6Ca + 2Mg + Al$. Abbreviations for TAS and QAPF classification: Gran = granite, Mon = quartz-monzonite, GranDio = granodiorite; abbreviations for de la Roche: Qtz rich = outside the classification fields, Alk Gran = alkali granite, Syeno Gran = syenogranite, Monzo Gran = monzogranite; LOI = loss on ignition.

(wt. %)	KB G1	KB G2	KB G3	KB G4	KB G5	KB G6	KB G7	KB G8
SiO_2	66.00	66.90	70.90	70.80	75.00	68.60	66.60	68.40
Al_2O_3	10.65	10.15	11.55	10.05	11.35	5.79	5.99	10.10
Fe_2O_3	2.80	3.18	4.90	2.89	3.58	3.51	4.10	1.72
MgO	0.03	0.02	0.04	0.03	0.04	0.00	0.03	0.01
CaO	3.69	2.99	0.34	1.90	0.42	8.65	6.41	4.48
Na_2O	4.53	4.63	4.19	3.37	3.96	0.08	1.24	3.32
K_2O	3.61	3.29	3.51	3.63	3.74	4.79	5.13	3.49
Cr_2O_3	<0.01	<0.01	<0.01	<0.01	0.00	<0.01	<0.01	<0.01
TiO_2	0.14	0.49	0.41	0.25	0.07	0.45	0.30	0.55
P_2O_5	0.04	0.02	0.05	0.02	0.03	<0.01	<0.01	0.03
MnO	0.16	0.19	0.06	0.23	0.01	0.05	0.47	0.04
BaO	0.03	0.03	<0.01	0.01	<0.01	0.02	0.04	0.01
LOI	1.75	1.82	1.7	1.57	1.48	2.26	2.39	2.35
Total	93.52	93.78	97.66	94.76	99.68	94.27	92.76	94.56
R1	1869	1950	2280	2590	2622	3320	2685	2503
R2	428	376	249	310	249	623	496	462
TAS	Mon	Mon	Gran	Gran	Gran	GranDio	GranDio	GranDio
De la Roche	Syeno Gran	Syeno Gran	Alk Gran	Alk Gran	Alk Gran	Qtz rich	GranDio	Monzo Gran
QAPF	Alk Gran	Alk Gran	Alk Gran	Alk Gran	Alk Gran	Alk Gran	Alk Gran	Qtz rich

As is indicated by the SiO_2 values, all rocks are silica oversaturated. The samples of KB S, especially, show an increased concentration of SiO_2 compared to KB G, which is in accordance to the increased quartz content indicated by the modal composition Figure 2. Other major elements are more variably distributed in the KBG sample suite, in which KBG 6 and 7 show elevated amounts of fluorite, but are in contrast to all other samples containing no albite in Figure 2. This leads to significantly decreased Na_2O and Al_2O_3 concentrations, but enhanced values of CaO and LOI (Figure 2, Tables 1 and 2). In addition to fluorite, gittinsite, which is present among the Zr-silicates in some samples (e.g., KB G1), is a carrier of CaO, too. Variations in major element composition are mirrored in element-based rock classification. Lower SiO_2 content in KB G samples but variable sums of $Na_2O + K_2O$ result in TAS classifications of monzonite, granodiorite, and granite (Table 1). In contrast, the de la Roche classification, which, in addition to Si and the alkalis, includes Fe, Ti, Ca, Mg, and Al, reflects element variability by assigning the classes syenogranite, alkali granite, granodiorite, and monzogranite to the KB G samples. While all KB S samples are entirely classified as granites in the TAS scheme, the de la

Roche scheme leads to classification as alkali granite and syeno granite in some cases or does not allow classification at all. The latter can be observed in samples with highest SiO_2 concentrations and R1 multication parameter > 3000. Therefore, these samples are classified as "Qtz rich" in this study.

Table 2. Major element composition of KB S. TAS classification for plutonic rocks according to [22]. QAPF classification according to [20]. Parameters R1 and R2 and resulting classification calculated according to [21]. Abbreviations for TAS, de la Roche, and QAPF classifications as defined in Table 1.

(wt. %)	KB S1	KB S2	KB S3	KB S4	KB S5	KB S6	KB S7	KB S8	KB S9	KB S10
SiO_2	79.90	77.20	72.70	74.20	71.70	76.50	77.40	72.50	75.00	73.30
Al_2O_3	9.04	10.55	12.90	10.75	9.77	8.91	10.50	9.97	11.35	10.80
Fe_2O_3	2.14	1.71	1.28	3.64	5.10	1.32	1.41	2.47	4.72	3.61
MgO	0.01	0.03	0.04	0.03	0.04	<0.01	0.03	0.03	0.02	0.03
CaO	0.04	0.06	0.20	0.18	0.47	0.52	0.10	2.58	1.87	2.30
Na_2O	1.02	2.27	4.21	3.62	3.43	2.93	3.04	2.69	3.95	2.59
K_2O	6.28	5.71	4.27	3.48	3.16	3.22	4.13	3.16	3.29	3.42
Cr_2O_3	<0.01	<0.01	<0.01	<0.01	<0.01	<0.01	<0.01	<0.01	<0.01	<0.01
TiO_2	0.09	0.09	0.38	0.07	0.18	0.22	1.32	0.34	0.16	0.92
P_2O_5	<0.01	<0.01	0.02	0.02	0.02	0.03	<0.01	0.01	<0.01	<0.01
MnO	0.01	0.06	0.01	0.04	0.43	0.02	0.01	0.01	0.01	0.01
BaO	<0.01	0.01	<0.01	0.01	0.05	0.01	0.01	0.01	<0.01	0.01
LOI	0.64	0.86	1.56	1.36	2.13	0.8	1.2	1.96	1.43	2.25
Total	99.17	98.55	97.58	97.41	96.49	94.48	99.15	95.74	101.8	99.25
R1	3434	2954	2306	2749	2685	3262	3040	3063	2700	3048
R2	180	212	267	223	221	205	213	349	334	349
TAS	Gran	Gran	Gran	Gran	Gran	Gran	Gran	Gran	Gran	Gran
De la Roche	Qtz rich	Alk Gran	Alk Gran	Alk Gran	Alk Gran	Qtz rich	Qtz rich	Qtz rich	Syeno Gran	Qtz rich
QAPF	Alk Gran	Alk Gran	Alk Gran	Alk Gran	Alk Gran	Alk Gran	Alk Gran	Alk Gran	Alk Gran	Alk Gran

The results of the REE analyses of the bulk rock samples are given in Tables 3 and 4. Chondrite normalized distribution patterns are given in Figure 7. All samples show REE concentrations strongly enriched in comparison to chondritic composition [23] and they have negative Eu anomalies, which are more pronounced in KB G samples (Eu/Eu* < 0.3) in contrast to KB S (Eu/Eu* < 0.2), see Tables 3 and 4. The HREE (heavy rare earth elements Gd-Lu) elements show a significantly higher enrichment in the KB G samples in comparison to KB S. The latter samples, on the other hand, show a steeper pattern, with elevated enrichment of LREE (light rare earth-elements La-Eu) ratios up to 7000 but a generally lower enrichment of HREE ≤ 1000 (Figure 7). KB G samples, instead, are less enriched in LREE < 3000, which leads to flat patterns.

Table 3. Rare earth elements (REEs) + Zr and Nb composition of KB G. Zirconium data from handheld X-ray fluorescence (XRF) measurement. Europium anomaly Eu/Eu* calculated based on the geometric mean of normalized Sm and Gd values.

(mg/kg)	KB G1	KB G2	KB G3	KB G4	KB G5	KB G6	KB G7	KB G8
La	224	231	1120	223	694	307	263	763
Ce	582	659	2480	547	1250	787	1100	1700
Pr	59	71	297	58	163	85	67	212
Nd	211	260	1080	207	642	299	239	768
Sm	61	83	295	60	172	61	52	160
Eu	6	8	20	5	11	4	5	12
Gd	65	92	298	79	186	52	56	137
Tb	19	25	61	24	26	11	13	29
Dy	175	203	357	207	129	88	115	188
Ho	49	54	60	50	10	24	33	41
Y	1285	1460	1575	1355	532	705	898	1090
Er	200	216	155	183	44	91	130	142
Tm	39	40	19	33	5	17	24	25
Yb	269	277	99	213	21	125	166	166
Lu	38	40	10	29	2	19	26	24
Eu/Eu*	0.28	0.21	0.21	0.21	0.19	0.19	0.29	0.25
Nb	1095	1140	>2500	1660	>2500	1905	1495	1445
Zr (wt. %)	3.27	2.66	0.08	1.91	1.61	2.68	3.28	2.46

Table 4. REE + Zr and Nb composition of KB S. Zirconium data from handheld XRF measurement.

(mg/kg)	KB S1	KB S2	KB S3	KB S4	KB S5	KB S6	KB S7	KB S8	KB S9	KB S10
La	460	510	1130	899	682	458	1290	505	385	1700
Ce	934	968	2280	1685	1230	902	2900	964	857	4480
Pr	108	125	281	206	159	112	373	113	115	620
Nd	399	475	1010	769	581	421	1395	424	456	2430
Sm	69	93	203	176	155	98	324	115	154	569
Eu	3	5	11	10	9	5	17	7	10	29
Gd	54	84	160	169	178	87	251	108	161	406
Tb	10	15	24	35	39	18	39	21	28	63
Dy	73	104	142	244	292	143	187	157	149	329
Ho	17	24	30	56	73	38	33	41	25	56
Y	436	617	838	1500	1775	965	805	975	650	1100
Er	52	72	87	172	238	137	86	139	57	134
Tm	7	10	13	24	33	22	11	22	6	15
Yb	45	65	79	150	205	159	78	162	32	92
Lu	6	9	10	19	26	22	11	24	3	11
Eu/Eu*	0.16	0.16	0.19	0.18	0.17	0.17	0.18	0.19	0.19	0.19
Nb	490	665	1945	1270	>2500	794	>2500	982	1410	1050
Zr (wt. %)	0.39	0.57	0.60	1.30	2.00	3.14	0.18	2.56	0.02	0.12

Zirconium has variable concentrations across the data set. KB G samples contain Zr up to 3.3 wt. % (Figure 7 orange range) whereas only two samples have concentrations < 1.9 wt. %. Within both sample suites, those of low Zr concentrations show decreasing enrichment of the HREE. In contrast, samples with high Zr concentration are characterized by increased HREE enrichment.

Niobium concentrations show a distribution similar to the Zr. maximum values given by the method's upper limit of quantification at 2500 mg/kg. Minimum concentrations of Nb in KB G samples generally are two times higher than the concentrations of KB S samples ranging between 490 and 790 mg/kg. Concentrations of Nb are generally higher in KB G and show a variation between 1000 mg/kg and 1900 mg/kg. In comparison, the concentration of Nb in KB S samples, excluding the samples exceeding the upper limit of quantification, cover the range from 490 mg/kg to 1945 mg/kg.

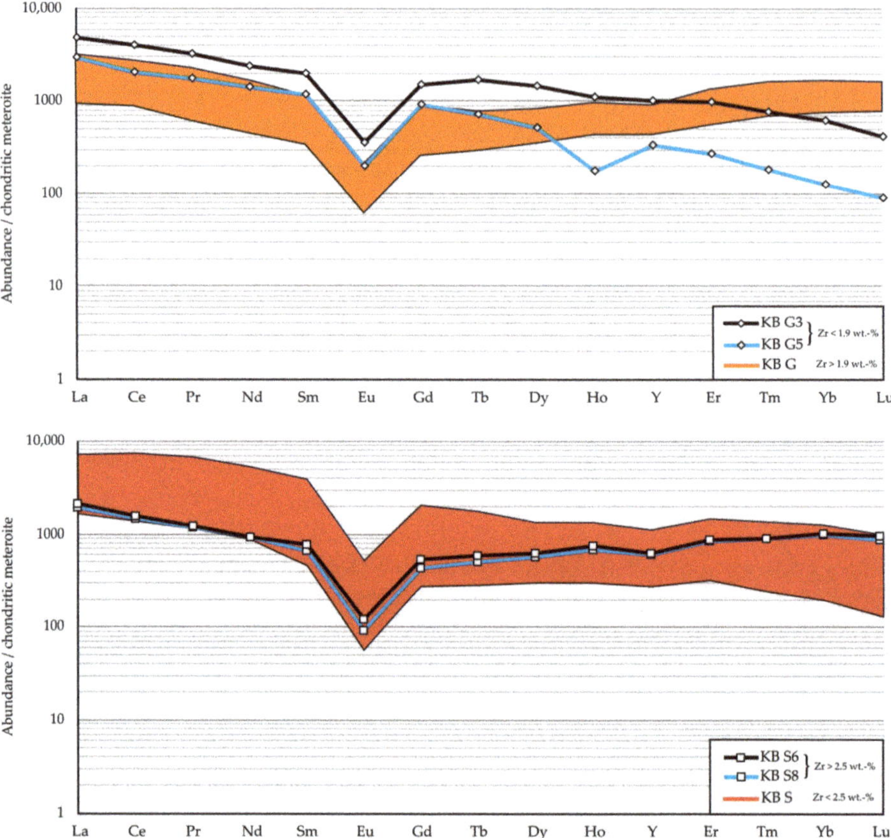

Figure 7. REE pattern for KB G (**upper**) and KB S (**lower**). Coloured range represents maximum and minimum values of individual sample patterns (orange and red) of parallel evolution for high content of Zr in KB G and low Zr content in KB S. The samples KB G3 and G5 were identified as having the lowest enrichment of Zr in the KB G samples and showed a decreasing tendency within the HREE. The same behaviour is shown for the KB S dataset where, in contrast, the samples KB S6 and S8 contain significantly higher amounts of Zr than the other samples (red range) representing the increasing tendency of the HREE. REE concentrations were normalized against chondritic composition according to [24].

4.3. Elemental Deportment

Petrographic and geochemical investigation points out that the major ore mineral classes are Zr-silicates, REE-carbonates, and Ti/Nb oxides that occur within the metasomatic rocks forming the centre of the Khalzan Buregtei deposit. Zircon is the most abundant ore mineral. It can be detected in all samples of ore metasomatites. Further, it is identified as the major carrier for the rare metals Zr, Hf, Y, and HREE due to the high concentrations of Y, Dy, and Er across the analysed zircon crystals, as demonstrated by the representative analyses of different zircon and gittinsite grains in Table 5.

Table 5. Exemplary analyses of gittinsite and zircon crystals of the metasomatic samples.

(wt. %)	Gtt	Gtt	Zrn	Zrn	Zrn	Zrn	Zrn	Zrn
SiO_2	36.32	40.01	32.49	30.89	30.90	27.25	27.43	27.98
ZrO_2	34.75	35.45	55.94	57.85	56.24	58.96	54.73	54.62
HfO_2	0.64	0.71	1.11	1.00	1.18	1.28	0.92	1.00
Fe_2O_3	0.89	0.41	1.23	1.04	0.76	1.03	0.26	2.49
Ce_2O_3	0.34	0.07	0.41	0.06	0.19	0.07	0.04	0.24
Dy_2O_3	0.22	0.25	0.27	0.12	0.26	0.25	0.81	0.38
Er_2O_3	0.08	0.10	0.31	0.15	0.31	0.36	0.78	0.41
Y_2O_3	0.42	0.54	2.07	0.78	2.01	1.99	5.98	3.12
Na_2O	0.05	0.05	0.06	0.04	0.07	0.21	0.40	0.42
CaO	18.41	17.45	1.03	2.30	1.84	0.45	0.07	0.34
MnO	0.58	0.67	0.11	0.11	0.14	0.02	0.01	0.05
Total	92.84	95.95	95.5	94.65	94.48	92.47	92.52	92.11

Table 6 summarizes EPMA analyses of bastnaesite-(Ce), parisite-(Ce), and synchysite-(Y), which are the most abundant minerals of the REE-carbonate group. It can be seen that the LREEs show a high concentration in the first two minerals, whereas synchysite-Y has got elevated concentrations of Y and is thereby suspected to contain unquantified amounts of HREE as well. Low total sums can be explained by unquantified amounts of HREE incorporated into the analysed minerals. The individual concentration levels of these elements are below the limit of quantification of the applied EPMA.

Table 6. Representative EPMA analysis of REE-carbonate minerals bastnaesite-(Ce), parisite-(Ce), and synchysite-(Y).

(wt. %)	Bst-(Ce)	Bst-(Ce)	Bst-(Ce)	Par-(Ce)	Par-(Ce)	Par-(Ce)	Syn-(Y)	Syn-(Y)
CaO	1.07	1.49	0.54	8.23	10.92	9.34	19.53	16.64
Y_2O_3	2.82	3.87	1.62	0.37	0.77	2.30	29.69	22.43
Ce_2O_3	32.52	31.87	34.60	32.87	29.61	31.26	1.29	2.36
La_2O_3	11.19	10.90	13.43	18.46	13.50	11.39	0.24	0.54
Pr_2O_3	3.79	3.73	3.88	2.54	2.85	3.36	0.23	0.30
Sm_2O_3	4.22	4.04	3.61	0.89	1.37	1.99	0.64	0.77
Nd_2O_3	15.42	14.74	15.14	6.60	9.19	12.25	0.52	0.85
SrO	0.11	0.14	0.16	nd	0.39	0.50	0.38	0.16
ThO_2	0.67	0.34	0.08	nd	0.06	0.05	0.18	0.45
SiO_2	0.48	0.67	0.08	1.38	1.42	0.24	0.484	4.37
F	8.10	9.03	4.56	6.96	7.92	7.25	5.01	4.99
CO_2	20.93	21.20	20.74	23.26	23.40	23.09	24.06	24.02
Total	101.32	102.02	98.44	101.56	101.40	103.02	82.24	77.88

Pyrochlore occurs as a major carrier for Nb in the Khalzan Buregtei deposit. Eight pyrochlore EPMA analyses are given in Table 7. They show that pyrochlore is enriched in the LREEs Ce, La, and Nd. Heavier REE elements were not detected. Due to the content of Ce, pyrochlore is identified as ceriopyrochlore-Ce, which contains only very small amounts of betafite [$Ca_2(Ti,Nb)_2O_6O$] and microlite [$(Na,Ca)_2Ta_2O_6(O,OH,F)$], as indicated by the low concentrations of Ti and Ta, respectively. Pyrochlore also shows a capability to enrich U to concentrations up to 1.16 wt. %. Semi-quantitative analyses indicate that, in addition to pyrochlore, rutile can be another carrier of Nb_2O_5 with variable concentrations from 2 to 5 wt. %.

Table 7. Elemental composition of representative pyrochlore grains from sampled parts of the Khalzan Buregtei deposit.

(wt. %)	Pcl 1	Pcl 2	Pcl 3	Pcl 4	Pcl 5	Pcl 6	Pcl 7	Pcl 8
Nb_2O_5	67.25	68.13	66.90	57.96	62.54	70.71	66.63	66.03
Fe_2O_3	0.94	0.97	0.92	2.51	1.04	0.91	0.80	1.05
TiO_2	4.81	4.85	5.30	8.16	3.61	2.44	4.33	4.62
Ta_2O_5	2.13	1.61	2.32	1.95	2.22	1.61	2.49	2.22
Na_2O	2.72	1.90	2.54	1.31	1.57	2.73	3.60	2.30
CaO	3.93	3.97	4.21	4.23	4.92	4.79	3.79	5.27
Ce_2O_3	11.21	10.96	10.99	11.92	9.31	10.10	10.79	4.16
La_2O_3	4.34	4.37	4.17	4.33	3.70	4.03	4.42	10.92
Nd_2O_3	2.54	2.58	2.54	3.07	2.21	2.38	2.63	2.51
Y_2O_3	0.07	0.12	0.12	0.20	0.72	0.18	0.07	0.09
UO_2	0.37	0.43	0.29	0.60	1.16	0.42	0.29	0.20
ThO_2	0.07	0.06	0.01	0.08	0.04	0.01	0.01	0.06
SiO_2	0.46	0.42	0.27	0.65	2.76	0.60	0.75	0.47
Total	100.83	100.38	100.58	96.99	95.80	100.90	100.60	99.89

5. Discussion

5.1. Classification Systematics

The classification of the mineralized rocks of the Khalzan Buregtei deposit applying traditional methods of magmatic petrology (QAPF and TAS, see Tables 1 and 2) leads to obviously inappropriate results due to a lack of consistency across the different classification schemes. Furthermore, the classification of some samples fails to unequivocally define rock types, as compositions do not match the compositional ranges for which the schemes were defined (R1-R2, see Tables 1 and 2). This mismatch is thought (see Section 4.2) to be caused by chemical variations due to post-magmatic mineral formation. In particular, the TAS classification for plutonic rocks, derived by [22] from the classic TAS scheme for volcanic rocks, is strongly based on Na_2O and SiO_2 concentrations and thus notably responds to post-magmatic processes. It is therefore concluded not to use the TAS or QAPF schemes for highly mineralized alkaline granitoid rocks such as those in the centre of the Khalzan Buregtei deposit. This is in accordance with [6], who proposed a more simple classification for the rock suite of the Khalzan Buregtei massif into three classes of barren and ore bearing metasomatites and unaltered rocks of the main intrusion phase, see Figure 1.

The multication parameters R1 and R2 combine concentration data of Si, Ti, Al, Fe, Ca, Mg, K, as well as Na and thus provide the opportunity to display the chemical composition of rocks and individual rock-forming minerals (Figure 8). Consequently, the effects of post-magmatic mineral reactions altering the intrusive alkaline magmatic rocks can be graphically visualized by these parameters. Figure 8 displays the R1-R2 plot of unaltered (blue mark) and metasomatic (red mark) rocks of the Khalzan Buregtei massif. The latter samples show significantly higher R1 values in contrast to the unaltered ones and show an elevated abundance of zircon within the ore cluster. The evolution towards the point representing quartz along the line connecting feldspars and quartz suggests silicification as one major alteration feature. Thus, the formation of post-magmatic quartz can be linked to the increase of the R1 parameter parallel to the pink vectors in Figure 8. Albitization shifts rock compositions towards the albite point. However, this effect is less pronounced in the diagram due to the proximity of the points representing albite and K-feldspar. The formation of other post-magmatic Ca-bearing minerals like fluorite, REE-carbonates, and gittinsite is reflected by increasing R2 values as indicated by the vertical shift along the purple vectors in Figure 8, which can be particularly well observed for the KB G samples. The influence of Mg on the R2 value can be neglected due to the very low concentration of this element in the entire samples series. Due to the R1 and R2 value of arfvedsonite and aegirine, decomposition of these minerals enhances the effect of the elevated R1 values described above.

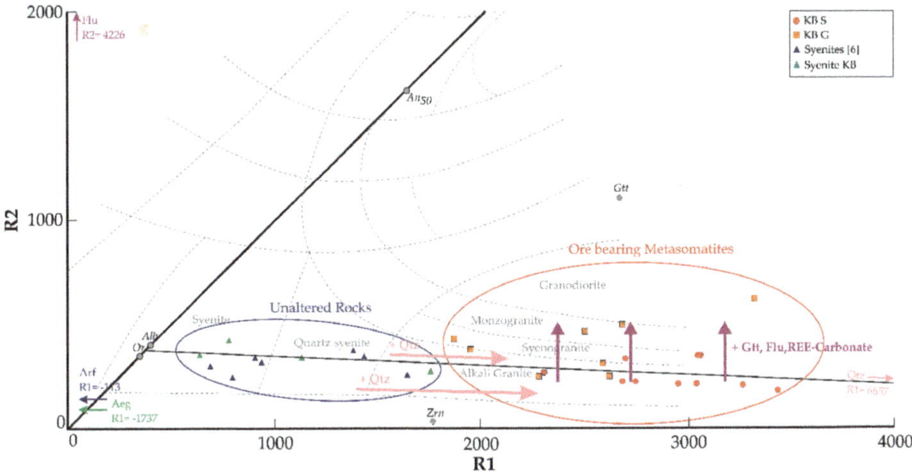

Figure 8. R1-R2 multication plot of the KB G and S samples, classification after [21] indicated by stippled grey lines. Fluorite (Flu) and quartz (Qtz) compositions plot outside the diagram and are indicated by their R1 and R2 value at the hosting axis. The blue mark indicates unaltered rocks of the Khalzan Buregtei Massif according to [5] (Syenites [5]) and Gronen (unpublished) (Syenite KB). Red mark highlights ore bearing metasomatic rocks of the KB S and KB G dataset. Purple vectors indicate the change of R2 value of samples containing CaO > 1.8 wt. %. Mineral abbreviations: Zrn = zircon, Gtt = gittinsite, Alb = albite, An_{50} = anorthite, Or = orthoclase K-feldspar, Arf = arfvedsonite, Aeg = aegirine. Composition of Arf and Aeg according to [24].

While traditional magmatic rock classification turns out to be inappropriate, which is also displayed in Figure 8 showing rocks with similar chemical and mineralogical compositions that plot into different fields, the pair of numeric R1-R2 values is nevertheless indicative of the petrographic character of each sample. Being based on whole rock major element data, these numeric values are appropriate for characterizing such complex metasomatically-altered rocks and therefore are considered as highly effective and unambiguous quantitative tools particularly suitable for distinction and visualization of altered or mineralized alkaline rocks and their rare metal ore bodies. Multication parameters are additive numeric values and consequently are a valuable base for modelling and geostatistics [25]. As whole rock chemical data are abundantly available already at pre-feasibility levels, multication parameters are also recommended for use in 3D modelling during exploration work or mine planning. While they fully characterize compositional rock properties, multication parameters are independent of the subjective impression or degree of skill of an individual petrographic investigator. Thus, they represent an important step to deposit characterization by quantitative parameters, which also is a prerequisite for any geometallurgical evaluation.

The geometallurgical approach considers mineralogical features of deposits in a quantitative manner with respect to their significance for mine planning and mineral processing [25]. The main focus lies on the most effective exploitation by accurate adjustment of the entire process chain from mining, mineral processing, and metallurgical procedures to mineralogical and textural ore properties [25–27].

5.2. Textural Properties of Ore Minerals and Their Use for Optimized Mineral Processing

Textural properties are crucial for the technology and costs required for ore processing. Therefore, identification and quantification of textural characteristics are as essential for ore evaluation as determination of ore mineral grade. Most ore minerals observed in the centre of the Khalzan Buregtei deposit crystallized in post-magmatic alteration reactions concomitant with the development of textures characterized by fine grain size, partial replacement, and intricate intergrowth between ore minerals and gangue (see Section 4.1, Figures 4 and 6). These textural properties significantly affect the ore mineral liberation size, indicating the coarsest size fraction produced in comminution that guarantees the maximum detachment of ore minerals and associated gangue [28]. For the ore-bearing rock samples of KB G and KB S at Khalzan Buregtei, the liberation size of the ore minerals was obtained by image analyses applied on the phase maps generated with the QEMSCAN© technique. The liberation size distribution of the bulk group of ore minerals comprising Zr-silicates and REE-carbonates are displayed in Figure 9A. The graphs for representative samples KB G 5, 8, and KB S 8 show that 80 wt. % of all ore minerals would be liberated in fractions varying between P80 < 45 μm and P80 < 85 μm. This is well in accordance with mineral processing test work on rare metal ores from Khalzan Buregtei conducted by [29]. Likewise, [30] suggested a comminution of <80 μm followed by scavenger flotation for REE-carbonate minerals for processing of a similar rare metal ore.

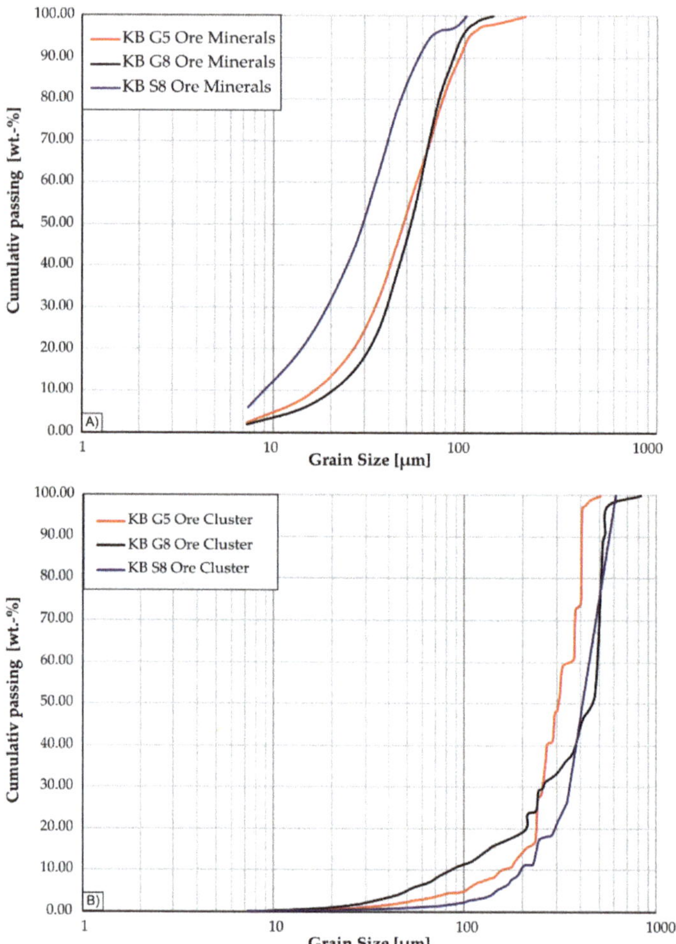

Figure 9. Particle size distributions of ore minerals (**A**) and ore clusters (**B**). The ore mineral groups Zr-silicates and REE-carbonates were combined into the class of ore minerals because these minerals are mostly observed in cluster assemblage. As can be obtained from the particle size distribution, the ore mineral size at P80 is significantly smaller (45–85 μm) in comparison to the particle size of the ore clusters (400–500 μm).

In general, comminution to low grain size is an energy and material-consuming step, belonging to the most expensive positions in processing operations [28,31]. In addition, although required for liberation of ore phases, low grain sizes also hamper processing activities, mainly leaving flotation as a suitable technique [28]. This enhances consumption of water and supplies (e.g., flotation chemicals) as well as of operation costs [28]. While flotation turned out to be effective for REE-carbonates, it cannot concentrate ore minerals with distinctly different physical properties, such as zircon or pyrochlore in the same step (e.g., [28,30]). Owing to the prices of Zr, Nb, and the HREE, the latter ore minerals, which remain in the flotation tailings together with a large mass of gangue minerals, would represent a higher value than the LREE predominantly extracted from carbonates. High process costs on one hand and limited concentration efficiency on the other hand would represent an obstacle for mining operations at the Khalzan Buregtei deposit as well as for other rare metal alkaline granites.

However, the aggregation of the ore minerals zircon, gittinsite, and REE-carbonate together with other post-magmatic phases, such as fluorite and hematite in clusters characteristic of the mineralized rocks of sampling sites KB G and KB S, allows an alternative approach to more efficient processing. Image analysis yields quantitative information on the grain size distribution of each individual ore mineral on the one hand as well as of the ore mineral clusters on the other hand. Figure 9B shows that diameters of 80 % of all ore clusters in representative samples KB G 5, 8, and KB S 8 are ≤400 to 500 μm. Consequently, liberation of ore clusters could already be achieved in size fractions, which are significantly coarser than those required for liberation of individual ore minerals (Figure 9A). It is worth mentioning that the increase in liberation size by a factor of, in part, >10 relative to the individual ore minerals, reflects the difference between magmatic and post-magmatic mineral grain sizes. The clusters, which predominantly formed as post-magmatic pseudomorphs after arfvedsonite and elpidite (see Section 4.1) mainly have the size of their former magmatic precursor.

The increased particle size of the ore clusters is a textural feature useable for optimization of mineral processing because the liberation of the ore clusters needs less intensive comminution, which is connected to a decrease in energy consumption (e.g., [31]). In a process chain, the separation of barren and ore bearing material needs to take place as early as possible to enhance the overall process effectiveness due to mass flow reductions and increased yield of target mineral phases. Extracting a pre-concentrate of ore mineral clusters from the coarse particle size fraction will achieve both goals.

For the extraction of ore mineral clusters, physical properties are needed, which distinguish the target phases from the barren material. One possible property is the increased density of the ore-bearing clusters, which is due to the locally-increased modal abundance of mineralogical phases of high specific gravity in contrast to the barren feldspar-quartz dominated rock domains. This makes particles amenable to density separation. Further, the close associations of clustered ore minerals with hematite (Figure 4A,B) represents a marked difference to the barren domains because the magnetic susceptibility of this Fe-bearing phase makes the ore clusters also amenable to magnetic separation.

Based on the detailed quantitative investigations of mineralogical and textural ore properties, a mineral-processing scheme was developed by [29]. The study demonstrates that the extraction of a pre-concentrate by dry magnetic separation on a particle fraction <250 μm was successfully conducted on a high volume ore sample taken from the Khalzan Buregtei deposit. Applying this optimized processing scheme, the mass flow was reduced by 55 wt. %, and an enrichment of zircon by a factor of 2.5 was achieved. As a consequence, following process steps like comminution of fractions rich in ore mineral clusters to liberation size of individual ore minerals and scavenger flotation, needed for the production of concentrates, can be conducted with the enhanced pre-concentrate having increased ore mineral content and decreased total mass.

This example shows that the mineralogical and textural properties "association of ore minerals to hematite" as well as "clustering of ore minerals" have a crucial impact on processing efficiency and thus finally on the profit of a mining operation. For a geometallurgical approach towards optimization of the process chain, it is important to delineate volumes of mineralized rock that are characterized by consistent mineral processing responses such as geometallurgical domains [25,32,33]. In the case of the Khalzan Buregtei deposit both properties are considered as key features for the definition of a geometallurgical domain.

The abundance of ore clusters is described for other alkaline massifs by several authors [34], for example, report clusters of REE rich apatite and allanite [(CaREE)(Al$_2$Fe^{2+})(Si$_2$O$_7$)(SiO$_4$)O(OH)] connected to Biotite-magnetite veins of the Loch Loyal syenite and conjugated REE-carbonate veins. Further, the replacement of arfvedsonite by hematite-quartz clusters is described by [35] for peralkaline arfvedsonite granites of the Amis complex in Namibia, which are connected to a high abundance of HFES silicates like zircon.

Clusters formed by gittinsite and zircon in association with secondary quartz replacing elpidite and arfvedsonite were also observed by [1] in the rare metal prospects of the alkaline granite at Strange Lake in Canada. These authors, too, describe post-magmatic minerals hematite, quartz and fluorite that formed in close association to ore minerals. The studies from different areas show that rare metal/HFSE enrichment in alkaline granitoid rocks is associated to the post-magmatic formation of ore-bearing clusters replacing precursor magmatic minerals. Consequently, application of the optimized beneficiation strategy, as outlined in this study, to those other rare metal alkaline granitoid deposits could be a step to improve economic mining operations.

6. Conclusion

Alkaline rock hosted rare metal deposits yield potential as future resource for commodities like Zr, Nb, Hf, Y, and the REE. The economic interest in this deposit type is underlined by various international studies of the last decade [1,7,9,10,30]. The complexity of the ore with respect to texture, composition, and variable types of ore minerals is often caused by post-magmatic alteration of the ore-bearing rock suite. The alteration is identified as one key process of the ore formation, because it enhances formation and concentration of ore minerals. However, post-magmatic alteration also causes rearrangement of the precursor magmatic rocks. As demonstrated in this study of Khalzan Buregtei, for Zr-REE-Nb deposits traditional used concepts of magmatic rock classification (e.g. TAS, QAPF) fail to properly characterize the altered alkaline rocks.) fail to properly characterize altered alkaline rocks. While not providing common magmatic-related names, multication R1-R2 parameters defined by [21] can serve as numeric values to characterize altered and mineralized rocks, as well as to visualize the chemical effect of alteration processes observed in petrographic inspection. Owing to their quantitative additive property, multication parameters are recommended to be implemented in numerical models to also display and analyse rare metal ore and their alkaline granitoid host rocks in geological 3D models.

The post-magmatic ore mineral assemblage at Khalzan Buregtei is characterized by overall fine grain size and intricate intergrowth. Consequently, in a mining operation, small liberation size, quantified in this study by QEMSCAN© image analysis, would require intensive comminution. High energy demand and lagged masses of finely-ground rock represent a significant obstacle to an economic use of this deposit. Petrographic inspection revealed that, except for the major Nb carrier pyrochlore, all ore minerals of Zr and REE (i.e., zircon, Zr-silicates, and REE-carbonates) occur in pseudomorph-like association mainly replacing precursor magmatic arfvedsonite. This leads to a concentration of ore minerals in clusters characterized by liberation sizes significantly exceeding those of individual ore minerals.

Furthermore, QEMSCAN© image analysis indicated that hematite is a characteristic constituent of clustered ore mineral aggregates. These textural properties allow pre-concentration of ore mineral clusters after clearly less energy consuming comminution to particle sizes < 250 μm and magmatic separation. Further comminution of the pre-concentrate and processing steps to concentrate zircon, Zr-silicates, and REE-carbonates then only have to be applied to 55% of the initial mass of the ore. This implies a significant reduction of energy consumption as well as of water and other supplies (e.g., flotation chemicals). Taking into account that clustering, as well as associations of ore minerals with hematite, are observed in other rare metal deposits formed in alkaline granitoids in different concentrations, the approaches developed for Khalzan Buregtei are considered as important steps towards more efficient and sustainable extraction of rare metal ores.

Author Contributions: Conceptualization, L.H.G. and S.S.; Investigation, L.H.G., J.L.K., and U.B.; Methodology, L.H.G. and S.S.; Project administration, H.W. and F.M.M.; Supervision, H.W. and F.M.M.; Validation, A.H.; Writing—original draft, L.H.G. and S.S.; Writing—review and editing, J.L.K., H.W., and F.M.M.

Funding: This publication was financially supported by the Aachen Know-How Center Resource Technology (AKR). The surrounding project OptiWiM is financially supported by the German Federal Ministry of Education and Research (BMBF, OptiWiM; FKZ: 033R162B).

Acknowledgments: The authors show gratitude to Chinzorekt of MONNIS International LLC and the staff of TD Bank of Mongolia, for granting access to Khalzan Buregtei license area and the permission to collect samples from outcrops. G. Stehr and the staff of the German-Mongolian Institute of Resource Technology (GMIT) and M. Bauer of Consulting, Business and Management GmbH (CBM GmbH, Bexbach) are gratefully acknowledged for administration and logistical support during the project. R. Klinghardt of technical staff of IML Klockmann Laboratory for Geometallurgy is gratefully thanked for his attendance during EPMA and QEMSCAN© analysis.

Conflicts of Interest: The authors declare no conflict of interest.

Appendix A

Table A1. Spectrometer setup for the EPMA high-resolution mappings.

Element	Line	Spectrometer	Position (mm)
Al	Kα	TAP	90.712
Ca	Kα	PETJ	107.636
Fe	Kα	LIFH	134.262
Si	Kα	TAP	77.208
Zr	Lα	PETJ	192.620

Table A2. Setup for EPMA measurements with focused electron beam on Zr-Silicates and pyrochlore. Sorted according to spectrometer optimization.

Element	Line	Spectrometer	Position (mm)	Bg +/− (s)	Peak (s)	Standard Matrix
Na	Kα	TAP	129.468	5	10	Jadeite
Y	Lα	PETJ	206.399	10	30	Glass
Zr	Lα	PETJ	194.628	5	10	Zircon
Th	Mα	PETJ	132.564	10	20	Th-Oxide
Hf	Lα	LIFH	109.114	15	30	Zircon
Al	Kα	TAP	90.509	5	10	Plagioclase
P	Kα	PETJ	196.998	5	10	Monazite
Nb	Lα	PETJ	183.547	15	30	Nb-Oxide
U	Mβ	PETJ	119.031	15	30	U-Si-Al-Ca-Gla
Er	Lα	LIFH	123.693	15	30	Glass
Si	Kα	TAP	77.225	5	10	Zircon
Ca	Kα	PETJ	107.636	5	10	Plagioclase
Dy	Lα	LIFH	132.284	15	30	Glass
Ti	Kα	PETJ	88.023	5	10	Rutile
Fe	Kα	LIFH	134.262	5	10	Faylaite
Mn	Kα	LIFH	145.742	5	10	Mn-Oxide
Ce	Lα	LIFH	177.487	15	30	Monazite

Table A3. Setup for EPMA measurements on REE-carbonate minerals with a beam diameter of 10 μm.

Element	Line	Spectrometer	Position (mm)	Bg +/− (s)	Peak (s)	Standard Matrix
Si	Kα	TAP	77.237	5	10	Plagioclase
F	Kα	TAP	199.162	5	10	Apatite
Y	Kα	PETJ	206.73	10	30	Glass
Ca	Lα	PETJ	107.647	5	10	Apatite
Ce	Lα	LIFH	177.503	15	30	Monazite
Sr	Lα	TAP	74.361	10	30	Celestine
Th	Mα	PETJ	132.609	10	30	Th-Oxide
La	Lα	LIFH	184.732	15	30	Monazite
Pr	Lβ	LIFH	156.531	15	30	Glass
Sm	Lα	LIFH	152.425	15	30	Glass
Nd	Lβ	LIFH	150.187	15	30	Glass

References

1. Salvi, S.; Williams-Jones, A.E. Alkaline granite-syenite deposits. In *Rare-Element Geochemistry and Mineral Deposits*; Linnen, L., Samson, I.M., Eds.; Geological Association of Canada: St. John's, NL, Canada, 2005; pp. 315–341.

2. Sørensen, H. Agpaitic nepheline syenites: A potential source of rare elements. *Appl. Geochem.* **1992**, *7*, 417–427. [CrossRef]

3. Semenov, E.I. Economic mineralogy of alkaline rocks. In *The Alkaline Rocks*; Sørensen, H., Ed.; Wiley-Interscience: Great Britain, UK, 1974; pp. 543–552.

4. European Commission. Report on critical raw materials and the circular economy. In *Commission Staff Working Document*; European Commission: Brussels, Belgium, 2018.

5. Kempe, U.; Möckel, R.; Graupner, T.; Kynicky, J.; Dombon, E. The genesis of Zr-Nb-REE mineralisation at Khalzan Buregte (Western Mongolia) reconsidered. *Ore Geol. Rev.* **2015**, *64*, 602–625. [CrossRef]

6. Richardson, D.G.; Birkett, T.C. Peralkaline rock-associated rare metals. In *Geology of Canadian Mineral Deposit Types*; Eckstrand, O.R., Sinclair, W.D., Thorpe, R.I., Eds.; Geological Survey of Canada: Ottawa, ON, Canada, 1995; Volume 8, pp. 523–540.

7. Motzfeldt Project—A Future World Class Multi-Element Project, RAM Resources Limited. Available online: https://www.govmin.gl/images/stories/minerals/events/perth_dec_2010/Ram_Resources.pdf (accessed on 24 January 2018).

8. Gysi, A.P.; Williams-Jones, A.E.; Collins, P. Lithogeochemical vectors for hydrothermal processes in the Strange Lake peralkaline granitic REE-Zr-Nb deposit. *Econ. Geol.* **2016**, *111*, 1241–1276. [CrossRef]

9. Chakrabarty, A.; Mitchell, R.H.; Ren, M.; Pal, S.; Pal, S.; Sen, A.K. Nb-Zr-REE re-mobilization and implications for transitional agpaitic rock formation: Insights from the Sushina Hill complex, India. *J. Petrol.* **2018**, *59*, 1899–1938. [CrossRef]

10. Melluso, L.; Guarino, V.; Lustrino, M.; Morra, V.; De'Gennaro, R. The REE- and HFSE-bearing phases in the Itatiaia alkaline complex (Brazil) and geochemical evolution of feldspar-rich felsic melts. *Mineral. Mag.* **2017**, *81*, 217–250. [CrossRef]

11. Sørensen, H. *The Alkaline Rocks*; Wiley-Interscience: Great Britain, UK, 1974; p. 622.

12. Kempe, U.; Götze, J.; Dandar, S.; Habermann, D. Magmatic and metasomatic processes during formation of the Nb-Zr-REE deposits Khaldzan Buregte and Taskhir (Mongolian Altai): Indications from a combined CL-SEM study. *Mineral. Mag.* **1999**, *63*, 165–177. [CrossRef]

13. Kovalenko, V.I.; Tsaryeva, G.M.; Goreglyad, A.V.; Yarmolyuk, V.V.; Troitsky, V.A.; Hervig, R.L.; Farmer, G.L. The peralkaline granite-related Khadzan-Buregtey rare metal (Zr, Nb, REE) deposit, Western Mongolia. *Econ. Geol.* **1995**, *90*, 530–574. [CrossRef]

14. Kovalenko, V.I.; Kozlovski, A.M.; Yarmolyuk, V.V. Trace element ratios as indicators of source mixing and magma differentiation of alkali granitoids and basites of the Haldzan-Buregtey massif and the Haldzan-Buregtey rare-metal deposit, western Mongolia. *Petrology* **2009**, *17*, 227–252. [CrossRef]

15. Andeeva, I.A. Genesis and mechanisms of formation of rare-metal peralkaline granites of the Khaldzan Buregtey massif, Mongolia: Evidence from melt inclusions. *Petrology* **2016**, *24*, 462–476. [CrossRef]

16. Google LLC. *Google Earth Pro Map Data: Image Digital Globe Landsat/Copernicus*; Google LLC: Mountain View, CA, USA, 2019.

17. Gonvindaraju, K. 1994 Compilation of working values and sample descriptions for 383 geostandards. *Geostandard. Geoanalytical Res.* **1994**, *18*. [CrossRef]
18. Sindern, S.; Meyer, F.M. Automated quantitative rare earth elements mineralogy by scanning electron microscopy. In *Handbook of Rare Earth Elements–Analytics*; Golloch, A., Ed.; De Gruyter: Berlin, Germany, 2016.
19. Heilbronner, R.; Barrett, S. *Image Analysis in the Earth Science—Microstructures and Textures of Earth Materials*; Springer: Berlin, Germany, 2014. [CrossRef]
20. Streckeisen, A.L. To each plutonic rock its proper name. *Earth Sci. Rev.* **1976**, *12*, 229–253. [CrossRef]
21. De la Roche, H.; Leterrier, J.; Grandclaude, P.; Marcal, M. A classification of volcanic and plutonic rocks using R1-R2 diagram and major element analyses—Its relationships and current nomenclature. *Chem. Geol.* **1980**, *29*, 183–210. [CrossRef]
22. Middlemost, E.A.K. Naming materials in the magma/igneous rock system. *Earth Sci. Rev.* **1994**, *37*, 215–224. [CrossRef]
23. McDonough, W.F.; Sun, S.-S. The composition of the Earth. *Chem. Geol.* **1995**, *120*, 229–253. [CrossRef]
24. Deer, W.A.; Howie, R.A.; Zussman, J. *An Introduction to the Rock Forming Minerals*, 3rd ed.; Mineralogical Socitey: London, UK, 2013.
25. Lund, C.; Lamberg, P. Geometallurgy—A tool for better resource efficiency. *Eur. Geol.* **2014**, *37*, 39–43.
26. Dominy, S.; O'Connor, L.; Parbhakar-Fox, A.; Glass, H.J.; Purevgerel, S. Geometallurgy—A route to more resilient mine operations. *Minerals* **2018**, *12*, 560. [CrossRef]
27. Compan, G.; Pizarro, E.; Videla, A. Geometallurgical model of a copper sulphide mine for long-term planning. *J. South. Afr. Inst. Min. Metall.* **2015**, *115*, 549–556.
28. Wills, B.A. *Mineral Processing*, 6th ed.; Butterworth Heinemann: Oxford, UK, 1992.
29. Katzmarzyk, J.; Gronen, L.; Wotruba, H.; Sindern, S.; Hellmann, A.; Meyer, F.M. Geometallurgical investigation of the processing of REE-Y-Nb-Zr complex ore. In Proceedings of the IMPC International Mineral Processing Congress, Moscow, Russia, 15–21 September 2018; Abstract No. 2009-66.
30. Yang, X.; Satur, J.V.; Sanematsu, K.; Laukkanen, J.; Saastamoinen, T. Beneficiation studies of a complex REE ore. *Miner. Eng.* **2015**, *71*, 55–64. [CrossRef]
31. Bond, F.C. Crushing & Grinding Calculations Part I. *Br. Chem. Eng.* **1951**, *622*, 378.
32. Gregory, M.J.; Lang, J.R.; Gilbert, S.; Hoal, K.O. Geometallurgy of the peppble porphyry copper-gold-molybdenum deposit, Alaska: Implications for gold distribution and paragenesis. *Econ. Geol.* **2013**, *108*, 463–482. [CrossRef]
33. Hoal, K.O. Getting the geo in geomet. *SEG Newslett.* **2008**, *73*, 11–15.
34. Walters, A.S.; Goodenough, K.M.; Hughes, H.S.R.; Roberts, N.M.W.; Gunn, A.G.; Rushton, J.; Lacinska, A. Enrichment of rare earth elements during magmatic and post-magmatic processes: A case study from Loch Loyal Syenite, northern Scotland. *Contrib. Miner. Petrol.* **2013**, *166*, 1177–1202. [CrossRef]
35. Schmitt, A.K.; Trumbull, R.B.; Dulski, P.; Emmermann, R. Zr-Nb-REE mineralization in peralkaline granites from the Amis complex, Brandberg (Namibia): Evidence for magmatic pre-enrichment from melt inclusions. *Econ. Geol.* **2002**, *97*, 399–413. [CrossRef]

Article

Automated Quantitative Mineralogy Optimized for Simultaneous Detection of (Precious/Critical) Rare Metals and Base Metals in A Production-Focused Environment

Mathis Warlo [1,*], Christina Wanhainen [1], Glenn Bark [1], Alan R. Butcher [2], Iris McElroy [3], Dominique Brising [3] and Gavyn K. Rollinson [4]

1   Division of Geosciences and Environmental Engineering, Luleå University of Technology, SE-971 87 Luleå, Sweden
2   Geological Survey of Finland/Geologian tutkimuskeskus, F1-02151 Espoo, Finland
3   Boliden Mineral AB, SE-936 81 Boliden, Sweden
4   Camborne School of Mines, University of Exeter, Cornwall Campus, Penryn, Cornwall TR10 9EZ, UK
*   Correspondence: mathis.warlo@ltu.se; Tel.: +46-920-493-940

Received: 17 June 2019; Accepted: 15 July 2019; Published: 18 July 2019

Abstract: Automated Scanning Electron Microscopy (ASEM) systems are applied in the mining industry to quantify the mineralogy of the ore feed and products. With society pushing towards sustainable mining, this quantification should be comprehensive and include trace minerals since they are often either deleterious or potential by-products. Systems like QEMSCAN® offer a mode for trace mineral analysis (TMS mode); However, it is unsuitable when all phases require analysis. Here, we investigate the potential of detecting micron-sized trace minerals in fieldscan mode using the QEMSCAN® system with analytical settings in line with the mining industry. For quality comparison, analysis was performed at a mining company and a research institution. This novel approach was done in full collaboration with both parties. Results show that the resolution of trace minerals at or below the scan resolution is difficult and not always reliable due to mixed X-ray signals. However, by modification of the species identification protocol (SIP), quantification is achievable, although verification by SEM-EDS is recommended. As an add-on to routine quantitative analysis focused on major ore minerals, this method can produce quantitative data and information on mineral association for trace minerals of precious and critical metals which may be potential by-products in a mining operation.

Keywords: automated scanning electron microscopy; QEMSCAN®; trace minerals; gold

1. Introduction

As the need for sustainability in mining is becoming increasingly important amongst the public, decision makers and the industry itself, detailed investigations into what ore deposits actually contain in terms of various minerals, potential by-products and industrial minerals are needed. By making use of a larger proportion of the mined ore (recovering also the trace metals) mining operations can be more sustainable and this will potentially also be beneficial in gaining public acceptance for mining (social license to operate). Precious metals such as Au and Ag are already readily produced as main- or by-products in many mining ventures even if they occur only in trace amounts and numerous metals could potentially follow the precious metals as by-products in a number of mining operations worldwide. Many European ore deposits contain various amounts of trace metals classified as "Critical Raw Material" (CRM) by the European Commission, i.e., they are of high economic importance for the EU but with a high risk associated with their supply [1].

While trace metal production may not be economically profitable at the moment this could change in the future as metal prices increase and the pressure is increasing for more sustainable mining. Hence, for sustainability as well as economic reasons, the trace mineral and metal characterization of an ore deposit should be considered when planning for a mining operation, and also in operating mines.

In order to predict trace metal deportment during processing of the ore, a good understanding of their mineralogical and textural characterization is necessary. Due to their low abundance in ore deposits and often fine-grained (<50 μm) nature, identification and quantification of trace minerals is difficult and requires the use of advanced micro-analytical techniques. Many capable techniques were developed over the last few decades, each with its own advantages and disadvantages [2], but especially, Automated Scanning Electron Microscopy (ASEM) systems found wide-spread acceptance and application in the mining industry. These systems provide fast and reliable quantification of the mineralogy and textures in a sample. Most prominent are the Quantitative Evaluation of Mineralogy by a SCANning electron microscope (QEMSCAN®) system and the Mineral Liberation Analyzer (MLA) system [3,4].

The QEMSCAN® system is the third generation of automated mineral analysis systems based on the then-named QEM*SEM® system. QEM*SEM®, was developed at the request of the mining industry by the Commonwealth Scientific and Industrial Research Organization (CSIRO) in Australia in the 1970s and marked the first automated mineral analyzer [5,6]. The MLA system is based on the concepts of Hall [7] and became commercially available in 2000 through FEI Company and JKTech, whilst QEMSCAN® was marketed by Intellection Pty Ltd. Both systems, QEMSCAN® and MLA, utilize backscattered electrons (BSE) and energy dispersive X-ray spectra (EDS) to create digital mineral images. In the QEMSCAN® system, low-count X-ray mapping is preferentially used for mineral classification. This is done by comparison of the X-ray element-spectra to existing mineral databases. BSE brightness is used to distinguish particles from the mounting media. A summary of the QEMSCAN® system and its various application modes is provided by Gottlieb et al. [3], Goodall et al. [8], and Pirrie and Rollinson [9]. By contrast, in the MLA system, particles are often defined through the BSE brightness and subsequently classified by one X-ray spectrum per particle. For particles of similar BSE brightness, X-ray mapping is used. The MLA system is described in detail by Gu [4]. While both systems are still widely applied in the industry and by research institutions, their commercial production has currently been terminated. This has given rise to new ASEM systems, most prominently the ZEISS Mineralogic Mining system and the TESCAN Integrated Mineral Analyzer (TIMA) [10]. These systems come with some improvements, such as faster speed of analysis. An introduction to the ZEISS Mineralogic Mining system is provided by Graham et al. [11] and references therein, and the principles and applications of the TIMA system are described by Hrstka et al. [10].

In the mining industry, ASEM systems are mostly applied for routine scans of particulate samples of ore and tailings concentrate to identify and quantify the mineralogy of the ore feed and products. Instruments are typically calibrated for fast acquisition rates to enable a high sample throughput. This comes at the expense of precision and resolution. As a consequence, trace minerals are often undetected due to their grain size being at or below the scan resolution. Most ASEM systems provide an analytical mode targeted towards the analysis of trace minerals (such as TMS mode for QEMSCAN®) and its usefulness for characterization of, e.g., Au has been demonstrated [8,12–14]. However, this analytical mode hardly finds application for deposits where trace minerals are only by-products, or of no current economic interest. Furthermore, since the TMS mode utilizes a BSE brightness threshold to filter for trace minerals, it is hardly applicable if the sample is enriched in Pb- and Bi-minerals due to their similar BSE brightness compared to Au-minerals. This forces the system to analyze many more particles than necessary (so-called false particles) and is thus more time-consuming. Hence, there is a need to improve detection of trace phases in general analyses.

Here, we compare results of analysis of a polished thin section from a Cu-(W-Au) ore between two QEMSCAN® systems; one is an industry-system (Boliden AB), the other a research-system (Camborne School of Mines). For the research-oriented scan, the setup of the analysis was thoroughly planned and much time was spent on the post-processing of the raw data, so this scan is assumed to be of the highest quality and used as a measure for the relative quality of the routine industrial scan. The goal was to determine the overall quality of general routine industrial scans and the possibility of detecting and quantifying trace phases at or below the scan resolution. Ideally, a scan should provide a basic idea of trace mineral mineralization in a sample and help the operator/decision maker decide if more detailed analysis is worth pursuing. In this case, the trace mineral Au was used to find an optimum methodology for detecting and quantifying trace minerals but the methodology presented applies to all trace minerals in an ore body. A guide towards analysis is provided. This analysis is novel in its collaborative inter-lab comparison between the industry and a scientific institution.

2. Materials and Methods

To promote analysis of uncrushed rock samples by ASEM systems prior to processing, a thin section sample of a drill core was chosen instead of a particulate sample for this analysis, despite the more common usage of particulate samples in the mining industry. Uncrushed samples have a higher uncertainty of representativeness, but they allow the study of original features like mineral distribution, structures and textures which carry important information for processing of the ore but are partly lost during crushing. Hence, for comprehensive ore characterization to aid in early mine planning, uncrushed rock samples are most suitable. To limit uncertainties, appropriate sampling and sufficient volume are necessary [15], in fact, possibilities to routinely scan drill core pieces are currently tested at Boliden AB. Furthermore, ideally, analysis should be performed on both crushed and uncrushed rock samples.

The sample selected was from the Liikavaara Cu-(W-Au) deposit, an intrusion-related vein-style deposit in Northern Sweden (Figure 1a,b), located close to the world-class Aitik Cu-Au deposit where the Liikavaara ore will be processed. Chalcopyrite, pyrite and pyrrhotite constitute the major ore minerals. Sphalerite, galena, scheelite, molybdenite, marcasite and magnetite are minor. The deposit also hosts several trace metals including Au, Ag, Bi and Sn which commonly occur in fine-grained minerals (<20 µm) [16]. The trace metal mineralogy is presented in Table 1, and a detailed description of the geology and mineralogy of the deposit is given by Zweifel [17] and Warlo et al. [16].

The deposit is currently in the pre-production stage, and production is estimated to start in 2023. Copper will be the primary commodity and Au and Ag will be by-products. Production of W, despite its enrichment and classification as a CRM, would require an additional processing step and is thus unprofitable at present. Bismuth is known for its potential to contaminate and lower the quality of the Cu concentrate, thus having good control over its mineralogy and distribution is beneficial.

The pre-production stage of the Liikavaara deposit, its enrichment in several trace metals of interest, a diverse fine-grained mineralogy, and previous studies, make the Liikavaara Cu-(W-Au) deposit an ideal candidate for this type of study.

Table 1. Trace metal mineralogy of the Liikavaara Cu-(W-Au) deposit.

Trace Metal	Mineralogy
Au	native Au, electrum (Au-Ag-alloy, Ag > 20%)
Ag	hessite (Ag_2Te), acanthite (Ag_2S), native Au, electrum
Bi	native Bi, pilsenite (Bi_4Te_3), tetradymite (Bi_2Te_2S), bismuthinite (Bi_2S_3)
Sn	cassiterite (SnO_2)

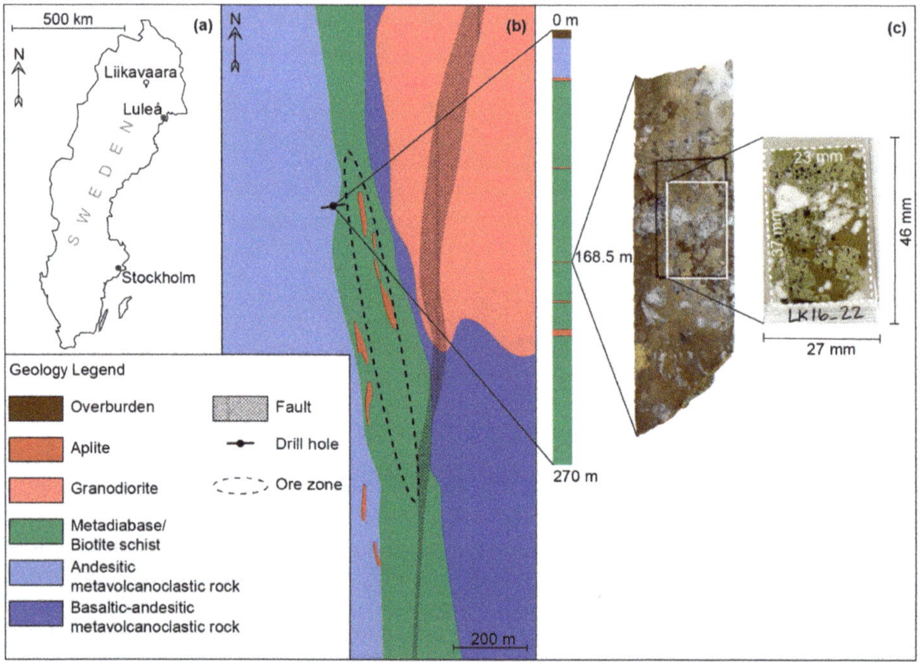

Figure 1. (**a**) Location of the Liikavaara Cu-(W-Au) deposit in Northern Sweden; (**b**) geological map of the Liikavaara deposit at 100 m below surface. Location of the drill hole for the sample analyzed in this study is shown; (**c**) thin section prepared from a drill core intersecting a mineralized quartz vein within an aplite dike. Modified from Warlo et al. [16].

The selected core sample (mineralized quartz vein from the proximal ore zone) was prepared into a polished thin section of 27 × 46 mm with a sample size of 23 × 37 mm (Figure 1c). In the corresponding other half of the drill core, an Au-grade of ca. 6 ppm was measured over a 1.3 m section.

The sampled vein is composed of quartz with minor tourmaline and scattered patches altered by fine-grained (<20 μm) calcite and chlorite (Figure 2a,e). It is strongly mineralized by pyrite and pyrrhotite, and by minor chalcopyrite and sphalerite (Figure 2b,f). Pyrite and pyrrhotite vary in grain sizes from a few microns to several centimeters in width. Grains are often fractured but pyrite retains a subhedral shape (Figure 2b,f). Chalcopyrite and sphalerite exist mostly as crack fillings and along grain boundaries in pyrite and quartz, but are also associated with tourmaline and disseminated (<50 μm) in areas altered by calcite and chlorite (Figure 2b,f). Several grains of scheelite (>1 cm), and one 400 μm grain of pilsenite, are observed (Figure 2c–e). SEM-BSE imaging coupled with EDS analysis revealed the occurrence of native bismuth, hessite, bismuthinite and electrum. Grains were mostly below 10 μm in size (Figure 3a–d).

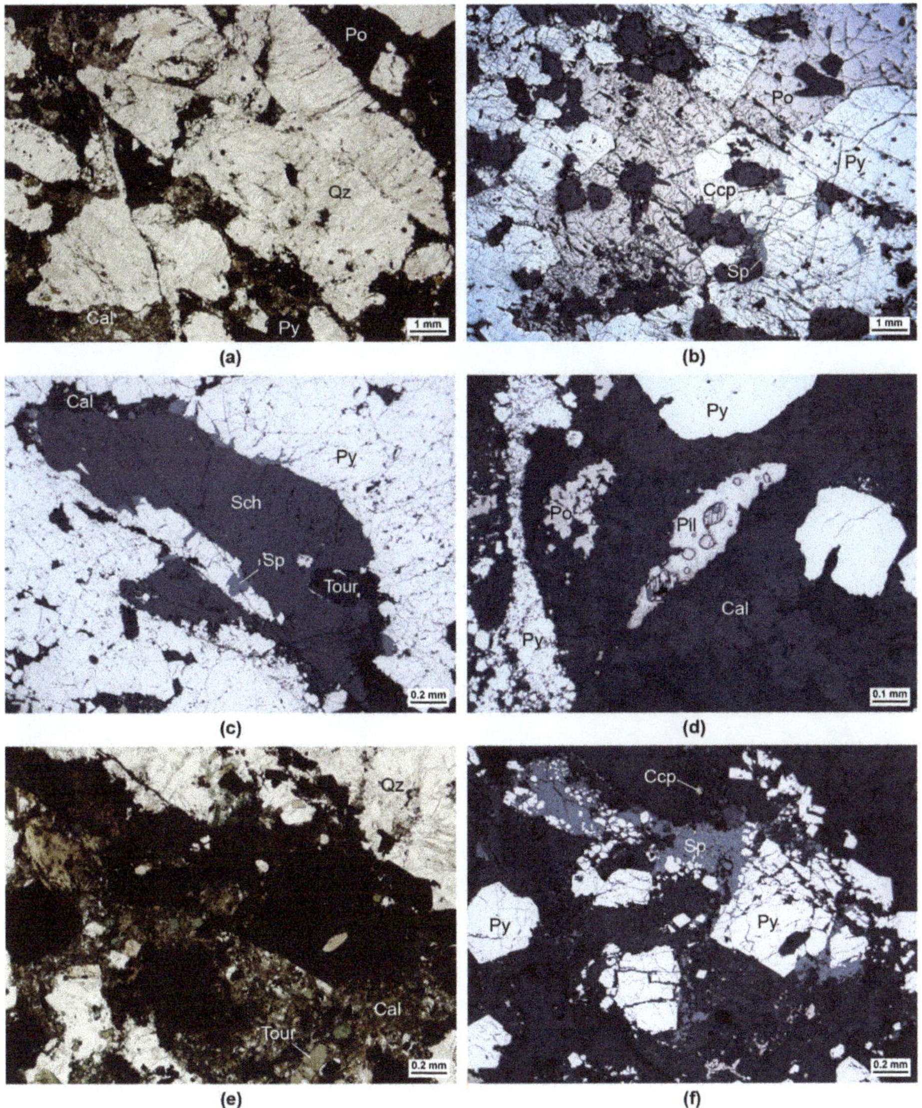

Figure 2. Petrographic images of the thin section analyzed by QEMSCAN® in this study. (**a**,**e**) are plane polarized light images, (**b**–**d**,**f**) are reflected light images; (**a**) grains of quartz surrounded by patches of calcite, and pyrrhotite and pyrite; (**b**) massive pyrrhotite and subhedral pyrite. Sphalerite and chalcopyrite occur along the edges and in cracks of pyrite; (**c**) scheelite grain within pyrite; (**d**) grain of pilsenite with patches of pyrrhotite surrounded by calcite; (**e**) and (**f**) assemblage of sphalerite, pyrite, calcite and tourmaline surrounded by quartz. Abbreviations: Cal—calcite, Ccp—chalcopyrite, Pil—pilsenite, Po—pyrrhotite, Py—pyrite, Qz—quartz, Sch—scheelite, Sp—sphalerite, Tour—tourmaline.

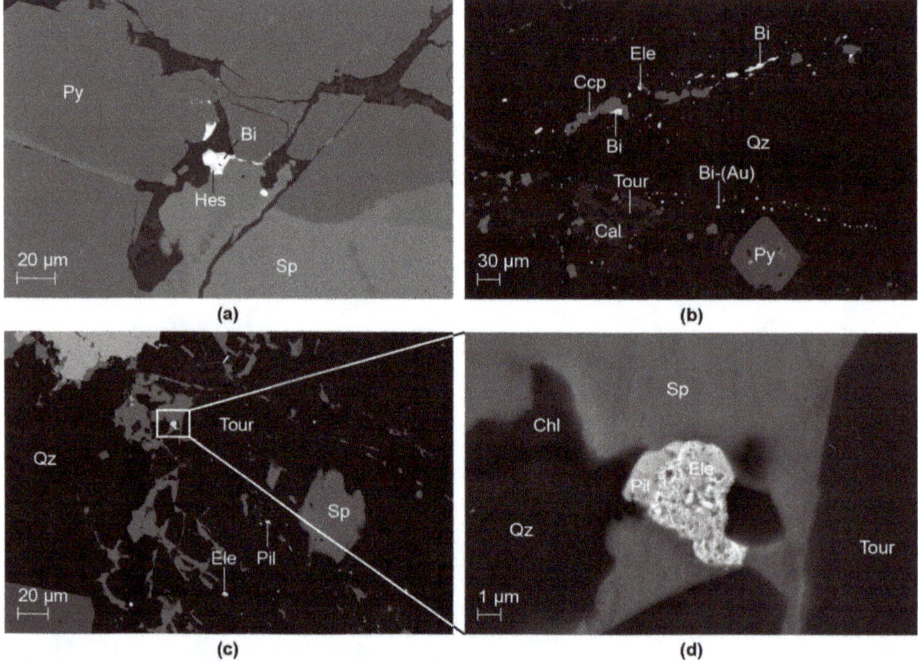

Figure 3. Backscattered (**a–c**) and secondary (**d**) electron images of the thin section prior to QEMSCAN®
analysis; (**a**) intergrowth of native Bi and hessite in a crack between grains of pyrite and sphalerite;
(**b**) droplet-shaped grains of native bismuth with Au in quartz; (**c**) grains of electrum at the border
of sphalerite and tourmaline, respectively; (**d**) magnified image of (**c**). Abbreviations: Au—gold,
Bi—native bismuth, Cal—calcite, Ccp—chalcopyrite, Chl—chlorite, Ele—electrum, Hes—hessite,
Pil—pilsenite, Py—Pyrite, Qz—quartz, Sp—sphalerite, Tour—tourmaline.

Petrography of the sample prior to QEMSCAN® analysis was carried out with a petrographic
microscope (Nikon ECLIPSE E600 POL) in transmitted and reflected light, and with a scanning electron
microscope (Zeiss Merlin FEG-SEM) at Luleå University of Technology. The same SEM was used for
verification of the trace minerals detected by the QEMSCAN® analyses.

The polished thin section was first analyzed with the QEMSCAN® system at Camborne School of
Mines (CSM), University of Exeter, Penryn, UK, to comprehensively characterize the mineralogy of
the sample with emphasis on the detection and identification of trace metal minerals. This consists
of a QEMSCAN® 4300 (Zeiss EVO®50 SEM with W-filament, four EDS, and an electron backscatter
detector) using iMeasure version 4.2 SR1 software for data collection, and iDiscover 4.2SR1 and 4.3
software for data processing. The sample was carbon coated to 25 nm at CSM prior to analysis. The
fieldscan measurement mode was performed at an X-ray resolution of 10 μm using a horizontal field
width of 1500 μm (150 × 150 analysis points per field), with a measurement area of approximately
19 mm × 35.5 mm (Figure 4), resulting in ~7 million analysis points and a scan time of 10:20 h. The
X-ray count per pixel used the default of 1000 counts. For mineral identification, a modified version of
the standard LCU5 Species Identification Protocol (SIP) was used, following the guidance in Section 7
of Rollinson et al. [18]. During data processing, particular emphasis was placed on the trace metal
minerals to enable identification of these and take into account their small size (some were at the single
pixel scale), which results in mixed spectra. This included electrum, bismuth minerals, molybdenite
and the silver minerals. However, the SIP (mineral database) was customized to the entire sample, to
ensure all the minerals in the sample were identified as accurately as possible, which involved checking

all the minerals present and developing the database entries as required. This, for example, involved improving existing entries, adding boundary categories for existing minerals caused by mixed spectra, and adding new entries for the trace metal minerals to ensure they were as accurately captured as possible given their small size.

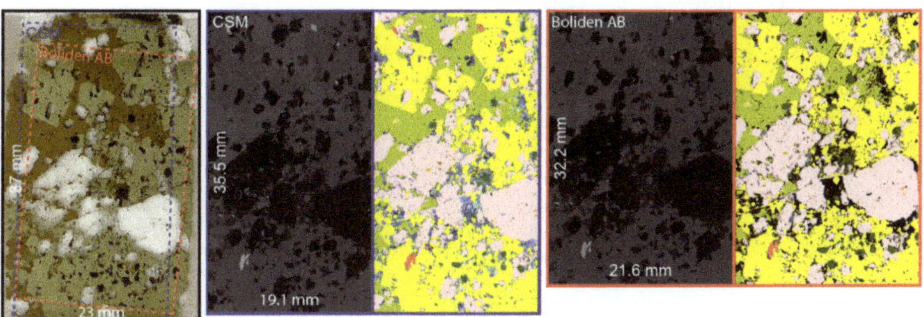

Figure 4. Optical scan of the sample and corresponding QEMSCAN® images (backscattered electrons (BSE) and mineral map) from Camborne School of Mines (CSM) and Boliden AB.

The same thin section was then measured at Boliden AB, Boliden, Sweden with a similar objective. However, settings were chosen to reflect a routine industrial application. At Boliden AB, a QEMSCAN® 650 (FEI with W-filament, two EDS, and an electron backscatter detector) was used with iMeasure version 5.4 software for data collection and iDiscover 5.4 software for data processing. The fieldscan measurement mode was performed at an X-ray resolution of 5 μm using a horizontal field width of 1500 μm (300 × 300 analysis points per field), with a measurement area of approximately 21.5 mm × 32 mm (Figure 4), resulting in ~24.6 million analysis points and a scan time of 23:50 h. The X-ray count per pixel used the default of 1000 counts. For mineral identification, a custom SIP for the Aitik deposit, based on several scientific and in-house mineralogical studies, was modified and adapted to the mineralogy of the Liikavaara Cu-(W-Au) deposit. After the measurement, an initial search for unknown phases was performed and corresponding minerals added to the SIP. This was followed by a data processing routine. Comparison of the results with the analysis at CSM led to application of the "boundary phase processor" and to several more additions to the mineral list (especially for Au-phases) to improve data quality (see Section 3).

3. Results

3.1. Results Prior to Optimisation

Results from both QEMSCAN® analyses, at CSM and Boliden AB, respectively, showed, overall, a good agreement with earlier detailed petrographic and SEM studies; however, there were some key differences between the data sets.

3.1.1. Measured Area

The areas scanned by QEMSCAN® differed in size between CSM and Boliden AB and did not cover the entire sample (Figure 4). Limitations were set by the design of the sample holders. Furthermore, it was not possible to match the produced BSE and mineral maps of Boliden AB and CSM through image manipulation such as cutting, rotating, and stretching. The reason for this is not certain, but could be an issue with magnification calibration during stitching of the individual fields to a unified image for either operator. These issues prevented full quantitative comparison of the modal mineralogy and mineral association between the two datasets.

3.1.2. BSE Map

Several fields of the BSE map from Boliden AB showed a shift in brightness and contrast to the surrounding fields which was not observed in the CSM map (Figure 5). This potentially indicates a poor vacuum condition in the SEM chamber during measurement, or beam fluctuation; however, no impact on the mineral map was observed.

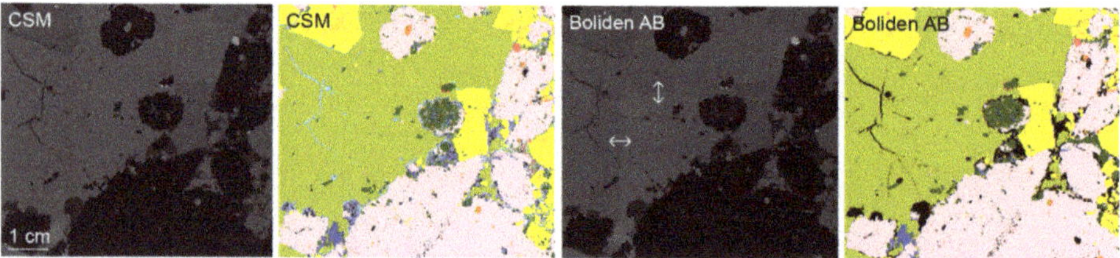

Figure 5. BSE image and corresponding mineral map of a part of the thin section analyzed in this study for CSM (left) and Boliden AB (right). A grey-level shift in the BSE image of Boliden AB can be observed (indicated by arrows). This shift was not observed for CSM and is not reflected in the mineral map.

3.1.3. Mineralogy

The QEMSCAN® analysis by CSM agreed with the mineralogy described by light optical microscopy and SEM of the selected sample (see Section 2). Major and minor sulfides and silicates were confirmed. Electrum, hessite, and Bi-minerals were all detected (Figures 6 and 7d). Additionally, traces of molybdenite, cassiterite, uraninite and Ag-minerals (other than hessite) were detected. Many of these phases mark single pixels in the mineral map. The fine-grained clusters of calcite observed during petrographic analysis were resolved to be complex intergrowths of calcite, ankerite and Fe-Ox/CO_3 (mostly goethite) (Figure 7d,e). Bi-minerals were categorized into Bi-tellurides and native bismuth/bismuthinite. No differentiation was made between Bi-tellurides (e.g., pilsenite and tetradymite). Similarly, native bismuth and bismuthinite were grouped together due to difficulties in separating Bi and S at the scan resolution in QEMSCAN® analysis. A total of seven pixels were identified as electrum (Figure 6). Follow-up work with the SEM-EDS confirmed these pixels correlated with six different grains of native Au and electrum in the sample. The grains were typically associated with quartz and often found within areas rich in Bi-minerals. The Au grains ranged in size from 6 × 12 μm to 2 × 3 μm (Figure 6). This means, despite their sub 10 μm grain sizes, a 10 μm resolution scan was sufficient to detect them. Further SEM-EDS studies revealed several more native gold and electrum grains (mostly <5 μm) in the immediate surrounding of the grains detected by the QEMSCAN® scan. Additionally, sub 5 μm droplet-shaped grains of native bismuth with partitioning of gold were frequently observed [19]. None of the pixels corresponding to these grains were identified as Au-minerals by the QEMSCAN® analysis of CSM.

Mineralogy was less detailed for the Boliden AB data set. Only 13 phases were distinguished by Boliden AB compared to 23 phases at CSM (which included all 13 phases from Boliden AB) (Figure 7). Minor sulfides were compiled under "other sulfides". Similarly, silicates were grouped together with the exception of quartz, tourmaline and chlorite. For trace metals, only "Bi-minerals" and "Au, Ag minerals" were distinguished but no pixels were identified for the latter. This means that the routine industry scan did not detect any gold grains from the analysis.

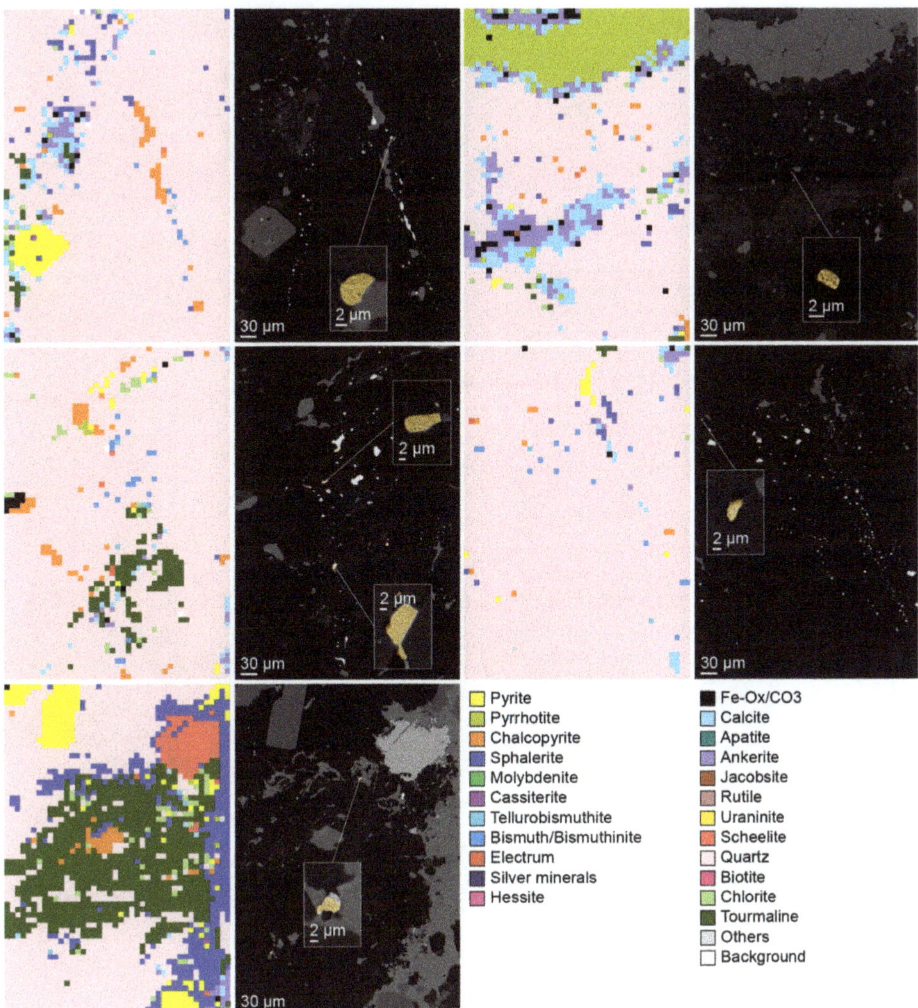

Figure 6. Areas of the QEMSCAN®mineral map by CSM (left) containing pixels identified as electrum (red) and corresponding backscattered electron images of the same areas by FEG-SEM (right). Gold grains detected by CSM and Boliden AB (after reprocessing) are false-colored in the BSE images. The Au grains detected by the CSM scan are also magnified. The magnified images are secondary electron images recorded by FEG-SEM.

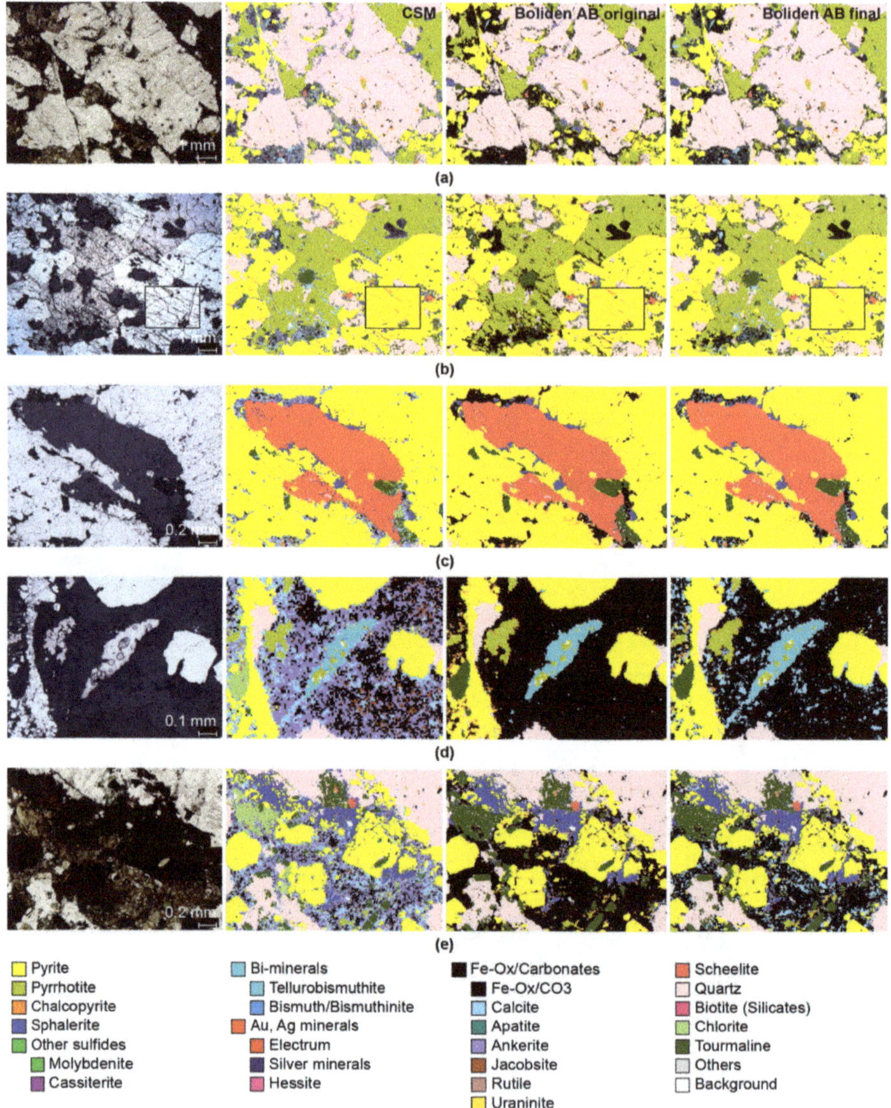

Figure 7. (**a–e**) Optical microscopy images and corresponding areas in the QEMSCAN® mineral maps of CSM and Boliden AB. In (b), the smaller squares show a magnified image of the underlying area to highlight differences in the micro fracture density between the optical images and the mineral maps. The CSM map includes all the minerals presented in the legend. The original Boliden AB map does not subdivide groups and uses "Silicates" instead of "Biotite". The final version by Boliden AB uses the same mineral list as CSM but lacks "Ankerite", "Jacobsite" and subdivision of "Au, Ag minerals". It includes application of the "boundary phase processor" on chalcopyrite and a single-element entry to the SIP (database) with a 25% intensity threshold for Au.

3.1.4. Textures

Texturally, mineral maps from both Boliden AB and CSM reflected the recorded BSE image. The scan resolution was higher for Boliden AB (5 μm pixel size compared to 10 μm for CSM) but in both maps shapes of grains, cracks, and inclusions were clearly visible (Figure 7). Only the fine-grained clusters of calcite, ankerite and goethite were not resolved but marked as a single phase for Boliden AB due to their less-detailed mineral list in which these minerals were simply categorized as Fe-Ox/Carbonates (Figure 7d,e). Micro cracks filled with chalcopyrite in pyrite and quartz were resolved for both data sets, but in higher detail in the CSM mineral map (Figure 7a,b). However, comparison to reflected light images showed that many fine cracks were lost in the BSE images and the mineral maps, for both data sets (Figure 7a,b).

3.1.5. Modal Mineralogy

A qualitative comparison between modal mineralogy calculated for the data sets from CSM and Boliden AB showed variation between the data sets (Table 2, columns "CSM" and "Boliden AB (original)"). For the major and minor ore minerals, a ~4 % higher mineral mass of pyrrhotite for CSM and half the mineral mass of chalcopyrite are notable. Other differences were the variations in silicates, especially chlorite and tourmaline, and a higher mass of unidentified phases for Boliden AB ("Others" 0.45 % compared to <0.01% for CSM). Also, as mentioned above, no Au- or Ag-minerals were detected for Boliden AB.

Table 2. Modal mineralogy based on mineral mass determined by QEMSCAN®. Data is normalized and quoted to two decimal places only (100 ppm). The columns show the results of the analyses by CSM and for Boliden AB before and after reprocessing. For the Boliden AB data, some minerals were grouped together. The 1st update of the Boliden data includes an extension of the mineral list. The 2nd and 3rd updates include the "boundary phase processor" and a specific SIP-entry for Au. Phases with "N.A." were not included in the Boliden AB SIP and hence, not analyzed.

Modal Mineralogy (Mineral Mass %)	CSM	Boliden AB (Original)	Boliden AB (1st Update)	Boliden AB (2nd + 3rd Update)
Chalcopyrite $CuFeS_2$	0.72	1.58	1.58	0.80
Sphalerite $(Zn,Fe)S$	0.91	0.82	0.82	0.79
Pyrite FeS_2	48.84	49.16	49.16	49.94
Pyrrhotite $Fe_{1-x}S$	19.08	15.42	15.42	15.42
Other sulfides		<0.01		
Molybdenite MoS_2	<0.01		<0.01	<0.01
Cassiterite SnO_2	0.01		<0.01	<0.01
Bi-minerals		0.02		
Tellurobismuthite Bi_2Te_3	0.02		0.02	0.02
Bismuth/Bismuthinite Bi/Bi_2S_3	<0.01		<0.01	<0.01
Au, Ag minerals		0	0	<0.01
Electrum Au-Ag alloy	<0.01			
Silver minerals	<0.01			
Hessite Ag_2Te	<0.01			
Scheelite $CaWO_4$	0.67	0.51	0.51	0.50
Quartz SiO_2	18.02	18.43	18.43	18.56
Biotite K-Mg-Fe-Silicate	0.13	*0.04	<0.01	<0.01
Chlorite Mg-Fe-Silicate	0.62	<0.01	<0.01	<0.01
Tourmaline Ca-K-Na-Silicate	2.82	3.98	3.98	3.97
Fe-Ox/Carbonates		9.58		
Uraninite UO_2	<0.01		<0.01	<0.01
Rutile TiO_2	0.01		<0.01	<0.01
Jacobsite $MnFe_2O_4$	0.14		N.A.	N.A.
Fe-Ox/CO3	1.96		3.28	7.54
Calcite $CaCO_3$	2.89		6.30	1.94
Ankerite $Ca(Fe,Mg,Mn)(CO_3)_2$	3.15		N.A.	N.A.
Apatite $Ca_5(PO_4)_3(F,Cl,OH)$	0.01		0.01	<0.01
Others	<0.01	0.45	0.49	0.51

* the category "Silicates" was used instead of "Biotite".

For the silicates, the variations were likely caused by differences in pixel classification, e.g., grains classified as chlorite by CSM were classified as tourmaline by Boliden AB (Figure 7d,e). This is a common issue caused by overlapping spectra of chlorite and tourmaline. To achieve good separation, calibration to a real tourmaline standard and cross-checking against real samples is required. Similar peaks are also an issue for pyrite and pyrrhotite. But neither the mineral maps nor the quantitative data suggested this to have been an issue in this study. Rather, the difference in the area analyzed was likely responsible for the variation of pyrrhotite.

3.2. Results past Optimisation

Based on the supposedly better-quality data from CSM, results from the Boliden AB analysis were reprocessed in several iterations in an attempt to achieve similar or higher quality data compared to those of CSM.

3.2.1. Mineral List

An attempt was made to reproduce the mineral list used at CSM (Figure 7). Ankerite and jacobsite could not be differentiated from the category "Fe-Ox/CO$_3$" and also, gold and silver minerals remained undifferentiated. The change of the phase of "silicates" to "biotite" resulted in a loss in mineral mass. An equivalent mass was gained in the "others" category (Table 2 column "Boliden AB (1st update)"). This shows, that most pixels initially classified as "silicates" did not meet the requirements to be classified as "biotite" in the Boliden AB data. While the total mass of carbonates, oxides and phosphates was similar between the data sets (9.58% for Boliden, 8.25% for CSM), separation showed a large variation between individual phases (Table 2, column "Boliden AB (1st update)"). Regarding the Bi-minerals, separation into "Tellurobismuthite" and "Bismuth/Bismuthinite" produced similar quantitative results. However, this was somewhat deceiving due to masses only being quoted to 100 ppm because of a significant risk of misidentification of grains at or below the scan resolution. In fact, for Boliden AB, hardly any "Bismuth/Bismuthinite" was found and most Bi-pixels were attributed to "Tellurobismuthite". Subsequent evaluation by SEM-EDS showed many of these pixels to actually be native Bi rather than Bi-tellurides. But even for the CSM data, the classification was not always correct.

3.2.2. Boundary Phase Processor

The Boliden AB data was reprocessed using the "boundary phase processor", which is a post-analysis processing tool that aims to counter false pixel identification caused by mixed spectra at grain boundaries and erroneous energy dispersive spectra. A spurious signal may be collected due to the electron beam being deflected from topography in the sample, e.g., holes, or due to sudden fluctuations in beam intensity. The method works by reclassifying individual pixels (to a mineral defined by the person processing the data), if the surrounding pixels are homogeneous and not of the same phase, e.g., a single pixel of chalcopyrite in a homogeneous pyrite grain. It also reclassifies pixels if they sit between two or more otherwise homogeneous phases, such as grain boundaries. The pixel is then either reclassified to the surrounding phase or to unknown. The method can be applied to individual phases. It is not possible for the method to distinguish between true and erroneous signals; therefore, there is a risk of wrongly reclassifying pixels. Thus, if and when the method is applied, it must be carefully assessed by the operator. Boliden AB applies this method frequently on a case by case basis if modification of the SIP is too time-intensive and/or yields little result. In their experience, the ratio of erroneous to true signals for such single pixels is mostly in favor of the error, hence warranting application of the "boundary phase processor" in most cases. While some errors could be fixed by improving the SIP, using the "boundary phase processor" is faster and easier.

Here, this method was applied on the chalcopyrite and trace phases. For chalcopyrite, it resulted in a drop in mineral mass from 1.58% to 0.80%, very close to the 0.72% from CSM (Table 2, column "Boliden AB (2nd + 3rd update)"). This means that about every second pixel that was originally identified as chalcopyrite was reclassified, mostly as pyrite and quartz. In the original mineral map, many pyrite grains appeared coated with a pixel-thick layer of chalcopyrite that was neither present in the mineral map of CSM nor in optical images. The chalcopyrite pixels were consequently reclassified as pyrite by the "boundary phase processor". However, many chalcopyrite-filled micro cracks within pyrite and quartz were also removed due to this method, despite them showing in the BSE and optical images (Figure 7a,b). For the trace phases, the change in bulk mineralogy was not noticeable due to most phases being <100 ppm. In the mineral map, the change was more apparent since many pixels of trace phases were reclassified to a major phase.

At CSM, issues with spurious signals were resolved through manual evaluation of the data and editing of the SIP. As a result, pyrite grains were uncoated and fine cracks filled with chalcopyrite were resolved. This shows the advantage of having an experienced operator spend time processing the data over relying on automated processing. Nevertheless, if either time or experience is lacking, the "boundary phase processor" is a helpful tool to improve the quality of analysis.

3.2.3. Gold Minerals

Next, the undetected Au-minerals in the Boliden AB analysis were targeted. In a first attempt, an entry in the SIP was created to identify any pixel with an Au-signal as Au. This entry was placed at the top of the SIP to guarantee that all pixels were filtered for Au. However, still, no pixels of Au were found. Manual inspection of some pixels that corresponded to Au-pixels in the CSM data revealed clear Au-signals, thus they should have been identified as Au based on the SIP. The problem was caused by the "boundary phase processor" used. Since Au-grains in the sample were dominantly below 5 µm in size, recorded Au-signals were mostly limited to single pixels. Consequently, processing of the data with the "boundary phase processor" reclassified all pixels that had been filtered as Au to their surrounding phase. Subsequent deactivation of the "boundary phase processor" for the Au-related SIP entries delivered a number of Au-pixels. However, evaluation by SEM-EDS revealed that many of the pixels were falsely classified, probably due to beam deflections. In order to exclude erroneous signals from being identified as Au, different threshold values for the Au-signal in the SIP were tested. Experimentally, the threshold was set to intensities of 20%, 30% and 40%, respectively, (compared to the Au elemental reference spectrum) and the results were verified by subsequent manual SEM analysis. The results are summarized in Table 3. At an intensity of 20%, 69 pixels were identified as Au, of which 20 were false. At an intensity of 30%, 30 pixels of Au were identified, with no errors. At an intensity of 40%, 20 pixels were identified as Au and no errors were found. Further tests with thresholds between 20% and 30% revealed that a threshold of 25% was the lowest possible intensity to avoid errors. At 25%, 39 pixels were classified as Au. A few pixels contained several <2 µm Au grains while some grains, >10 µm, were covered by more than one pixel.

Table 3. A range of Au signal intensity (SI) thresholds, defined in the SIP to identify pixels as Au, and their respective outcome after processing of the Boliden AB QEMSCAN® analysis. Pixels identified as Au include the errors.

Au SI Threshold (%)	Pixels Identified as Au	Errors	Missed Au Pixels (Compared to 20 % SI)
20	69	20	0
21	58	12	3
22	48	4	5
23	46	3	6
24	43	2	8
25	39	0	10
26	37	0	12
27	34	0	15
28	31	0	18
29	30	0	19
30	30	0	19
40	20	0	29

Figure 8 shows the Au grains detected by the Boliden AB analysis using different Au intensity threshold values (Table 3). As expected, there is a clear correlation between detectability and Au grain size. Generally, decreasing the threshold value allows detection of smaller grains because their excitation volume creates a weaker Au signal. However, it also increases the risk for errors. The detection limit is, on average, approximately 5%–10% per analysis point at 1000 counts and depends strongly on the element and the matrix [18,20]. At an optimal threshold value of 25% of the signal intensity (Au), several grains below 2 μm were detected. Compared to the six Au grains detected by CSM, four were detected even at a threshold of 40% in the Boliden AB data (Figure 7a,c,d), one at a threshold of 21% (Figure 7b), and one was not detected at all despite a grain size of ~5 μm (Figure 7e). It is possible that the undetected Au grain lay just between two beam spots and thus was not measured. Another explanation could be its close spatial association with pilsenite and sphalerite, whose signals may have overlaid the Au signal. In contrast, two grains detected by Boliden AB were large enough that the 10 μm scan from CSM could have picked them up too, but missed them. Manual SEM inspection of the CSM mineral map showed that one grain was classified as pyrite and the other as rutile, despite none of these minerals being in the vicinity of the respective pixels.

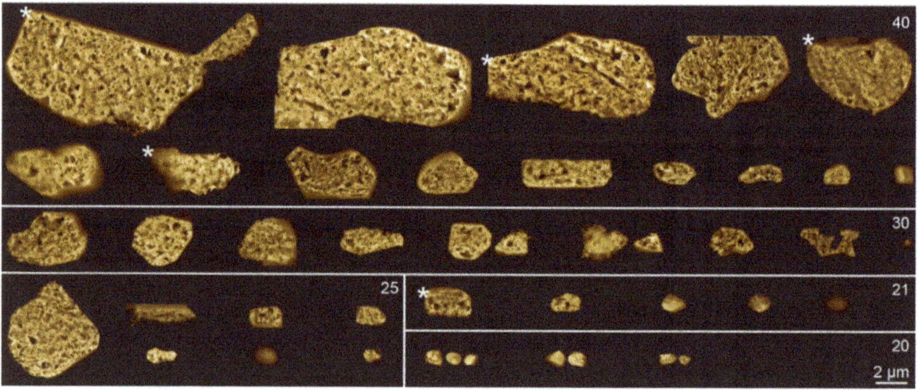

Figure 8. Secondary electron images of false-colored gold grains detected by the QEMSCAN® analysis of Boliden AB for a range of signal intensity thresholds (20%–40%) of Au defined in the SIP-entry for the identification of Au minerals. A clear correlation between detectability and size exists. For the grains close to each other, they were marked by a single pixel in the mineral map and it was impossible to discern which had been hit by the electron beam. The grains with an asterisk were also detected by the QEMSCAN® analysis of CSM.

The experiments with the SIP-entry for Au showed that detection and correct differentiation were largely controlled by the SIP and that the same procedure used for the detection of Au could be applied to other trace metals. It could, therefore, be treated as a proxy for all trace metals. To test this methodology, a quick test confirmed that the detection of Ag-minerals was indeed possible by creating a SIP-entry specific for the detection Ag and placing it just below the entry for Au at the top of the SIP. Several tens of Ag-minerals were identified, despite none having been detected in the sample previously.

4. Discussion

Scans by both CSM and Boliden AB covered only about 75 % of the sample due to limitations of their sample holders (Figure 4). Obtaining representative samples of the rock/ore is challenging and samples have to be selected carefully. Structures, textures and mineral distribution are often heterogeneous and some features may be observed only in a small area of a sample. Although the edges are usually of poorer polish quality, a 75% scan can result in a significant loss of information and hence, an effort should be made to analyze the entire sample area (or as large an area as possible), using for instance, properly designed sample holders.

For Boliden AB, the backscattered electron image of the sample area showed several fields with shifts in grey-levels likely caused by poor vacuum conditions or a fluctuating beam current. However, the shifts were not reflected in the mineral map; therefore, the X-ray yield was not significantly affected and/or within the tolerance of the SIP. It is possible that this artefact would have been more problematic in an MLA system, as it relies primarily on the BSE signal for particle distinction. There are a plethora of possible interactions of components within and outside the instrument that can affect vacuum and beam stability, such as vacuum pumps in need of repair and unexpected highs/lows in the power supply or an old filament. Although the results of this analysis were not affected, future complications are possible. Thus, troubleshooting to find the root of the problem is recommended, although it exceeds the scope of this paper.

Regarding the spatial resolution of the scan, CSM opted for a pixel size of 10 μm compared to 5 μm for Boliden AB. The 10 μm resolution was chosen based on the findings of unpublished work by one of the authors (G.K.R.) and a study by Boni et al. [21], which showed the difference between a 10 μm and a 1 μm scan to be marginal from a bulk mineralogical point of view for most samples. Furthermore, this study did not show any significant differences in bulk mineralogy directly attributable to the difference in resolution between Boliden AB and CSM. However, this is deceiving when it comes to the identification and quantification of trace phases. Due to accuracy issues with quantities <100 ppm, no more precise values were reported for many trace phases. In fact, two Au grains detected by Boliden AB (after re-processing the data several times) were misidentified by CSM despite a grain size at CSM scan resolution. Additionally, dependent on the Au signal threshold value in the SIP of Boliden AB, many pixels were erroneously identified as Au. However, all Au grains detected by CSM were confirmed by SEM-EDS. Furthermore, this study showed that with the right SIP, the 5 μm scan at Boliden AB was able to exclude errors and resolve many more Au grains compared to the 10 μm scan at CSM, although quantitatively, they both were below 100 ppm. This means, a scan resolution <10 μm can improve quantification of trace phases if the SIP is of sufficient quality and the data is verified by another method. In fact, for studies focused only on the quantification of Au, even higher resolutions of 1–2 μm are used [8,12,22]. However, this is not realistic with the QEMSCAN® technology for comprehensive routine analyses of uncrushed samples in the mining industry as the runtime would drastically increase. In this study, the 5 μm scan required already more than twice the amount of time compared to the 10 μm scan. Whether the benefit of a better trace mineral quantification outweighs the downside of a longer scan time is up to the mining company to decide. There is also always the option to follow up a fieldscan with a high-resolution scan in TMS mode or a high-resolution fieldscan of smaller areas of interest.

Concerning mineralogy, the differences between the QEMSCAN® analyses at Boliden AB and at CSM are apparent. Camborne School of Mines as a research institution has the ambition to achieve the highest level of detail with as little unknowns ("Others") as possible for every analysis. In this study, 23 phases were distinguished with less than 100 ppm mineral mass that was left unidentified. Most trace minerals were at or below the scan resolution in terms of grain size; however, they were often possible to separate from the surrounding phases and are marked as single pixels in the mineral map. While some single pixels were misclassified, all pixels of valuable trace minerals such as Au and Ag were confirmed by manual SEM-EDS. This was achieved by detailed work with the SIP using the SMART approach [23]. At CSM, the same SIP is used for every sample, regardless of type and origin (geology, archaeology, agriculture, forensics, etc.). However, with every analysis, the SIP is edited and adapted to account for natural compositional variations of minerals between samples. Unknown pixels are individually reviewed to try to deduct the mineral phase (or phases) responsible for the EDS signal and, if successful, a corresponding SIP-entry is added. With detailed mineralogical knowledge of a sample prior to QEMSCAN® analysis, through thorough optical microscopy and SEM work, the output data will contain much fewer uncertainties. For this study, previous mineralogical studies by Warlo et al. [16] were initially used to better constrain the SIP entries of some phases.

In contrast, for Boliden AB, a rough understanding of the mineralogy of a sample is often considered sufficient. Gangue phases such as silicates, carbonates and oxides have no economic value for the company and similarly, minor and trace ore minerals are often too low in abundance to be economic. Boliden AB, therefore, does not prioritize a detailed characterization and separation of these particular phases. Furthermore, due to the much higher required sample throughput compared to CSM, thorough manual editing of the SIP for every analysis is too time consuming and, therefore, economically unfeasible for Boliden AB. Instead, for each deposit or process-mineralogical type of ore, an individual SIP is developed and consequently, used for the quantitative mineral analysis of the whole deposit. The time it takes to develop a customized SIP is strongly dependent on the mineralogical complexity of the deposit. These customized SIPs are commonly based on prior analysis by optical microscopy. In this case, the SIP for the nearby Aitik Cu-Au deposit was used as a basis due to its somewhat similar mineralogy to the Liikavaara deposit and the SIP being supported by several mineralogical studies even though the two deposits are genetically different. The SIP was then slightly adapted for this particular study, based on previous mineralogical studies by Warlo et al. [16]. The SIP is then typically used for every sample from the same deposit with editing focused mostly on adjusting for mineralogical variations between samples. This saves time (editing of ~14 samples per day) and commonly delivers data of sufficient quality for the mining operation. Nevertheless, the quality of analysis is dependent on how well mineralogy and chemical composition of the minerals in the sample fit with their definitions in the SIP. Major ore minerals are usually well-constrained but especially, mineralogy of gangue and minor and trace phases is not always fully studied/understood and consequently, their SIP-entries are vague or missing.

Furthermore, fine textures with phases smaller than the excitation volume of the electron beam (e.g., trace minerals) commonly produce mixed X-ray signals and thus, are not identified by conventional SIP entries. This explains the shorter mineral list of Boliden AB compared to CSM in this study and the larger variations in modal mineralogy for gangue and trace phases compared to major ore phases. It is also the reason for the amount of unidentified phases. However, although not of economic value, there is definitely a benefit in distinguishing the various gangue phases and trace ore minerals in a sample. The hardness of the gangue phases, for example, dictates crushing of the ore, sheet silicates affect the flotation, and some trace metals are deleterious to primary metals (main commodity). The importance of understanding the mineralogy of trace minerals and gangue minerals especially in Cu-Au ores is also highlighted by Agorhom et al. [24] in their review on trace element recovery in copper flotation. Hence, recognizing these potential problems should be of interest in a mining venture. Boliden AB showed the potential of their QEMSCAN® system to separate between different gangue phases and minor ore minerals by expanding the mineral list to match CSM. However, it also showcased their limitations caused by a less-developed SIP. Ankerite and jacobsite, for example, could not be differentiated from "Fe-Ox/CO3" since no SIP-entries existed for these phases and no reference material was available to create new entries. To compensate for this less comprehensive SIP compared to CSM and its inability to handle mixed signals and also to deal with signal errors caused by a deflected beam, Boliden AB often applies the "boundary phase processor". The results showed that it helped to increase similarity in the bulk mineralogy for chalcopyrite between Boliden AB and CSM and to remove falsely classified rims of chalcopyrite around pyrite grains, but at the expense of also removing previously resolved micro-cracks of chalcopyrite. Hence, a comprehensive SIP is a key requirement to high quality and meaningful data. This is, however, not limited to the QEMSCAN® system. Although the means of mineral identification may differ between ASEM systems, all rely on a comparison of the recorded signal with a database for classification. If minerals are defined by grey-scale values, X-ray intensities, or stoichiometry is marginal. In fact, a study by Kern et al. [25], using the MLA system showed improvements in calculating Sn deportment in a skarn deposit by including mixed phases in their mineral reference list in order to resolve mixed spectra at grain boundaries rather than relying on so-called touchups (similar to a "boundary phase processor").

Generally, the "scientific" and the "industrial" approach by CSM and Boliden AB, respectively, are both justified for their respective purpose. However, with the rising economic importance of many trace metals and their implications on ore processing and the environment, control over their occurrence and distribution should be of interest to the mining industry and consequently, aimed for with the use of some advanced automated quantitative mineralogical type of analysis. This study explored the potential of routinely identifying economic trace minerals in rocks prior to processing with industrial QEMSCAN® settings. It was shown that by including single-element SIP-entries as filters at the top of the SIP detection, at best quantification of trace minerals is indeed possible, albeit without being able to distinguish between minerals of similar element composition (e.g., native Au and electrum). One challenge is erroneous signals that cause the misidentification of pixels. While for major ore minerals, Boliden AB utilizes the "boundary phase processor" to correct for these errors, it cannot be applied to trace minerals as they are themselves adversely affected by this method. Instead, a threshold value for the X-ray signal intensity of the trace metal mineral must be added to the single-element SIP-entry. The optimal threshold value to exclude all erroneous signals while including as many true signals as possible may differ between trace metals. To determine the optimal threshold value, QEMSCAN® data has to be reviewed by other analytical methods, e.g., SEM-EDS to separate true from erroneous pixels. It is not plausible to fully implement this in an industrial routine. However, in this study, even with a threshold value 60% higher than the ideal value (40% compared to 25%), around half of the Au-pixels were captured (20 of 39 pixels). Thus, implementing SIP-entries with conservative threshold values for all economic trace metals in a deposit would already be beneficial with a minimum amount of work. While this, without follow-up analysis, will not provide reliable quantitative mineralogical information and data on grain size and shape, it should provide a basic overview of trace mineral association and distribution and allow for targeted follow-up studies.

5. Conclusions

This study investigated the potential of comprehensive routine quantitative mineralogical characterization of uncrushed rock samples by QEMSCAN® (as an example of ASEM) in the mining industry, with emphasis on trace mineral quantification. Analytical quality and methodology between an industrial and a scientific application of the QEMSCAN® system were compared. It was shown that in comparison to a scientific application, the quality of the industry data was largely reliant on the quality of the species identification protocol (SIP) or mineral library used. Especially, the capability to identify different gangue minerals and trace phases and to resolve mixed spectra was inferior for the analysis with settings for an industrial application. The resolution of mixed spectra was achieved through the "boundary phase processor" after modification of the SIP (the preferred method for the scientific analysis) proved too time intensive. It was demonstrated that by modification of the SIP for the analysis using industrial settings, gangue mineral differentiation could be improved. Additionally, the identification and quantification of trace minerals (in this case, Au-minerals) was significantly improved by the addition of single-element entries to the top of the SIP. Due to a potential of erroneous spectra caused by, e.g., a deflected electron beam, a threshold value had to be added to the single-element SIP. The lowest possible threshold value to avoid errors had to be determined experimentally (25% signal intensity for Au) and by verification with another analytical method (SEM-EDS). For a routine application, continuous verification is time consuming and thereby implausible, but a conservative threshold value could be implemented at the expense of missing some pixels of trace minerals. With this method, a 5 µm field scan was able to identify Au grains of less than 2 µm. It was also successfully tested for Ag. However, no information on trace metal mineralogy, grain size, and shape was collected. It thus cannot be compared to the data quality achievable with a 1 µm phase-specific search. However, as an add-on to routine quantitative mineralogical analysis focused on major ore minerals this method can also produce quantitative data and information on mineral association for trace minerals whose metals may be potential by-products in a mining operation. This method will then lay the foundation for further targeted analysis of, for instance, precious- and critical trace metals.

While this study was performed on a single thin section sample only, the method developed to quantify trace minerals should be reproducible for other samples as well. In general, the more complex the mineralogy and textures of a sample and the finer the trace minerals, the more challenging an analysis will be. Additionally, the quality of the analysis is dependent on the quality of polishing (issues with beam deflections). Although this may impact the threshold value necessary to exclude errors, it should not affect the usability of the method itself.

Author Contributions: Conceptualization, M.W., C.W., G.B. and A.R.B.; Data curation, M.W., D.B. and G.K.R.; Formal analysis, M.W., I.M., D.B. and G.K.R.; Investigation, M.W., D.B. and G.K.R.; Methodology, M.W., I.M., D.B. and G.K.R.; Supervision, M.W., C.W., G.B. and A.R.B.; Validation, M.W.; Writing—original draft, M.W.; Writing—review & editing, M.W., C.W., G.B., A.R.B., I.M., D.B. and G.K.R.

Funding: This research received no external funding.

Acknowledgments: We kindly thank Boliden AB for letting us test their methodology, to develop their operation, and for permission to publish the findings. This work was carried out as part of a doctoral project (MW) financed by CAMM[2] (Centre for Advanced Mining & Metallurgy, Luleå University of Technology) and Boliden Mineral AB. The authors appreciate the in-depth discussions with Thomas Riegler on earlier versions of the manuscript, and three anonymous reviewers are thanked for improving the manuscript.

Conflicts of Interest: The authors declare no conflict of interest.

References

1. *Study on the Review of the List of Critical Raw Materials*; Criticality Assessments; European Commission: Brussels, Belgium, 2017. [CrossRef]
2. Goodall, W.R.; Scales, P.J. An Overview of the Advantages and Disadvantages of the Determination of Gold Mineralogy by Automated Mineralogy. *Miner. Eng.* **2007**, *20*, 506–517. [CrossRef]
3. Gottlieb, P.; Wilkie, G.; Sutherland, D.; Ho-Tun, E.; Suthers, S.; Perera, K.; Spencer, S.; Butcher, A.; Rayner, J. Using Quantitative Electron Microscopy for Process Mineralogy Applications. *Jom* **2000**, *52*, 24–25. [CrossRef]
4. Gu, Y. Automated Scanning Electron Microscope Based Mineral Liberation Analysis an Introduction to JKMRC/FEI Mineral Liberation Analyser. *J. Miner. Mater. Charact. Eng.* **2003**, *2*, 33–41. [CrossRef]
5. Miller, P.; Reid, A.; Zuiderwyk, M. QEM* SEM Image Analysis in the Determination of Modal Assays, Mineral Associations and Mineral Liberation. In Proceedings of the XIV International Mineral Processing Congress, Toronto, ON, Canada, 17–23 October 1982; pp. 1–20.
6. Reid, A.; Gottlieb, P.; MacDonald, K.; Miller, P. QEM* SEM Image Analysis of Ore Minerals: Volume Fraction, Liberation, and Observational Variances. In Proceedings of the Second International Congress on Applied Mineralogy in the Minerals Industry, Los Angeles, LA, USA, 22–25 February 1984; Won, C.P., Donald, M.H., Richard, D.H., Eds.; The Metallurgical Society of AIME: Englewood, CO, USA, 1984; pp. 191–204.
7. Hall, J. Composite Mineral Particles: Analysis by Automated Scanning Electron Microscopy. Ph.D. Thesis, University of Queensland, Brisbane, Austrilia, 1978.
8. Goodall, W.; Scales, P.; Butcher, A.R. The Use of QEMSCAN and Diagnostic Leaching in the Characterisation of Visible Gold in Complex Ores. *Miner. Eng.* **2005**, *18*, 877–886. [CrossRef]
9. Pirrie, D.; Rollinson, G.K. Unlocking the Applications of Automated Mineral Analysis. *Geol. Today* **2011**, *27*, 226–235. [CrossRef]
10. Hrstka, T.; Gottlieb, P.; Skála, R.; Breiter, K.; Motl, D. Automated Mineralogy and Petrology-Applications of TESCAN Integrated Mineral Analyzer (TIMA). *J. Geosci.* **2018**, *63*, 47–63. [CrossRef]
11. Graham, S.D.; Brough, C.; Cropp, A. An Introduction to ZEISS Mineralogic Mining and the Correlation of Light Microscopy with Automated Mineralogy: A Case Study using BMS and PGM Analysis of Samples from a PGE-bearing Chromite Prospect. *Precious Met.* **2015**, 1–12.
12. Butcher, A.R.; Helms, T.A.; Gottlieb, P.; Bateman, R.; Ellis, S.; Johnson, N.W. Advances in the Quantification of Gold Deportment by QEMSCAN. In Proceedings of the Seventh Mill Operators Conference, Kalgoorlie, Australia, 12–14 October 2000; pp. 267–271.
13. Goodall, W.R. Characterisation of Mineralogy and Gold Deportment for Complex Tailings Deposits Using QEMSCAN®. *Miner. Eng.* **2008**, *21*, 518–523. [CrossRef]

14. Goodall, W.R.; Butcher, A.R. The Use of QEMSCAN in Practical Gold Deportment Studies. *Miner. Process. Extr. Metall.* **2012**, *121*, 199–204. [CrossRef]
15. Gy, P. *Sampling for Analytical Purposes*; John Wiley: New Jersey, NJ, USA, 1998; p. 153.
16. Warlo, M.; Wanhainen, C.; Martinsson, O.; Karlsson, P. Mineralogy and Origin of the Intrusion-Related Liikavaara Cu-(W-Au) Deposit, Northern Sweden. *GFF* **2019**. (under review).
17. Zweifel, H. Aitik-Geological Documentation of a Disseminated Copper Deposit: A Preliminary Investigation. Ph.D. Thesis, Geological Survey of Sweden, Stockholm, Sweden, 1976.
18. Rollinson, G.K.; Andersen, J.C.Ø.; Stickland, R.J.; Boni, M.; Fairhurst, R. Characterisation of Non-Sulphide Zinc Deposits Using QEMSCAN®. *Miner. Eng.* **2011**, *24*, 778–787. [CrossRef]
19. Warlo, M.; Wanhainen, C.; Bark, G.; Karlsson, P. Gold and Silver Mineralogy of the Liikavaara Cu-(W-Au) Deposit, Northern Sweden. In Proceedings of the 15th SGA Biennial Meeting, Glasgow, UK, 27–30 August 2019.
20. Andersen, J.C.Ø.; Rollinson, G.K.; Snook, B.; Herrington, R.; Fairhurst, R. Use of QEMSCAN® for the Characterization of Ni-Rich and Ni-Poor Goethite in Laterite Ores. *Miner. Eng.* **2009**, *22*, 1119–1129. [CrossRef]
21. Boni, M.; Rollinson, G.; Mondillo, N.; Balassone, G.; Santoro, L. Quantitative Mineralogical Characterization of Karst Bauxite Deposits in the Southern Apennines, Italy. *Econ. Geol.* **2013**, *108*(4), 813–833. [CrossRef]
22. Agorhom, E.A.; Swierczek, Z.; Skinner, W.; Zanin, M. Combined QXRD-QEMSCAN Mineralogical Analysis of a Porphyry Copper-Gold Ore for the Optimization of the Flotation Strategy. In Proceedings of the XXVI International Mineral Processing Congress (IMPC), New Delhi, India, 23–28 September 2012.
23. Haberlah, D.; Owen, M.; Botha, P.W.S.K.; Gottlieb, P. Sem-Eds-Based Protocol for Subsurface Drilling Mineral Identification and Petrological Classification. In Proceedings of the 10th International Congress for Applied Mineralogy (ICAM), Trondheim, Norway, 1–5 August 2011; Springer: Berlin/Heidelberg, Germany, 2012; pp. 265–273.
24. Agorhom, E.A.; Lem, J.P.; Skinner, W.; Zanin, M. Challenges and Opportunities in the Recovery/Rejection of Trace Elements in Copper Flotation—A Review. *Miner. Eng.* **2015**, *78*, 45–57. [CrossRef]
25. Kern, M.; Möckel, R.; Krause, J.; Teichmann, J.; Gutzmer, J. Calculating the Deportment of a Fine-Grained and Compositionally Complex Sn Skarn with a Modified Approach for Automated Mineralogy. *Miner. Eng.* **2018**, *116*, 213–225. [CrossRef]

 *minerals*

Article

Automated SEM Mineral Liberation Analysis (MLA) with Generically Labelled EDX Spectra in the Mineral Processing of Rare Earth Element Ores

Bernhard Schulz [1],*, Gerhard Merker [2] and Jens Gutzmer [3]

[1] Institut für Mineralogie, Lagerstättenlehre und Petrologie, TU Bergakademie Freiberg, Brennhausgasse 14, D-09596 Freiberg/Sachsen, Germany

[2] Merker Mineral Processing, Oertenroeder Str. 21, D-35329 Gemuenden, Germany

[3] Helmholtz-Zentrum Dresden-Rossendorf, Helmholtz-Institut Freiberg für Ressourcentechnologie, Chemnitzer Str. 40, D-09599 Freiberg/Sachsen, Germany

* Correspondence: bernhard.schulz@mineral.tu-freiberg.de; Tel.: +49-3731-39-2668

Received: 9 August 2019; Accepted: 29 August 2019; Published: 30 August 2019

Abstract: Many rare earth element (REE) deposits have experienced multistage geological enrichment processes resulting in REE bearing mineral assemblages of considerable complexity and variability. Automated scanning electron microscopy (SEM) mineral liberation analysis of such REE ores is confronted by the difficult assignment of energy-dispersive X-ray (EDX) spectra to REE mineral names. To overcome and bypass this problem, a generic and reliable labelling of EDX reference spectra obtained from REE-bearing minerals based on their contents of Si, Ca, F and P in a bulk normalised analysis is proposed. The labelled spectra are then combined into groups of REE-P (~monazite), REE-Ca-Si-P (~britholite), REE-Ca-F (~synchysite) and REE-F (~bastnaesite, parisite, fluocerite). Mixed spectra with low counts for REE from minute REE mineral grains are combined into a separate group. This classification scheme is applied to automated SEM mineral liberation analysis (MLA) data from beneficiation products by comminution and multistage flotation of REE carbonatite ores. Mineral modes, mineral grain size distribution, mineral liberation, mineral locking and mineral grade versus recovery curves based on the analysis of >200,000 particles in a sample can be recognised and interpreted in virtual grain size fractions. The approach as proposed here will allow future process mineralogical studies of REE deposits to be robust and comparable.

Keywords: REE minerals; REE carbonatite ore; comminution; multi-stage flotation; EDX spectra

---

1. Introduction

Rare earth element (REE) ore deposits occur in a wide variety of geological contexts and are hosted by a considerable diversity of host rocks [1]. Igneous host rocks appear mainly carbonatites and peralkaline plutonic rocks as nepheline syenites [2–5]. Although pegmatites may also contain significant amounts of REE-bearing minerals [6], they are usually too small in volume to be of economic significance. REE mineralisation also occurs in hydrothermal veins and stockworks [7,8]. Recent and fossil mineral placers can be sedimentary REE mineral deposits. Significant enrichment of REE is also possible by chemical weathering with recoverable REE concentrations occurring in lateritic clays formed at the expense of magmatic REE deposits. Some of the largest known REE deposits owe their origin to a sequence of natural enrichment processes. Primary igneous occurrences in alkali syenites and carbonatites as well as sedimentary heavy mineral accumulations and weathering crusts underwent metamorphic and hydrothermal overprint in depth, followed by weathering. Due to their possible complex geological evolution, the deportment of the elements in such REE mineral deposits are often ambiguous and the subject of scientific discussions.

REE ores are not only marked by geological complexity, but REE ore mineralogy is also very complex, with different minerals having complex chemical and crystallographic properties. More than 200 REE minerals are known [9]. Actually, in the largest REE mineral deposit of Bayan Obo in Mongolia one can distinguish three ore types with Fe-REE-, dolomite-REE- and silicate-REE-ores. The economic REE minerals are bastnaesite and monazite [10]. Other important REE mineral deposits are associated with carbonatite intrusions, as Mountain Pass (USA) and Mount Weld (Australia). The ore minerals in these deposits are bastnaesite, allanite (part of the epidote mineral group), monazite, apatite and pyrochlor. In lateritic clay deposits, REE released during chemical weathering of igneous host rocks may occur adsorbed in clay minerals [11–14].

The beneficiation of REE ores poses major technological challenges [15–18]. These may be understood—and then overcome—by applying modern and quantitative analytical methods that yield not only chemical but also mineralogical and microfabric data. Bulk chemical analysis of ore and processing products by X-ray fluorescence (XRF) and inductively coupled plasma mass spectrometry (ICP-MS) will readily provide elemental concentrations, especially of the REE [19]. X-ray diffraction (XRD) analysis, in turn, allows the identification of minerals, but is afflicted with considerable uncertainty and error when mineral modal abundance is below ~1 wt %. However, both analytical methods require powdered samples at grainsizes <2 μm and provide no tangible information on particle and mineral grain sizes, particle compositions, mineral intergrowths and liberation. However, such particle-related parameters are essential during the beneficiation of REE ores. Therefore, non-destructive, element sensitive methods based on scanning electron microscopy (SEM) with energy dispersive spectroscopy (EDS) and involving automated backscattered electron (BSE) image analysis, referred to as automated mineralogy, are widely applied [20–23]. Automated mineralogical studies of REE ores are confronted with the following mineralogical challenges:

1. REE-bearing minerals occur in many mineral classes, including REE-phosphates, REE-carbonates, REE-halogenides, REE-oxides, REE-silicates, REE-arsenates, as well as combinations of it as REE-fluoro-carbonates [24].
2. Many REE minerals have a complex mineral chemistry with light REE (LREE, elements La to Eu), heavy REE (HREE, elements Gd to Lu), Th, U, Si, Al, Ca, F, CO_3, PO_4, Nb, Y, As, S and others.
3. Most of the REE minerals are solid solutions with single and coupled substitution involving LREE, HREE, Y, Si, Al, Ca, F, P, Nb, Th, U and others. Compositional variations are widespread.
4. Many REE minerals are hydrated phases. This considerably hinders their identification by chemical analysis.
5. The often complex and multistage geological processes of REE enrichment lead to mineral intergrowths, pseudomorphs, partial and complete replacement, hydration and dehydration. This often results in a complex REE mineral assemblage.

Mineral phases are defined first by the crystallographic parameters (structure) and second by the chemical composition [24]. For the REE-bearing minerals as outlined above, there arises the consequence that even when crystallographic parameters by XRD are available it is often very hard to define a correct mineral name based on energy-dispersive X-ray (EDX) spectra and element compositions. Also, the conventional method of labelling EDX spectra from REE-bearing minerals by mineral names is severely hampered by the fact that the X-ray energy peaks and sub-peaks of LREE and HREE display considerable overlap and interference along the keV scale, which cannot be resolved by an analysis with the EDX, even when interference corrections are applied. This requires dedicated routines during the application of automated SEM methods [25,26]. Our study introduces an approach of applying generic labelling of a list of EDX reference spectra obtained from REE-bearing minerals in nepheline syenites and carbonatites, which is followed by a distinct mineral grouping. This allows robust classification and the extraction of mineralogical data from automated scanning electron microscopy-mineral liberation analysis (SEM-MLA) measurements. Many of the problems of the nomenclature and assignment of the REE-bearing minerals are thus avoided.

2. Approach and Analytical Methods

At the TU Bergakademie Freiberg/Saxony a scanning electron microscope FEI Quanta 600 (FEI, Hillsboro, OR, USA) equipped with a field emission gun (FEG) as electron source, two Bruker energy dispersive X-ray (EDX) SDD detectors (Bruker Quantax 200 with two Dual XFlash 5030 EDX detectors, (Bruker, Berlin, Germany), and backscattered electron (BSE) and SE detectors is applied for Mineral Liberation Analysis [20]. For the measurements presented here, the commercial MLA 2.9 software package (JKTech, Brisbane, Australia) has been used [21].

The analytical routine proposed here commences with the survey of a sample by a BSE image, labelled as a frame, at 25 kV acceleration voltage and 10 nA beam current. The instrument-specific working distance was at 12 mm. REE-bearing mineral grains with high average atomic numbers and molecular masses appear in light grey in the BSE image compared to gangue minerals as quartz and feldspar with darker grey colour (Figure 1a–c). The calibration of the BSE greyscale with contrast and brightness was performed with gold reference. After automated image analysis (Figure 1d–f), the electron beam is directed into the barycentres of contiguous mineral grains characterised by their BSE grey values, and a single EDX spectrum is obtained (XBSE measurement mode, [21]). In the case of thin sections or polished blocks of ore, the image analyser performs only the particle segmentation within a frame, and a grid of single EDX spectra is gained from each contiguous domain with distinct BSE grey values (GXMAP measurement mode). Each EDX spectrum is normalised by the counting rates $(cts/s)_N$ of the coupled EDX detectors and plotted against the keV scale (Figure 1g–i). These EDX spectra have characteristic peaks at distinct positions in the keV scale and distinct relative cts/s allowing identification of major elements present, giving a semiquantitative indication of their concentrations. Both measurement routines allow a later examination of both geometrical and mineralogical particle and grain parameters, as various size and shape parameters, mineral locking and mineral liberation [21]. The classification of the measured EDX spectra from the sample is performed by a comparison to a list of labelled reference EDX spectra. The list of reference spectra can incorporate up to 250 different reference EDX spectra which were gained under similar measurement conditions from the analysed samples and/or from related reference samples.

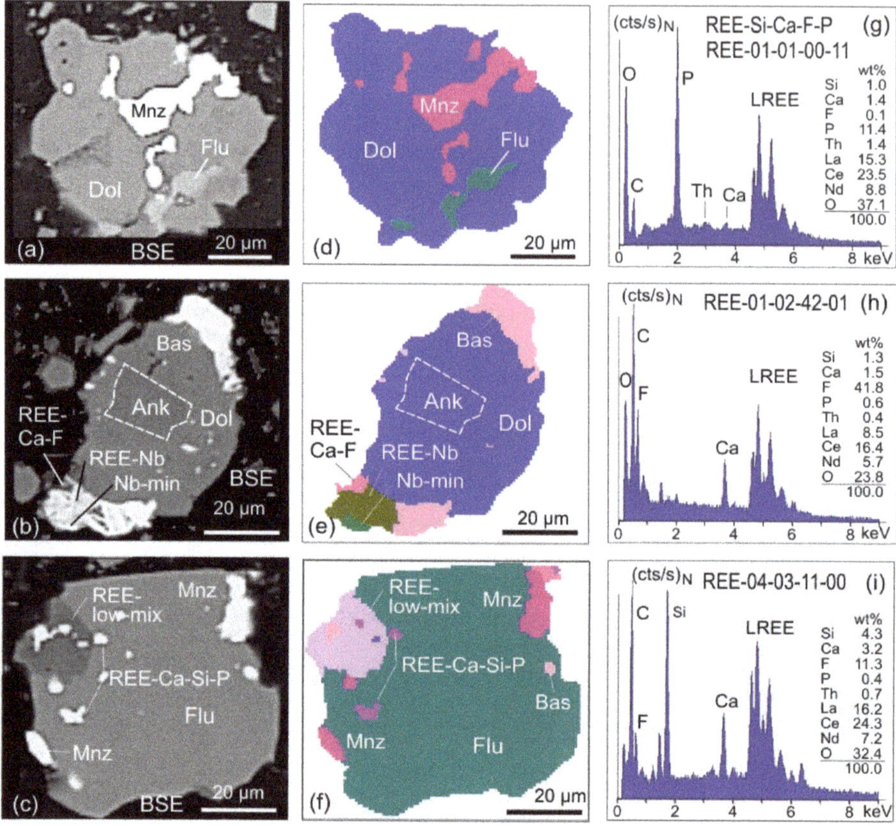

Figure 1. (**a–c**) Particles of grinded rare earth element (REE) carbonatite ore in backscattered electron images (BSE). The REE minerals monazite (Mnz), bastnaesite (Bas), and the REE-Ca-F minerals (synchysite) are in light grey to white; the REE-Nb- and Nb-bearing minerals (Nb-min) are also in light grey; the gangue minerals as ankerite (Ank), dolomite (Dol) and fluorite (Flu) are dark grey. (**d–f**) REE carbonatite ore particles after classification of automated scanning electron microscopy (SEM) measurement in the mineral liberation analysis (MLA)-XBSE routine. Classification with list of generically labelled energy-dispersive X-ray (EDX) spectra and corresponding grouping of the spectra (see text). REE minerals in the particles display only poor and partial liberation. (**g–i**) EDX spectra of some REE minerals in keV versus normalised counting rate $(cts/s)_N$. Positions of maxima in the spectra are labelled by the corresponding elements. The generic labelling of the spectra in REE-Si-Ca-F-P is according to a quantitative EDX element analysis of the corresponding REE mineral grain, shown on the right side.

3. Energy-Dispersive X-ray (EDX) Spectra of Rare Earth Element (REE)-Bearing Minerals

As outlined above, REE-bearing minerals have rather complex chemical compositions [24]. This mineralogical and chemical diversity is also evident from the currently most important economic REE bearing minerals (Table 1). The REE phosphate monazite $(LREE,Y,Th,Si,Ca)PO_4$ shows highly variable concentrations of the LREE elements La, Ce, Nd in solid solutions toward cheralite $(REE,Ca,Th)(P,Si)O_4$, huttonite $(ThSiO_4)$ and the xenotime group of minerals $(Y,HREE)PO_4$ involving coupled substitution of Y, Ca, Si, Th and P [24]. Hydrated species such as rhabdophane $(REE)PO_4(H_2O)$ and florencite $REE)Al_3(PO_4)_2(OH)_6$ occur as well. In the britholite group $(REE,Ca,Th)_5(SiO_4,PO_4)_3(OH,F)$ variable REE and Y contents occur together with Ca, Si, P and F. The synchysite group $Ca,REE(CO_3)_2F$ contains

Ca, F and C together with REE. Parisite, with more F and C but less Ca can be considered there as a subspecies. Bastnaesite REE(CO$_3$)F is a hydrated halogene-bearing carbonate mineral with variable REE, Y, F, C concentrations. In contrast, fluocerite (REE)F$_3$ is a simple fluor-bearing mineral (Table 1). A systematic search for REE-bearing minerals in available databases, such as *MinIdent* [27,28] and websites (http://webmineral.com/chem/Chem; http://rruff.info.ima) by using the mineral chemical compositions leads to long and desperately confusing list of mineral names. This complexity of mineral compositions and mineral names renders correct identification of individual REE minerals difficult. This pertains in particular to automated mineralogy studies where mineral identification is based on EDX spectra with minute X-ray counts, as ~10,000 cts exemplified in this study. Even when pertinent expert knowledge upon the corresponding mineral groups is available, an assignment of the EDX spectra to mineral names remains biased. To overcome this particular problem of reference EDX spectra denomination for SEM-based image analysis measurements, the following workflow is proposed. Although the proposed workflow is based here on the MLA 2.9 or MLA 3.1 software versions, it is easily transferable to similar instrumental and software platforms.

1. During a first automated SEM-MLA measurement, all EDX spectra in a given sample are captured (XBSE-STD measurement mode). The discrimination of these spectra is provided by a high reliability value of $1e^{-10}$ which means a high degree of conformance. The spectra that fall within this limit of conformance receive consecutive numbers (Mineral 1, Mineral 2, …). Dependent on the mineralogical diversity of the ore, ca. 100–150 different EDX spectra can be collected from REE-bearing and gangue minerals (e.g., carbonates and silicates). A certain fraction of these spectra will be from grain boundaries or artefacts of preparation effects. These can be ignored during further assessment by a tentative classification of the measurement.

2. The MLA software functions allow driving the SEM stage to the mineral grains (e.g., Mineral 1, Mineral 2, …) where the consecutively numbered spectra were obtained for the first time during the measurement. A quantitative element analysis by EDS is performed from these grains. The EDX spectra from gangue minerals can then be labelled by corresponding mineral names (e.g., calcite, dolomite, fluorite, …). It is recommended labelling several slightly differing spectra from the same gangue minerals (e.g., calcite1, calcite-mix).

3. The EDX spectra from REE-bearing mineral grains receive a generic label that matches the normalised results of EDS element analysis. An EDX spectrum from an REE-bearing mineral suggesting e.g., 3.9 wt % Si, 4.7 wt % Ca, 14.3 wt % F and 2.4 wt % P (when totals are normalised to 100) will be labelled as REE-04-05-14-02 (Figure 1g–i). The range of the labelled elements should be the same for all REE-bearing minerals in a particular study, e.g., REE-Si-Ca-F-P, to assure consistency and comparability. When P and F are purposefully positioned at the end of the label, this will facilitate the subsequent step of spectra grouping (see below). Due to the carbon coating and the analytical uncertainty related to peak interferences, the elements C and the REE are not used in this generic labelling process. In a similar manner EDX spectra from REE-bearing minerals with Y, Nb or further elements such as As can also be labelled. A similar approach has previously been applied to automated SEM-MLA measurements of sewage sludge ashes [29], soils [30] or to zoned metamorphic garnets [31].

4. A labelling of the EDX spectra based on quantitative EDS analysis of the REE is not reasonable because the relative REE concentrations have only secondary relevance for the distinction of mineral classes. Also, a considerable analytical uncertainty is caused by the REE peak interferences, which could yield erroneous absolute concentrations of REE. Therefore, when the totals are normalised to 100 wt %, the analysis of Si, Ca, F and P will provide at best the relative proportions of these elements in a REE bearing mineral grain.

Table 1. List of REE-bearing minerals and their assignment to mineral groups as applied in the presented MLA studies of REE ores. The given mineral compositions (in wt %; density Dens.) are not representative and refer to the analysed REE ores and/or are partly taken from databases (e.g., www.webmineral.com).

MLA-Group	Mineral	General Formula	Dens.	$\sum$REE	P	Ca	Si	C	F	O
REE-P-monazite	monazite	(LREE,Y,Th,Si,Ca)PO4	5.10	59.49	13.21	0.00	0.00	0.00	0.00	27.29
REE-Al-P-phases	florencite	(LREE)Al3(PO4)2(OH)6	3.58	27.31	12.07	0.00	0.00	0.00	0.00	43.66
REE-Ca-Si-P-phases	britholite	(LREE,Ca)5(SiO4,PO4)3(OH,F)	4.45	46.44	2.00	14.70	9.40	0.00	0.50	26.85
REE-Ca-F-phases	synchysite	(LREE,Ca)(CO3)2F	4.02	43.89	0.00	12.56	0.00	7.53	5.95	30.07
REE-F-phases	bastnaesite	(LREE)(CO3)F	4.97	63.94	0.00	0.00	0.00	5.48	8.67	21.90
REE-F-phases	fluocerite	(LREE)F3	6.13	71.07	0.00	0.00	0.00	0.00	28.93	0.00

The SEM-MLA measurements of various REE ores will provide a list of reference EDX spectra. Given a possible maximum of 250 reference EDX spectra for a measurement classification, then ~50 spectra should be sufficient for REE-bearing minerals. The mineral xenotime will be readily distinguished by the abundance of Y and will not require more than ~5 spectra [32]. The reference spectra list further encloses minerals with Nb (~15 spectra) and also minerals with both Y and Nb (~5 spectra). The other spectra in the list concern gangue minerals, i.e. feldspars (~15 spectra for K-feldspar, albite, plagioclase), quartz (5 spectra), carbonate minerals (~20 spectra for calcite, dolomite, ankerite, siderite), fluorite (~5 spectra), as well as accessory minerals such as Ti-bearing minerals (~15 spectra for rutile, ilmenite, titanite), apatite (5 spectra) and zircon (3 spectra).

The classification of an SEM-MLA measurement against the labelled reference EDX spectra list provides the proportion in area% of the grains that are classified by a distinct spectrum. However, the evaluation of area proportions dedicated to up to 250 EDX spectra is at best unmanageable. Therefore, the area proportions of several EDX spectra have to be integrated or summarised into groups. An educated grouping of spectra is usually sufficient to address important issues of the mineral processing [32]. A re-grouping of existing data is possible at any time without performing a new measurement. Depending on the MLA software version, a mineral formula, an exact or approximated element composition, and a specific weight can be assigned to each reference EDX spectrum (MLA 3.1 version) or to a group of spectra (MLA 2.9 version). The REE ores analysed in this study were grouped as follows (Table 1):

1. EDX spectra from REE-bearing minerals with high content of P are summarised under the group REE-P-monazite. A typical mineral of this group would be monazite.
2. EDX spectra from REE-bearing minerals with low content of P but elevated contents of Si and Ca are summarised as the REE-Ca-Si-P group. A typical mineral of this group would be britholite.
3. The EDX spectra from REE-bearing minerals without P but high contents of Ca and F, and intermediate to low contents of Si are summarised as the REE-Ca-F group. A typical mineral of this group would be synchysite.
4. The EDX spectra from REE-bearing minerals with dominant F at low Ca and Si concentrations are combined as the REE-F group. Typical minerals of this group are bastnaesite and parisite. The element carbon cannot be used for labelling due to the carbon-coating of the samples.
5. EDX spectra from grains with detectable but low REE contents are merged to form the REE-Low-Mix group. In contrast to the groups 1–4 with high cts/s on the numerous lines of REE and corresponding elevated element contents, the REE-Low-Mix group integrates spectra with low counts on the REE lines and apparently low REE contents. Minerals containing low REE concentrations (for example allanite) can generate such spectra, but similar may be true for small acicular and fibrous crystals of REE minerals enclosed in gangue minerals. In the latter case, the excitation bulb of the electron beam captures gangue minerals beside and below, which will led to such mixed spectra, with low counts on the REE.

A further group of EDX spectra includes Y-HREE and P-rich xenotime and associated Nb-Ta minerals. Dependent on the mineralogy of the deposit, particular attention during spectra grouping is recommended to minerals that contain Y and Nb, as well as Nb and REE, such as the aeschynite mineral group. For the presented case studies, the REE-Nb group and the Nb-Y groups were established. For the discrimination of spectra from Y and Nb the secondary lines have to be considered. After a tentative grouping of the EDX spectra from REE-bearing minerals from an ore sample, the corresponding area proportions should be examined. In the following, the convenience and suitability of the classification workflow introduced above is presented in some case studies dealing with automated SEM of complex REE ores and process samples from three different deposits.

4. Case Studies

Beneficiation test work requires the investigation of comminution and separation characteristics of a given raw material. The first task should always be a thorough study of the unprocessed feed material, i.e. the ore itself in its pristine state [23]. Characterisation by SEM-based image analysis will provide important clues towards the development of a suitable beneficiation strategy. It will inform to the determination of the optimal grinding conditions and time. This is done to achieve a high degree of mineral liberation, whilst minimising the generation of fines. SEM-based image analysis data will provide important insight to assess the success of comminution. In the case of REE ores, comminution is often followed by a multi-stage flotation process [33]. The success of flotation test work can be monitored by SEM-based image analysis. Our case studies illustrate that the mineralogical properties, intergrowths, locking relationships and grain sizes of the REE minerals, as captured by automated mineralogy, are crucial in the critical assessment of the performance of beneficiation processes of fine grained and complex REE ores.

4.1. Case Study 1: Run-of-Mine Ore

The proper identification of the REE-bearing minerals, their mineral grain size and intergrowths is crucial to select suitable technologies and machine parameters for comminution and mineral separation. This is exemplified by the analysis of a polished thin section of a syenite from the Thor Lake Intrusion in Canada, Northwest Territories [34,35]. The sample has been taken from the mineralised T-Zone at the northern margin of the syenite intrusion. The abundance of gangue minerals such as fluorite and quartz in the sample attest to an intense hydrothermal overprint of the pluton. The SEM-MLA analysis (at 25 kV, 10 nA) has been performed in the GXMAP mode at 175 times magnification and with a greyscale trigger (25–255) that includes all minerals but excludes epoxy resin. The analysis took 15 h and is composed of 300 square frames of 1500 µm edge length, covering an area of 2.25 mm^2. The analysed area is covered by 2,158,015 single EDX spectra, which means ~7,200 spectra per frame or 3,200 spectra per mm^2 with a stepsize of 18 × 18 µm. EDX spectra obtained from the REE-bearing minerals can be subdivided into 3 groups (Figure 2). Most abundant are the minerals of the REE-Ca-F group (resembling synchysite) with 3.58 area% (Figure 2b). REE-P-monazite is present, but of such low abundance that it has been ignored for the purpose of this study. A comparably large area proportion (4.16 area %) of the spectra related to the REE-Low-Mix group is obvious. This can be explained by the intense intergrowths of very fine-grained REE-bearing mineral grains with gangue minerals. Indeed, BSE images and GXMAPs suggest that REE minerals are concentrated in 1 × 1.5 mm large aggregates composed of countless miniscule REE bearing mineral grains (Figure 2b). These minute REE mineral grains (0.2–5 µm) are tightly intergrown with phyllosilicates, feldspar and Fe-Ti-oxides. This leads to the conclusion that physical treatment of the ore for separation of REE mineral grains will require a very fine grind size, and that REE mineral grains—even if they are liberated—are unlikely to be recovered by a conventional flotation process properly.

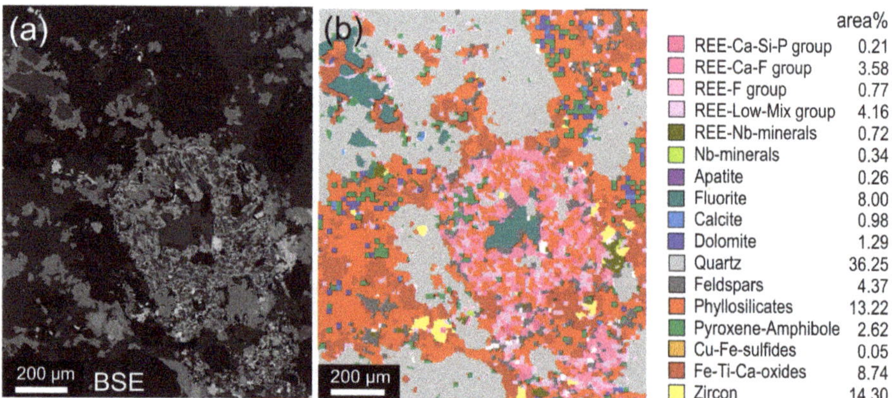

Figure 2. (**a**) Backscattered electron image (BSE) of one frame (magnification 175 times) of an automated scanning electron microscopy-mineral liberation analysis (SEM-MLA) measurement of a polished thin section from a hydrothermally overprinted alkali plutonite. A complete measurement of the 25 × 40 mm sized thin section is composed of ~300 frames. The BSE image displays only a comparably low resolution due to technical reason. (**b**) Classified, grouped and color-coded presentation of the frame in (a) in an automated SEM-MLA measurement in the GXMAP routine; the modes in the mineral legend are in area percent and are related to the whole sample area. Stepsize is 18 μm. The images in (a) and (b) display fine-grained and heterogeneously composed parts and REE mineral grains in intimate intergrowth with phyllosilicates and Fe-oxides.

4.2. Case Study 2: Comminution

This case study concerns a carbonatite REE ore, with monazite as the most abundant REE mineral. Monazite is part of the REE-P-monazite group of EDX spectra with 3.5 wt % modal proportion (please note, in this case study we report wt %, different to the previous case study where data was reported as area %). All other groups of REE mineral spectra attain a total of only 0.7 wt % whilst the REE-Low-Mix group accounts for 0.5 wt %. The prevalent gangue minerals are dolomite (~70 wt %) and Fe-Mg carbonates (~13 wt %). Fluorite reaches ~7.0 wt % in abundance, whilst phyllosilicates, other silicates and quartz all together amount to a maximum of 2.5 wt % (Figure 3a).

Dry grinding experiments on the REE carbonatite ore were performed with a laboratory rodmill, starting with crushed (<2 mm particle size) feed material. For the determination of the optimal grinding fineness, two experiments at 45 min (sample M45) and at 90 min (sample M90) grinding time were conducted. The products (~10 g) were thoroughly mixed with an adequate amount (~10 g) of powdered graphite of pure and fine quality as a parting agent, and stirred into ~2 cm³ of fast-hardening epoxy resin for the production of grain mount blocks of 30 mm in diameter [36,37]. The thickness of the epoxy layer containing sample grains is <5 mm to prevent severe gravity segregation effects. The horizontal block surfaces were polished after a thickness reduction of ~1 mm by grinding. MLA measurements in the XBSE analysis routine included 200,000 particles per sample which were examined within 3–4 h. The XBSE analysis routine is based on a single EDX spectrum within the barycentre of each mineral grain as identified by its grey colour in the BSE image. The cumulative bulk particle size distribution curves for both grinding tests display similar shapes. At 45 min grinding time the P50 (corresponds to cumulative 50 wt % of the distribution curve) is at ~19 μm, and at 90 min grinding time at ~16 μm. For the REE mineral monazite, the most important ore mineral in this case study, the corresponding grainsizes at P50 are 7.5 μm (M45) and 7.2 μm (M90), respectively. When compared at P50 the grainsizes of the carbonates are reduced from 20 to 17 μm, and those of the fluorite from 15 to 12 μm at the longer grinding time (not shown). This illustrates significant effects of selective comminution.

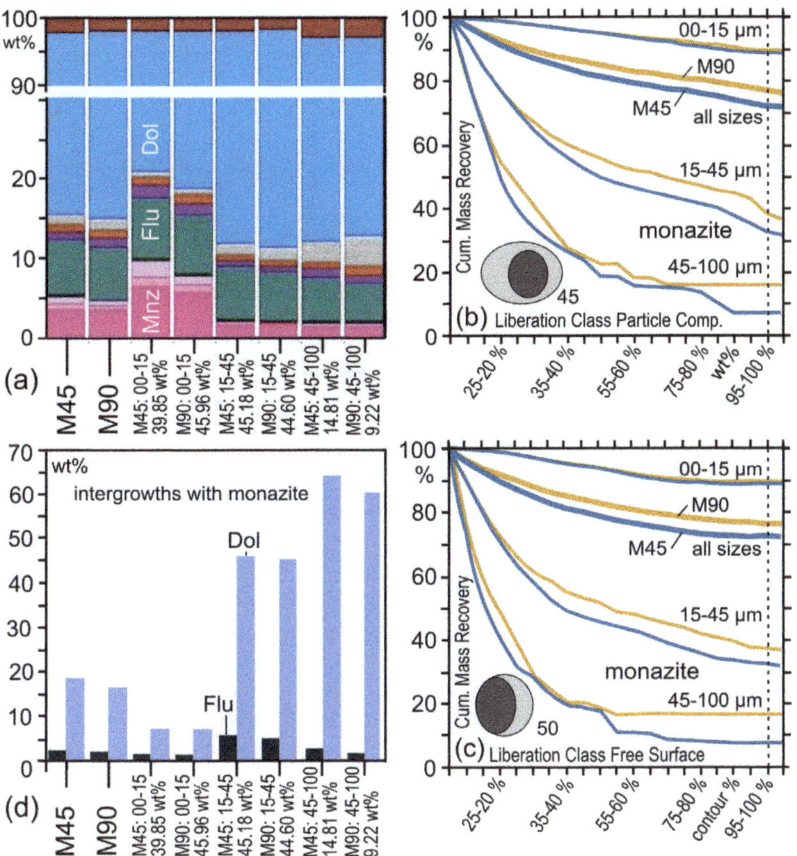

Figure 3. Results of comminution tests of REE carbonatite ore with 45 and 90 min length of time. (**a**) Modal mineralogy (in wt %, y-axis) of complete samples and of particle size classes by virtual sieving with the filter routine of equivalent circle (EC) diameter (see text). Proportion of the corresponding particle size class in wt % (x-axis). Dolomite (Dol); fluorite (Flu) and REE-P-monazite (Mnz) are labelled. (**b**) Mineral liberation of REE-P-monazite in terms of proportion in wt % of particle composition. Inset sketch displays the liberation class particle composition of 45 wt % REE-P-monazite in a schematic view. (**c**) Mineral liberation of REE-P-monazite in terms of proportion in contour% of free surface. Inset sketch displays the liberation class free surface of 50% REE-P-monazite in a schematic view. (**d**) Presentation of the intergrowths of non-liberated REE-P-monazite with fluorite (Flu) and dolomite (Dol). Proportions of fluorite and dolomite in wt % in the complete samples and in particle size fractions by virtual sieving with the filter routine EC diameter (see text).

With a longer grinding time one expects to achieve better mineral liberation of REE mineral grains. However, the problem of over-grinding of the REE mineral grains also increases with a longer grinding time. Over-grinding leads to a large proportion of very fine grains at <<10 μm that are known to usually float poorly [38], and will thus hamper separation by flotation [39]. Cumulative grain size distribution curves provide a first control of potential over-grinding. A subsequent sieve classification of the ground ore with subsequent study of further parameters (e.g., mode, mineral liberation, mineral locking) in the distinct grain size fractions would be of great interest [40]. At the given particle sizes, a mechanical sieve classification is only reliable at particle sizes >20 μm. An alternative is the virtual sieving by an electronic method. The shape classification parameter of the equal circle (EC) diameter turned out to give reasonable results for the 2-dimensional image, however, dependent on the overall particle shapes (e.g., rounded, cubic, platy, elongated, acicular, fibrous), distinct differences to the results of mechanical sieving are noted [23]. In the studied samples the rounded and cubic particles prevail. Virtual sieving of the XBSE data sets was performed in the particle size fractions 0–15 μm, 15–40 μm und 40–100 μm. As expected, increasing grinding time from 45 to 90 min resulted in a larger proportion of the smallest sieve size fraction from ~40 to 46 wt % (Figure 3a). It is obvious from the modal mineralogy that the REE mineral monazite is prominently enriched in the smallest sieve grain size fractions. In the smallest sieve size fraction 0–15 μm the mode of monazite slightly decreases with the longer grinding time while the mode of carbonates increases (Figure 3a). This is a consequence of the lower mechanical stability of the carbonates due to their cleavage planes, when compared to monazite.

The dataset of the XBSE measurement also allows the extraction of parameters of mineral liberation as (1) mineral liberation by particle composition, and (2) mineral liberation by free surface. For both parameters the particles are examined in liberation classes ranging from 0–100%. The liberation class 95–100% (fully liberated mineral grains) for mineral liberation by particle composition for the mineral group (REE-P-)monazite includes all particles that comprise of 95–100 wt % of (REE-P-)monazite. Correspondingly, the parameter mineral liberation by free surface, includes all particles with monazite where the (2D)-contour of the monazite grain is 95–100% free of inherent other mineral phases. The cumulative proportions of each liberation class in wt % are plotted along the Y axis as mass recovery (Figure 3b,c). For our case study, more than 90 wt % of the monazite of the grain size fraction 0–15 μm appear fully liberated. This is the best liberation among all (virtual) sieve grain size classes, as monazite is often locked by carbonates and fluorite in coarser particle size fractions (Figure 3d). Interestingly, the locking of monazite with fluorite is highest in the (virtual) sieve size fraction of 15–40 μm (Figure 3d). A longer comminution at 90 minutes resulted in no further improvement of monazite liberation in the size fraction 0–15 μm. For the larger sieve size fractions and the complete sample, the longer grinding time results in a moderate increase of the cumulative mass recovery of about 5% for the liberation class 95–100% (Figure 3b,c).

In the complete samples, the REE-P-monazite in the particles are often locked by carbonates (at 15–18 wt %), and with fluorite (at ~2 wt %, Figure 3d). With increasing sieve grain size fraction, the proportion of inherent carbonate minerals also increases, but with slightly lower values for the long comminution test at 90 min. At the locking of REE-P-monazite with fluorite the highest proportions are observed in the (virtual) sieve grain size fraction of 15–40 μm (Figure 3d). As a consequence of the results presented above, a multi-phase grinding process with only short periods of milling and intermittent classification has been established. This prevented the undesirable formation of fines and associated losses of REE-P-monazite. In addition, a sizing step by hydrocyclones was introduced to reject slimes (<5 μm particle size) prior to flotation.

4.3. Case Study 3: Flotation

The presented method for REE mineral classification was deployed in the evaluation of mineral processing tests for a further REE carbonatite ore. The studied samples were taken from multi-stage open cycle flotation tests (Figure 4). Previous to flotation, multi-stage comminution was performed with an interim classification step at 40 μm and recirculation of the >40 μm grain size fraction, followed by de-sliming with removal of the fraction <5 μm using a hydrocyclone. The de-slimed material was the feed to multi-step flotation including rougher flotation, scavenger flotation, two rougher-cleaner steps and one scavenger-cleaner flotation. Wet re-grinding was applied to the rougher concentrate prior to the cleaner stages, and also to the middling concentrate from the scavenger flotation prior to an additional scavenger-cleaner stage (Figure 4). This approach was chosen to accomplish a further improvement of the REE-liberation for boosting REE- and Y-recovery.

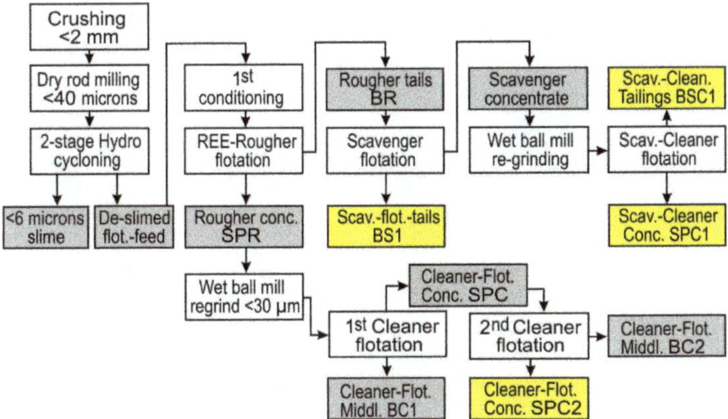

Figure 4. Flow scheme of multi-stage open cycle flotation tests with REE carbonatite ore, containing monazite and bastnaesite as principal REE minerals. Positions of analysed samples along the flow scheme are marked in yellow, flotation products which are not analysed are marked in grey.

Polished epoxy grain mounts with 30 mm in diameter were prepared from four flotation process samples (Figure 4) and analysed by automated MLA in the XBSE mode. Between 317,000 and 340,000 particles were analysed in each block during 4–5 h. Sample SPC2 represents the final concentrate after two cleaner flotation steps. Sample SPC1 is a middling concentrate produced by scavenger-cleaner flotation of a rougher-scavenger concentrate (following re-grinding). The sample BSC1 is a middling from the scavenger-cleaner stage that still contains liberated REE-mineral fines. The sample BS1 is the final tailing (Figure 4).

After EDX spectra classification and grouping, the REE-P-monazite and the REE-F-phases are the dominant groups among the REE-bearing minerals, and are denoted as monazite and bastnaesite for simplification (Figure 5a). The gangue minerals are the carbonate minerals dolomite, siderite and ankerite (regrouped as siderite-ankerite) but also apatite and quartz. The mineral grain sizes for monazite and bastnaesite in the pristine ore do not exceed 30 μm. In the cumulative grain size distribution curves the P50 values are 8 μm for monazite and 11 μm for bastnaesite. In all samples from this ore type, the P50 grain sizes for monazite are lower than those for bastnaesite (Figure 5b).

Virtual sieving based on the parameter equivalent circle (EC) diameter was performed at several sieve grain size classes. Coarse sieve grain size fractions (>40 μm) are not further considered here, as they account only for ~5.2 wt % in the sample BS1 and less than 0.5 wt % in the other three samples. In the final concentrate SPC2 a grade of 33.28 wt % of monazite and of 18.47 wt % of bastnaesite is achieved (Figure 5a). In contrast, only 0.59 wt % monazite and 0.29 wt % of bastnaesite report to the final tailings (sample BS1). The contents of the REE-Low-Mix spectra group in sample BSC1 is at 2.24 wt % fairly elevated when compared to the other samples with modes markedly below 1.0 wt %.

The REE minerals in the flotation tests reach their highest grades in the virtual sieve grain size class of 0–15 μm (Figure 5a,c). As expected, the multi-stage grinding and de-sliming process produced a narrow range of particle sizes in the virtual sieve grain size classes 0–15 μm and 15–40 μm. The cumulative grain size distribution curves display P50 particle sizes between 9 μm in sample BSC1 and 18 μm in sample BS1 (Figure 5b). The concentrates SPC2 and SPC1 display intermediate P50 particle sizes of ~12 μm.

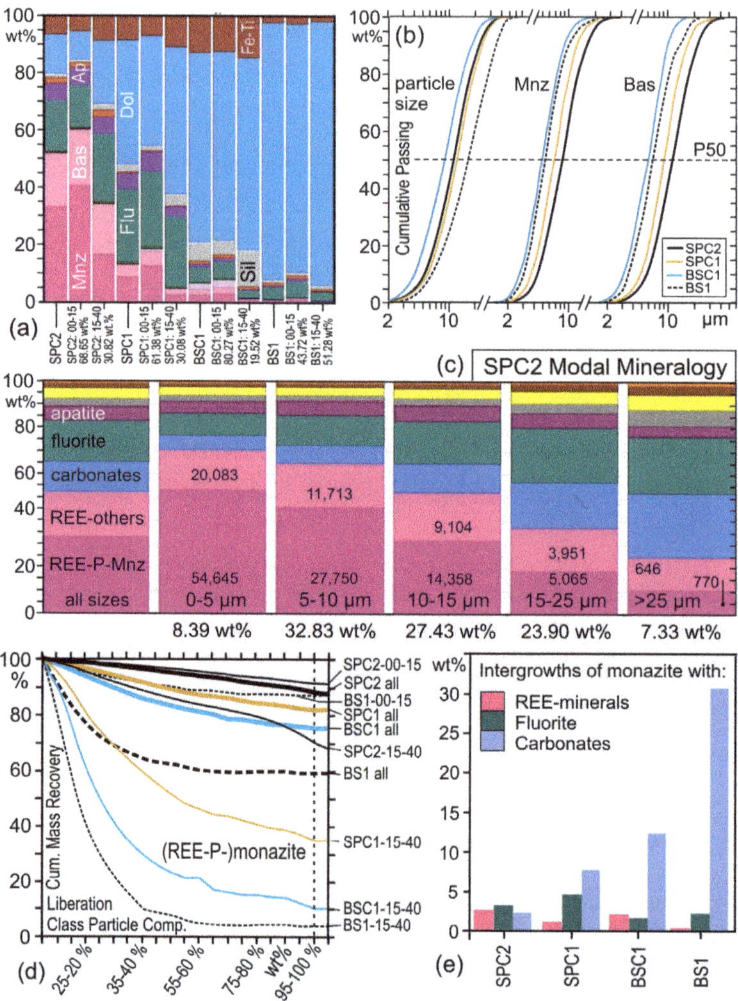

Figure 5. Results of mineral processing tests with the steps SPC2-SPC1-BSC1-BS1 during multistage flotation of REE carbonatite ore (see Figure 4 for positions of samples in the flow scheme. (**a**) Modal mineralogy (in wt %) of complete samples and selected particle size fractions after virtual sieving with the filter mode equivalent circle diameter. Ap—apatite; Bas—bastnaesite; Dol—dolomite; Fe-Ti—Fe-Ti-minerals; Flu—fluorite; Mnz—REE-P-monazite; Sil—silicate minerals. SPC2—second cleaner concentrate; SPC1—scavenger cleaner concentrate; BSC1—scavenger cleaner middlings; BS1—tailings. (**b**) Cumulative particle size distributions (cumulative passing) of the step samples of the mineral processing tests in (a), for all particles, for (REE-P)-monazite (Mnz) and bastnaesite (Bas). (**c**) Modal mineralogy of virtual sieve fractions of concentrate SPC2. Proportions of fractions are listed below the columns. Maximum modes of REE minerals are found in the fraction 0–5 µm. Numbers are grain counts. (**d**) Mineral liberation of (REE-P)-monazite in terms of proportion in wt % of particle composition. Data for complete samples (labelled as all) in thick lines; data for selected particle size classes in microns after virtual sieving in thin lines. Same legend as in (b). (**e**) Intergrowth relationships of non-liberated (REE-P)-monazite with other REE-minerals, fluorite and carbonates.

SEM-based image analysis is currently the only available routine analytical method to quantify parameters such grain sizes or liberation of distinct minerals in fine-grained material without a previous mechanical mineral separation. Thus, the effects of selective comminution in mineral processing can be critically assessed [32]. In this case study, the monazite has a P50 value of 8.5 μm, whereas the bastnaesite has a P50 of 12 μm in the concentrate sample SPC2 (Figure 5b). Furthermore, it is noted that the P50 grain sizes of the REE minerals in the concentrate samples are always higher than in the tailings (Figure 5b).

Unsurprisingly, the liberation of monazite in the parameter particle composition is for all samples always best in the (virtual) sieve grain size class 0–15 μm. In this grain size class, the liberation class 95–100% in the final concentrate sample SPC2 has a value of cumulative mass recovery of 93% (Figure 5d). For the complete sample SPC2 the cumulative mass recovery of the monazite and the other REE minerals has a very high value of 88% for monazite and 82% for bastnaesite for the liberation class 95–100% (Figure 5d). For sample BS1 these values are minimal at 60% monazite and 56% bastnaesite for the liberation class 95–100%. This illustrates the need for an efficient comminution and liberation.

Grain size dependent trends are exemplified for three particle size fractions (Figure 5a). At grain size fractions below 15 μm these trends are continued, as exemplified for sample SPC2, so that at grain size fractions below 5 μm, more than 70 wt % mode of REE minerals are observed (Figure 5c). In the grain size fraction <15 μm the 95–100% liberation of monazite is well above 90% cumulative mass recovery (Figure 5d). This emphasises that high modes in distinct grain size fractions *in combination* with a high degree of liberation are important for a later enrichment of REE minerals. However, this positive effect is partly counteracted by the very fine grain size of the liberated grains.

In the concentrate sample SPC2 the monazite is mostly in contact with fluorite (3.1 wt %) and carbonates (2.4 wt %) but also often intergrown with other REE minerals (2.5 wt %). In sample SPC1 the intergrowth with carbonates has the highest value at 7.7 wt %. In the tailings BS1 the target mineral group monazite is mostly locked by carbonates (31.2 wt %) and rarely locked within fluorite (Figure 5e). Liberation data can be used to determine recovery curves, also known as mineral grade vs. recovery curves [41,42]. These curves allow the comparison of efficiencies in mineral-processing schemes. The values for the curves are defined by the proportions of the given mineral in wt % in the various liberation classes. As can be expected, the final concentrate SPC2 displays the best curve for monazite, whereas the tailings BS1 illustrate the worst case (Figure 6a). This is confirmed by the (virtual) sieve grain size fractions, where curves for the class 0–15 μm are more favourable than those observed for bulk (i.e. unsieved) samples. For monazite, the larger (virtual) sieve size fraction 15–40 μm shows a more advantageous curve for the concentrate SPC2 when compared to SPC1. In the coarse (virtual) sieve grain class in sample SPC2 one can recognise a potential for a partition of more monazite. The curve for bastnaesite displays a more advantageous trend than that for monazite (Figure 6a).

This assessment of potential recoveries is especially interesting for the scavenger and cleaner sample BSC1 and the tailings in sample BS1. For these samples, the (virtual) sieve grain size fraction 0–15 μm displays a quite favourable recovery curve for the REE-bearing minerals (Figure 6b,c). In contrast, the recovery curve for the (virtual) sieve grain size fraction 15–40 μm displays a potential for further recovery of REE minerals by improving their liberation, possibly by re-grinding. This potential can be also evaluated and visualised by a simple line-up of the particles with REE-bearing minerals (Figure 6d).

The line-up function is an important tool within the MLA processing software packages. Even in the tailings sample BS1 numerous well-liberated REE mineral grains remain. Most of these have grain sizes <10 μm, i.e. they can be expected to not float well. The loss of such REE mineral grains is undesirable, but technically induced. The reasons of non-floating fines may be: (1) insufficiently adapted hydrodynamics for fines flotation; (2) a too short flotation time (kinetic problem); (3) insufficiently adapted bubble size distribution; and (4) an insufficiently optimized reagent dosage. Another important observation in the line-up view is the presence of REE mineral grains locked in coarse-grained carbonate particles. In this case the liberation of REE mineral grains may be improved by finer grinding, so that these grains will also float. However, further grinding will inevitably also result in more fines, a typical trade-off when developing a process flow sheet.

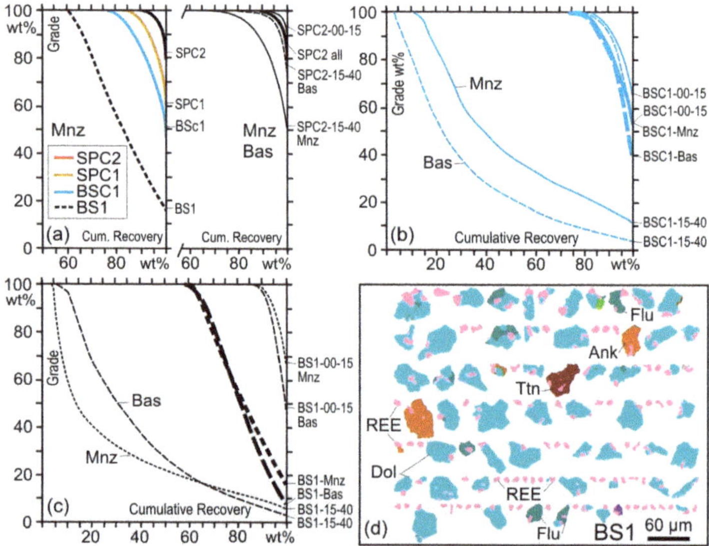

Figure 6. (**a–c**) Mineral grade vs recovery curves of REE carbonatite ore samples in mineral processing tests involving multistage flotation. Data of 100 wt % cumulative recovery is given by the mode of bastnaesite (Bas—broken lines) and REE-P-monazite (Mnz). Data of 100 wt % mineral grade is given by the proportion in wt% of fully liberated grains, extracted from the mineral liberation data in Figure 5d. SPC2—second cleaner concentrate; SPC1—scavenger cleaner concentrate; BSC1—scavenger cleaner middlings; BS1—scavenger tailings. Data for the complete samples in thick lines; data for particle size classes (in microns, virtual sieving) in thin lines. (**d**) Particle line-up of monazite and bastnaesite (REE) in the BS1 scavenger tailings sample. Note that many REE mineral particles are fully liberated but did not float. REE mineral grains of same size are mostly enclosed in carbonate (ankerite Ank, dolomite Dol), fluorite (Flu) and titanite (Ttn) particles.

5. Analytical Uncertainties

The presented data from grain mounts of processing products are based on the analysis of 336,000–599,000 individual particles in a single sample. When the analytical uncertainties are evaluated, the mineral mode appears as the most prominent parameter in comparison to the particle and grain sizes and their shape geometries. An MLA measurement starts with the acquisition of the first frame in the centre of a round epoxy grain mount block. The subsequent frames are then arranged in a single spiral toward the margin of the block [16]. During the preparation of the grain mount block, particle separation may be induced by the stirring of the particles into the liquid epoxy and subsequent gravitational subsidence of high-density particles during epoxy hardening. These effects can lead to heterogeneous particle distributions in the surface of a polished block [42,43], an effect that is particularly prominent for sample materials (mineral mixtures) with large differences in particle sizes and/or densities. For evaluating this potential uncertainty induced by heterogeneous particle distribution in the polished block surface, the full datasets with >300.000 particles were compared to the datasets from the inner part of the spiral. At a scale of 0–35 wt % the data align almost perfectly (Figure 7a). As a consequence the measurement of the inner spiral appears representative of the complete sample. However, when the inner spirals and the outer spirals are compared at the scale 0–5 wt % of mode, it is obvious that minerals with a mode below 1 wt % may be heterogeneously distributed within a sample, as is the case for the REE-Low-Mix group in sample M45 (Figure 7b). This apparent heterogeneity is attributed to a nugget effect for minor and trace minerals induced during sample preparation. Yet, it is encouraging that the modes of the other mineral groups match very well between the different data sets.

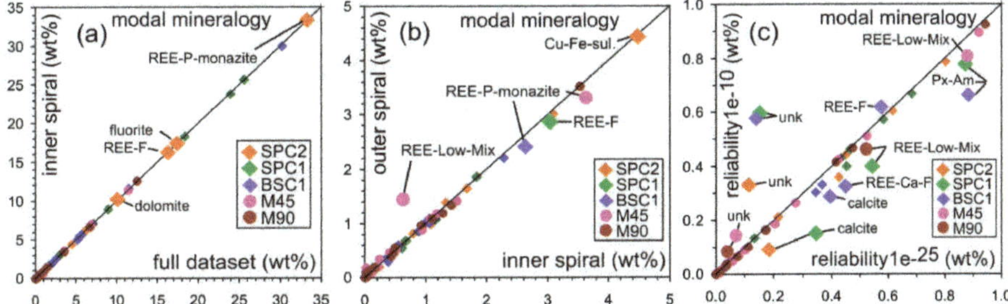

Figure 7. Influence of various parameters on the modal mineralogy of processed REE carbonatite ores. Datasets from single samples are based on the analysis of 300,000–600,000 particles in epoxy grain mount blocks of 30 mm in diameter. Samples are: SPC2—second cleaner concentrate; SPC1—scavenger cleaner concentrate; BSC1—scavenger cleaner middlings; BS1—scavenger tailings; M45, M90 are samples of comminution test with 45 and 90 min of milling. See legend insets. (**a**) Comparison of full dataset composed by inner and outer spiral of measurement frames vs. inner spiral, in the scale 0–35 wt %. At this scale a measurement of the inner spiral appears as sufficient. (**b**) Comparison of the inner spiral vs. the outer spiral of measurement frames, in the scale 0–5 wt %. Although some outliers with low modes exist, the epoxy block displays homogeneous distribution of particles. (**c**) Influence of the parameter reliability of conformity of EDX spectra during measurement classification against reference spectra set. The values of reliability of conformity are intermediate ($1e^{-25}$) and high ($1e^{-10}$) and refer to grouped spectra. A high reliability of conformity leads to a significant increase of "unknown" (unk) which is only relevant for modes between 0 and 1 wt %.

There are also uncertainties related to the spectra classification. The EDX spectra classification algorithm for the MLA software packages follows the principle of *best match* along a scale of reliability between $1e^{-10}$ (absolute conformance) to $1e^{-100}$ (no conformance), as outlined by [16]. REE-bearing minerals display a comparably complex pattern of X-ray emission lines, with many peaks and sub-peaks that are marked by considerable interference (Figure 1g–i). Due to the complex X-ray spectra characteristic for REE-bearing there is considerable risk that EDX spectra are not at all classified, if classification is carried out at a high reliability value. The classification algorithm allows no alternative assignment to another EDX reference spectrum or to another mineral in the list. Due to this principle, the spectra which cannot be classified by the higher reliability scale value will remain as unknown and increase the mode of *unknown* grains. For the study presented here the sample EDX spectra were thus classified by the reliability values of $1e^{-10}$ (high degree of conformance) and $1e^{-25}$ (fair degree of conformance). The latter reliability value is applied to process samples, in an effort to reduce the amount of unknowns below 0.1 wt % mode (by assigning the specific weight of carbon to the unknown spectra).

Applying a reliability value of $1e^{-25}$ to the samples of the third case study, the modes of unknowns remained low, ranging between 0.04 and 0.15 wt %. These values increase to 0.21–0.59 wt % when a reliability value of $1e^{-10}$ is applied to the same data sets. For the modes above 1 wt %, the differences between the two classification schemes are negligible and far below 1 wt %, except for dolomite (−1.2 wt %) in sample BSC1. The largest differences are observed for minor and trace minerals (modal abundance <1.0 wt %; Figure 7c). It is obvious that the increase of the unknowns leads to a reduction of the modes of REE mineral groups. This is the case for the REE-Low-Mix and the REE-Ca-F groups in the samples M90, SPC1 and BSC1 (Figure 7c). For the modes above 1 wt % (not shown), the differences are marginal. The samples of the grinding tests display similar trends. Here, one also observes a reduction of the REE-Low-Mix group, presumably induced by the low intensities along the LREE spectra.

6. Conclusions

Many REE mineral deposits are marked by a diverse set of REE bearing minerals. These REE-bearing minerals are distributed across several mineral classes, representing complex solid solutions with diverse substitutions, and crystallise in numerous hydrous species. All of these factors result in considerable problems in labelling the energy-dispersive X-ray spectra and quantitative element data from REE bearing minerals with the corresponding mineral names.

For the efficient examination of automated SEM mineral liberation analysis data of REE ores, a generic system of labelling the reference EDX spectra from REE-bearing minerals is proposed. This generic labelling is based on quantitative elemental EDS analyses of the REE-bearing minerals, placing particular emphasis on the elements Si, Ca, F and P. EDS spectra are assigned to the following groups (1) REE-P-monazite; (2) REE-Ca-Si-P (represented by britholithe); (3) REE-Ca-F (represented by synchysite); (4) REE-F (represented by bastnaesite and parisite), and a further group (5) REE-Low-Mix. The latter comprises of spectra with low counts in the energy range of the LREE. In case studies, this classification approach has been applied to classify automated SEM-MLA measurements on REE syenite and REE carbonatite ores.

In a REE syenite ore the fine-grained REE minerals in aggregates up to 1 mm in size are closely intergrown with Fe-Ti phases and phyllosilicates. This causes a hard mineability of the REE ores, although the bulk rock REE concentrations are convenient. In comminution tests of a REE carbonatite ore, a longer grinding time of 90 min, compared to 45 min, leads to no significant improvement of the liberation of the dominant REE mineral monazite. A successful concentration of monazite and bastnaesite to >50 wt % in a REE carbonatite ore requires a multi-stage flotation process with regrinding and de-sliming. A considerable proportion of well liberated yet fine REE mineral particles did not float and were lost in the tailings caused by insufficiently optimized flotation conditions and generally known problems with flotation slimes.

The datasets were classified with the spectra list involving generically labelled REE spectra with high ($1e^{-10}$) and fair ($1e^{-25}$) reliability of match. When classified at a reliability value of $1e^{-10}$, one can state an increase of the *unknowns* by a factor 2, however, the mode proportions of *unknowns* still remains below 1.0 wt %. A high reliability of match ($1e^{-10}$) induces a reduction of the mode proportions of the REE-Low-Mix group and other REE spectra groups in favour of the *unknowns*. At mode proportions of >5 wt% the effects of the reliability values are marginal and almost negligible.

The case studies illustrate the generic characters of the classification approach, as it is found to be highly applicable to different types of REE ores and mineral-processing products.

Author Contributions: Conceptualization, B.S., G.M., J.G.; methodology, B.S.; investigation, G.M., B.S.; resources, J.G.; data curation, B.S., G.M.; Writing—Original Draft preparation, B.S.; Writing—Review and Editing, B.S., G.M., J.G.; visualization, B.S. and G.M.; project administration, G.M.; funding acquisition, G.M. and J.G.

Funding: This research was supported by TU Bergakademie Freiberg and the Helmholtz Institute Freiberg for Resource Technology.

Acknowledgments: The authors acknowledge the great expertise of A. Bartzsch, R. Wuerkert and M. Stoll at the difficult preparation of numerous epoxy grain mount blocks from REE carbonite ores at Helmholtz Institute Freiberg for Resource Technology. S. Gilbricht is thanked for her untiring support during the automated SEM-MLA measurements at the instruments of the Geometallurgy Laboratory at the TU Bergakademie Freiberg. The authors acknowledge also the constructive comments of three reviewers to the manuscript.

Conflicts of Interest: The authors declare no conflict of interest.

References

1. Chakhmouradian, A.R.; Wall, F. Rare earth elements: Minerals, mines, magnets (and more). *Elements* **2012**, *8*, 333–340. [CrossRef]
2. Guarino, V.; de' Gennaro, R.; Melluso, L.; Ruberti, E.; Azzone, R.G. The transition from miaskitic to agpaitic rocks as marked by the accessory mineral assemblages, in the Passa Quatro alkaline complex (southeastern Brazil). *Can. Mineral.* **2019**, *57*, 1–23. [CrossRef]

3. Guarino, V.; Wu, F.Y.; Melluso, L.; Gomes, C.B.; Tassinari, C.C.G.; Ruberti, E.; Brilli, M. U-Pb ages, geochemistry, C-O-Nd-Sr-Hf isotopes and petrogenesis of the Catalão II carbonatitic complex (Alto Paranaíba Igneous Province, Brazil): Implications for regional-scale heterogeneities in the Brazilian carbonatite associations. *Int. J. Earth Sci.* **2017**, *106*, 1963–1989. [CrossRef]

4. Melluso, L.; Guarino, V.; Lustrino, M.; Morra, V.; de' Gennaro, R. The REE- and HFSE-bearing phases in the Itatiaia alkaline complex (Brazil), and geochemical evolution of feldspar-rich felsic melts. *Mineral. Mag.* **2017**, *81*, 217–250. [CrossRef]

5. Marks, M.A.W.; Markl, G. A global review of agpaitic rocks. *Earth-Sci. Rev.* **2017**, *173*, 229–258. [CrossRef]

6. Chakhmouradian, A.R.; Zaitsev, A.N. Rare earth mineralization in igneous rocks: Sources and processes. *Elements* **2012**, *8*, 347–353. [CrossRef]

7. Williams-Jones, A.E.; Migdisov, A.A.; Samson, I.M. Hydrothermal mobilisation of the rare earth elements—a tale of "Ceria" and "Yttria". *Elements* **2012**, *8*, 355–360. [CrossRef]

8. Richter, L.; Diamond, L.W.; Atanasova, P.; Banks, D.A.; Gutzmer, J. Hydrothermal formation of heavy rare earth element (HREE)–xenotime deposits at 100 °C in a sedimentary basin. *Geology* **2018**, *46*, 263–266. [CrossRef]

9. Jones, A.P.; Wall, F.; Williams, C.T. *Rare Earth Minerals: Chemistry, Origin and Ore Deposits*; The Mineralogical Society Series 7; Chapmann & Hall: London, UK, 1996; p. 372.

10. Yang, K.F.; Fan, H.R.; Santosh, M.; Hu, F.F.; Wang, K.Y. Mesoproterozoic carbonatitic magmatism in the Bayan Obo deposit, Inner Mongolia, North China: Constraints for the mechanism of super accumulation of rare earth elements. *Ore Geol. Rev.* **2011**, *40*, 122–131. [CrossRef]

11. Long, K.R.; Van Gosen, B.R.; Foley, N.K.; Cordier, D. *The Principal Rare Earth Elements Deposits of The UNITED STATES—A Summary of Domestic Deposits and a Global Perspective*; USGS Scientific Investigation Report 2010–5220; USGS: Reston, VA, USA, 2010; p. 96. Available online: https://pubs.usgs.gov/sir/2010/5220/ (accessed on 30 August 2019).

12. Hatch, G.P. Rare Earth Elements: Dynamics in the Global Market for Rare Earths. *Elements* **2012**, *8*, 341–346. [CrossRef]

13. Mariano, A.N.; Mariano, A., Jr. Rare earth elements: Rare earth mining and exploration in North America. *Elements* **2012**, *8*, 369–376. [CrossRef]

14. Kynicky, J.; Smith, M.P.; Xu, C. Rare earth elements: Diversity of rare earth deposits: The key example of China. *Elements* **2012**, *8*, 361–367. [CrossRef]

15. Reisman, D.; Weber, R.; McKernan, J.; Northeim, C. *Rare Earth Elements: A Review of Production, Processing, Recycling, and Associated Environmental Issues*; EPA Report; EPA/600/R-12/572; U.S. Environmental Protection Agency (EPA): Washington, DC, USA, 2013. Available online: https://nepis.epa.gov/Adobe/PDF/P100EUBC.pdf (accessed on 30 August 2019).

16. Weng, Z.H.; Jowitt, S.M.; Mudd, G.M.; Haque, N. Assessing rare earth element mineral deposit types and links to environmental impacts. *Appl. Earth Sci.* **2013**, *122*, 83–96. [CrossRef]

17. Dutta, T.; Kim, K.-H.; Uchimiya, M.; Kwonc, E.E.; Jeon, B.-H.; Deep, A.; Yun, S.-T. Global demand for rare earth resources and strategies for green mining. *Environ. Res.* **2016**, *150*, 182–190. [CrossRef] [PubMed]

18. Wall, F.; Rollat, A.; Pell, R.S. Responsible sourcing of critical metals. *Elements* **2017**, *13*, 313–318. [CrossRef]

19. Rollinson, H. *Using Geochemical Data: Evaluation, Presentation Interpretation*; Longman Group: Harlow, UK, 1993; p. 354.

20. Gu, Y. Automated scanning electron microscope based mineral liberation analysis. An introduction to JKMRC/FEI Mineral Liberation Analyser. *J. Miner. Mater. Character. Eng.* **2003**, *2*, 33–41. [CrossRef]

21. Fandrich, R.; Gu, Y.; Burrows, D.; Moeller, K. Modern SEM-based mineral liberation analysis. *Int. J. Miner. Process.* **2007**, *84*, 310–320. [CrossRef]

22. Lastra, R. Seven practical application cases of liberation analysis. *Int. J. Miner. Process.* **2007**, *84*, 337–347. [CrossRef]

23. Petruk, W. *Applied Mineralogy in the Mining Industry*, 1st ed.; Elsevier Science: Amsterdam, The Netherlands, 2000; p. 288.

24. Chang, L.L.A.; Howie, R.A.; Zussman, J. *Rock-Forming Minerals*, 2nd ed.; Non-silicates: Sulphates, carbonates, phosphates, halides; Geological Society London: London, UK, 1998; Volume 5B, p. 383.

25. Thompson, W.; Lombard, A.; Santiago, E.; Singh, A. Mineralogical studies in assisting benefication of rare earth element minerals from carbonatite deposits. In Proceedings of the 10th International Congress for Applied Mineralogy (ICAM), Trondheim, Norway, 1–5 August 2011; Springer: Berlin/Heidelberg, Germany, 2012; pp. 665–672, ISBN 9788273851390.

26. Smythe, D.M.; Lombard, A.; Coetzee, L.L. Rare earth element deportment studies utilising QEMSCAN technology. *Miner. Eng.* **2013**, *52*, 52–61. [CrossRef]

27. Smith, D.G.W. Identification of rare earth and yttrium minerals using the minident database. In Russel, P.E. Microbeam Analysis. In Proceedings of the 24th Annual Conference of the Microbeam Analysis Society, Asheville, NC, USA, 16–21 July 1989; pp. 559–562.

28. Smith, D.G.W.; de St. Jorre, L. The MinIdent Data Base-examples of applications to the Thor Lake rare metal deposits. In Proceedings of the 14th General meeting of International Mineralogical Association, Stanford, CA, USA, 13–18 July 1986; p. 234.

29. Greb, V.G.; Guhl, A.; Weigand, H.; Schulz, B.; Bertau, M. Understanding phosphorous phases in sewage sludge ashes: A wet-process investigation coupled with automated mineralogy analysis. *Miner. Eng.* **2016**, *99*, 30–39. [CrossRef]

30. Pietranik, A.; Kierczak, J.; Tyszka, R.; Schulz, B. Understanding heterogeneity of a slag derived technosol: The role of automated SEM-EDS analyses. *Minerals* **2018**, *8*, 513. [CrossRef]

31. Schulz, B. Polymetamorphism in garnet micaschists of the Saualpe Eclogite Unit (Eastern Alps, Austria), resolved by automated SEM methods and EMP-Th-U-Pb monazite dating. *J. Metamorph. Geol.* **2017**, *35*, 141–163. [CrossRef]

32. Merker, R.G.; Leissner, T.; Schulz, B. Use of virtual fractions for MLA of Y-bearing REE ores. In Proceedings of the 28th International Mineral Processing Congress 2016, Québec, QC, Canada, 11–15 September 2016. Conference Proceedings Paper ID 894.

33. Merker, R.G.; Smith, D.L. Rare earths—espects of market and beneficiation. *World Metall.-Erzmetall.* **2018**, *71*, 189–197.

34. Pinckston, D.R.; Smith, D.G.W. Mineralogy of the Lake zone, Thor Lake rare-metals deposit, N.W.T., Canada. *Can. J. Earth Sci.* **1995**, *32*, 516–532. [CrossRef]

35. Sheard, E.R.; Williams-Jones, A.E.; Heiligmann, M.; Pederson, C.; Trueman, D.L. Controls on the concentration of zirconium, niobium, and the rare earth elements in the Thor Lake rare metal deposit, Northwest Territories, Canada. *Econ. Geol.* **2012**, *107*, 81–104. [CrossRef]

36. Jackson, B.R.; Reid, A.F.; Wittemberg, J.C. Rapid production of high quality polished sections for automated image analysis of minerals. *Proc. Australas. Inst. Miner. Metall.* **1984**, *289*, 93–97.

37. Echlin, P. *Handbook of Sample Preparation for Scanning Electron Microscopy and X-ray Microanalysis*, 1st ed.; Springer Science: Berlin/Heidelberg, Germany, 2009; p. 326. [CrossRef]

38. Finch, J.A.; Gomez, C.O. Separability curves from image analysis data (Technical Note). *Miner. Eng.* **1989**, *2*, 565–568. [CrossRef]

39. Leistner, T.; Peuker, U.A.; Rudolph, M. How gangue particle size can affect the recovery of ultrafine particles during froth flotation. *Miner. Eng.* **2017**, *9*, 1–9. [CrossRef]

40. Lastra, R.; Petruk, W. Mineralogical characterization of sieved and un-sieved samples. *J. Miner. Mater. Charact. Eng.* **2014**, *2*, 40–48. [CrossRef]

41. Neethling, S.J.; Cilliers, J.J. Grade-recovery curves: A new approach for analysis of and predicting from plant data. *Miner. Eng.* **2012**, *36*, 105–110. [CrossRef]

42. Sandmann, D.; Schulz, B.; Gutzmer, J.; Eickhoff, A. Controls on the Industrial Regrind Milling Process of Coarse-Grained Ores by SEM-Based Image Analysis; Abstracts 89. In Proceedings of the DGK, DMG & +MG, Salzburg, Austria, 20–24 September 2011; p. 131.

43. Leigh, G.M.; Sutherland, D.N.; Gottlieb, P. Confidence limits for liberation measurements. *Miner. Eng.* **1993**, *2*, 155–161. [CrossRef]

Article

Mineralogical Imaging for Characterization of the Per Geijer Apatite Iron Ores in the Kiruna District, Northern Sweden: A Comparative Study of Mineral Liberation Analysis and Raman Imaging

Patrick Krolop [1,*], Anne Jantschke [2], Sabine Gilbricht [1], Kari Niiranen [3] and Thomas Seifert [1]

[1] Division of Economic Geology and Petrology, Technische Universität Bergakademie Freiberg, Akademiestraße 6, 09599 Freiberg, Germany
[2] Chair of Bioanalytical Chemistry, Dresden University of Technology, 01069 Dresden, Germany
[3] Luossavaara Kiirunavaara AB (publ.), 98186 Kiruna, Sweden
* Correspondence: patrick.krolop@mineral.tu-freiberg.de; Tel.: +49-373-139-3516

Received: 14 August 2019; Accepted: 6 September 2019; Published: 10 September 2019

Abstract: The Per Geijer iron oxide apatite deposits are important potential future resources for Luossavaara-Kiirunavaara Aktiebolag (LKAB) which has been continuously mining magnetite/hematite ores in northern Sweden for over 125 years. Reliable and quantitative characterization of the mineralization is required as these ores inherit complex mineralogical and textural features. Scanning electron microscopy-based analyses software, such as mineral liberation analyzer (MLA) provide significant, time-efficient analyses. Similar elemental compositions of Fe-oxides and, therefore, almost identical backscattered electron (BSE) intensities complicate their discrimination. In this study, MLA and Raman imaging are compared to acquire mineralogical data for better characterization of magnetite and hematite-bearing ores. The different approaches demonstrate advantages and disadvantages in classification, imaging, discrimination of iron oxides, and time consumption of measurement and processing. The obtained precise mineralogical information improves the characterization of ore types and will benefit future processing strategies for this complex mineralization.

Keywords: MLA; mineral processing; iron ore; Kiruna; Raman spectroscopy; magnetite; hematite

---

1. Introduction

The mineralogical characterization of ore deposits using scanning electron microscopy (SEM)-based automated mineralogy systems such as mineral liberation analyzer (MLA) or QEMSCAN® are of major importance for mineral processing. Declining ore grades, the reconnaissance of new targets, and demands for improved energy efficiency require fast, reliable, and high-quality information on mineral resources. Highly variable and complex mineralogy affects mineral processing and extractive metallurgy, as has been shown by authors in the past [1–6]. Collecting quantitative data of mineral distributions, associations and textures are one of its key abilities. Using this technique, information gained by traditional bulk mineralogical analysis, chemical assay, and optical microscopy can be supported and expanded. Automated mineralogy is widely used in the mining business as it allows rapid identification and characterization of key minerals in ore and rock samples. Furthermore, it can be applied to all sorts of different mineral deposit types [1,7–11]. Mineral liberation analysis (MLA) software is one of the widely applied SEM-based applications offering a rapid quantitative characterization of mineral species and their relations in polished samples [12]. It is commonly used in conjunction with other micro-compositional or microstructural techniques such as electron-probe microanalyses (EPMA) or X-ray diffraction (XRD). However, SEM-based techniques cannot precisely distinguish mineral polymorphs (e.g., rutile-anatase-brookite) or discriminate between minerals having a similar elemental composition (e.g., magnetite and hematite). Especially the discrimination of magnetite and hematite becomes an important factor in future processing stages at the Luossavaara-Kiirunavaara Aktiebolag (LKAB) enterprise. Almost pure magnetite mineralization is only prevalent in the Kiirunavaara deposit. Increasing hematite contents in the other deposits in the Kiruna district (Malmberget and Leveäniemi), and most predominantly in Per Geijer will significantly influence the beneficiation and pelletizing process. In order to sustain efficient comminution and magnetic separation processes evaluation by automated mineralogy or similar methods will add significant value in the future.

The effect that similar average atomic numbers result in similar BSE intensities (BSE grey values) requires modification of the MLA technique, especially in the data processing stage. A technique for distinguishing between hematite and magnetite by MLA has been established by Figueroa et al. [13]. However, stable beam currents and even sample surfaces are obligatory for maintaining constant BSE grey values. As this often proves problematic, there is demand for additional analytic solutions.

In this study, Raman imaging is used as an alternative tool to highlight advantages and disadvantages of both methods with the focus on magnetite-hematite mineralization of the Per Geijer iron ore deposits close to the well-known Kiirunavaara deposit. This comparative approach for ore characterization provides detailed information by mapping mineral distributions of a potential new iron ore resource.

2. Geology of the Per Geijer Deposits

The iron oxide apatite ores (IOA) in the Kiruna area are hosted by the Svecofennian Kiirunavaara Group. Svecofennian rocks are represented by volcanic and sedimentary units generated by reworking of older crust and subduction and accretion of several volcanic arc complexes [14–17]. Three formations account for that group: (1) Hopukka Formation, (2) Luossavaara Formation, and (3) Matojärvi Formation. The giant Kiirunavaara ore body with pre-mining resources of more than 2000 Mt [18], is a massive tabular almost purely magnetite body with generally low phosphorus content. The smaller, magnetite-rich Luossavaara deposit north of Kiirunavaara is situated in the same stratigraphy. Closely associated with the east and northeast reside the Per Geijer orebodies at the stratigraphically upper contact of the Luossavaara Formation or within the lower part of the Matojärvi Formation [19]. These deposits are located north of the town of Kiruna and consist of five orebodies that have been mined in intervals during the 20th century as open pits. The Nukutus, Henry, and Rektorn mineralization occur stratigraphically at the upper contact of the Luossavaara Formation, while the Haukivaara and Lappmalmen deposits, at least partly, are located within the overlying Matojärvi Formation. Lappmalmen is a blind and, so far, unexploited ore body only known from exploration drilling. The ores exhibit large variations in texture, mineral composition, and relation to wall rocks [18]. Depending on the orebody and the vertical position of the mineralization in the deposit, magnetite and hematite occur in varying textures and proportions leading to different ore types (e.g., hematite-dominated, magnetite-dominated, magnetite/hematite-mixed), as proposed by [20]. Generally, hematite occurs together with magnetite often replacing it, especially close to the upper part of the deposit. The overall phosphorus content is high and occurs as apatite. Other main gangue minerals are carbonate and quartz. The Per Geijer deposits have an average composition of 40–50% Fe and 3–5% P but higher and lower iron and phosphorus contents can locally occur.

3. Materials and Methods

3.1. Samples and Preparation

Sixty-five samples were selected from drill cores representing all five Per Geijer ore bodies after specific pre-defined ore type sections and macroscopic petrographic evaluation. For this study, 18 polished thin sections were prepared by the preparation laboratory of the Helmholtz Institute Freiberg for Resource Technology. Sample HA-81155-2 was selected for comparative SEM-MLA and

Raman measurements because of the occurrence of magnetite, hematite, and Ti-bearing hematite in satisfying abundances.

3.2. Scanning Electron Microscope (SEM) and Mineral Liberation Analyzer (MLA)

Samples were studied with the scanning electron microscope (SEM) including mineral liberation analyzer (MLA) (FEI, Brisbane, Australia) software on a Quanta 650 FEG-MLA650F (FEI©, Brno, Czech Republic). The analyses were carried out at the geometallurgical laboratory of the Department of Mineralogy at TU Bergakademie Freiberg, Germany. Energy-dispersive X-ray spectroscopy (EDS) was performed with two Bruker Nano Dual X-Flash 5030 detectors (Bruker, Berlin, Germany), using an accelerating voltage of 25 kV and a working distance of 12 mm. The workflow of the SEM-MLA analysis described in the following sections is shown in Figure 1.

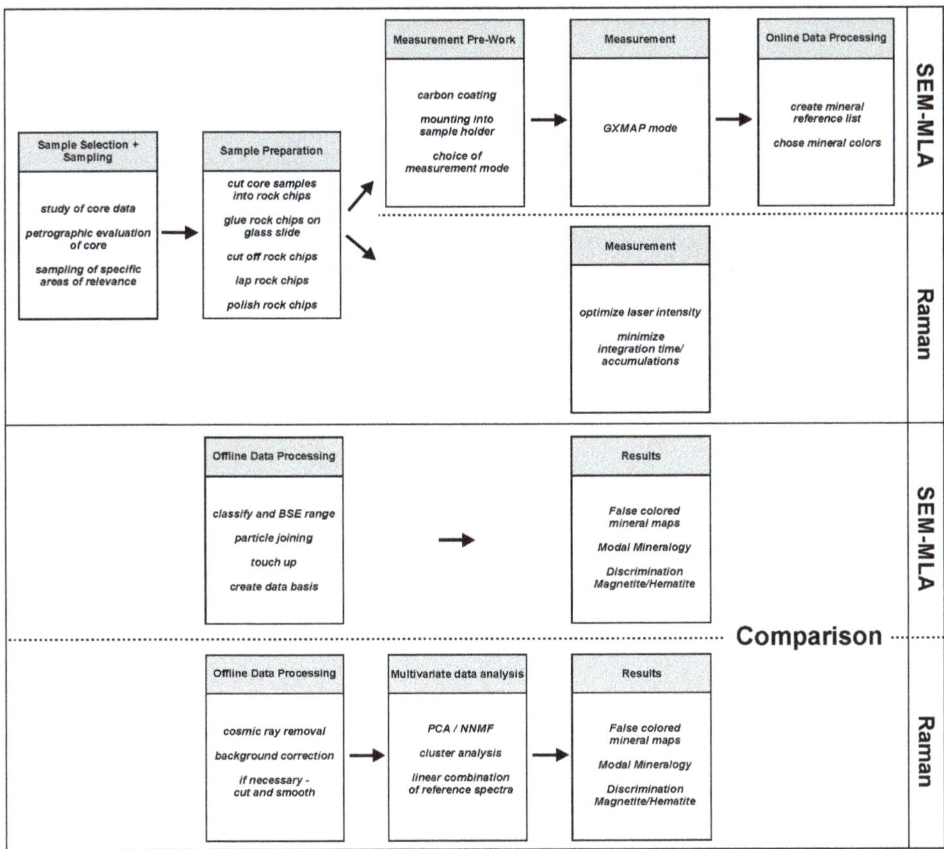

Figure 1. Workflow of the SEM-MLA and Raman analysis.

3.2.1. Measurement Mode and Setup

The grain-based X-ray mapping (GXMAP) mode was selected for this study. The GXMAP technique analyses selected grains (in this case, the whole thin section) with a closely spaced grid of X-ray measurements resulting in false-colored mineral maps [12,21]. Here, the whole sample was continuously mapped with an EDS analysis every 12 μm (slower mapping) or 36 μm (faster mapping). In both measurements, the frame resolution was 500 × 500 pixel with a magnification of either 3 or

2 µm/pixel, respectively. Based on backscattered images, the minerals were first separated according to their grey values and then mapped in a definite grid of EDS analysis. The basic setup included the definition of a BSE grey value range (BSE trigger). As element contents are responsible for the BSE contrast, chemically similar minerals have identical BSE grey levels. Thus, for hematite and magnetite classification, reference BSE values had to be manually defined. Furthermore, they were compared with optical reflected light images of the sample in which the two iron oxides have characteristic reflection properties. The details of the SEM image acquisition are listed in Table 1. A detailed description of the functionality of MLA and the measuring modes can be found in the literature [12,22–24].

Table 1. Details of the scanning electron microscope (SEM) + mineral liberation analyzer (MLA) and Raman measurements.

Device	MLA GXMAP (f)	MLA GXMAP (s)	Raman
Sample type Excitation energy	ca. 20 mm × 40 mm ts 25 kV	20 mm × 40 mm ts 25 kV	20 mm × 40 mm ts 532 nm, 5–20 mW laser power
Step size	36 µm	12 µm	30 µm
Acquisition time	7 ms	7 ms	275 ms/200 ms
Measurement time	0.75 h/cm^2	1.5 h/cm^2	9 h/cm^2 (30 µm Renishaw) 45 h/cm^2 (12 µm WITEC) 42 h/mm^2 (1 µm WITEC)

(s)—slow measurement, (f)—fast measurement, (ts)—thin section.

3.2.2. Mineral Reference List

A preliminary mineral reference list was generated by collecting mineral standards manually. Chemical compositions were obtained with the Bruker Esprit 1.9 software (Bruker, Berlin, Germany), mineral names, and formulae, as well as physical properties (e.g., density), were added.

3.2.3. Online Data Processing

The preliminary mineral list had to be adjusted after first classification due to larger unknowns. Unknowns are detected when the comparison between the spectra of the mineral reference list and the measured X-ray spectra fails. This error in consistency occurs when the limits of a spectrum matching threshold (80%) and/or a low count limit (800) exceed the predetermined values. Thus, an assignation to a suitable mineral phase was not possible. However, some unknowns can be the result of mixed spectra of two different phases.

3.2.4. Offline Data Processing

For off-line image and data processing collected X-ray spectra were classified. As for online data processing, comparison with the spectra of the mineral reference list is necessary. Therefore, the spectrum matching threshold and low count limit were set to 80% and 800 counts, respectively. However, the X-ray spectra and the BSE grey level of the different iron oxides (e.g., hematite and magnetite) are very similar. Consequently, the results from MLA were checked against the results from optical microscopy and manual correction was applied. For most samples, BSE grey value ranges were set at 150 to 190 for hematite and 191 to 250 for magnetite. These values were checked for every sample after classification and adjusted, if necessary. However, single frames in some samples yielded wrong classification of magnetite and hematite based on local BSE grey value variations. Manual correction was obtained by using the edit selected particles mode. By re-defining wrongly classified areas manually, the validity for the samples was improved.

To remove small unknowns and other noise from the image, the touch-up tool was used. All unknowns with a size smaller than 50 pixels were turned into the containing host (unknown to any host), which mainly affected pixels with mixed spectra. The remaining unknowns generated by holes and brackets of the thin section holder were turned into background (size < 2,000,000 pixels). Minerals, which feature a pixel size smaller than 50 or 70 that are located in another mineral, were transformed to the containing host mineral (mineral to any host). Finally, databases were created comprising information about modal mineralogy, mineral associations, as well as grain properties. For this study, only information of the modal mineral abundance was taken into account.

3.3. Raman Spectroscopy and Imaging

Raman maps were collected using an alpha300 M+ Raman Microscope System (WITec GmbH, Ulm, Germany) or an inVia Qontor Raman Confocal Microscope (Renishaw, Pliezhausen, Germany). The workflow of the Raman analysis described in the following sections is shown in Figure 1.

3.3.1. Measurement Mode and Setup

The measurements were carried out at Papiertechnische Stiftung (PTS) Heidenau and Center for Molecular Bioengineering at TU Dresden (BCUBE). Both systems were equipped with an upright microscope using a 532 nm laser excitation and a Charge Coupled Device (CCD) detector with 1024 pixel. The Renishaw system was used with a 100× objective (Leica Objective N plan EPI 100×/NA 0.85) using 10% laser power (corresponding to approximately 5 mW) and a 1800 gr/mm grating (spectral range: 60–1837 cm^{-1}). The Raman map of Figure 2 (2.127 cm × 2.232 cm) was collected with 30 μm step size and 275 ms exposure time using the Renishaw system.

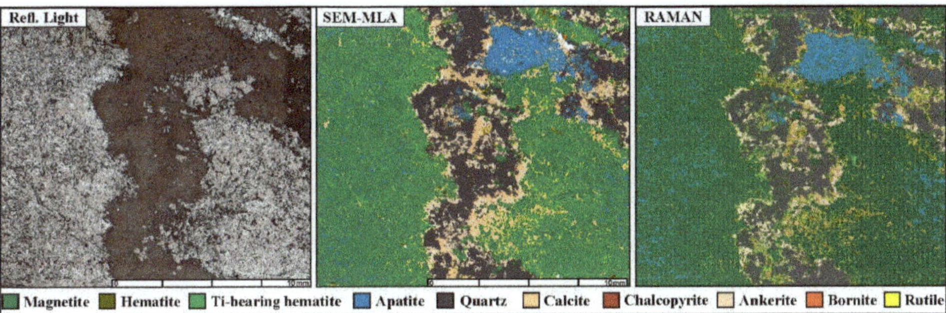

Figure 2. Mineral distribution maps from reflected light (optical microscopy) and processed SEM-MLA and Raman analysis.

The WITec confocal Raman system is based on a Zeiss microscope and was used with a 100× objective (Zeiss EC Epiplan NA 0.9) using 20 mW laser power and a 600 gr/mm grating (spectral range: −123–3777 cm^{-1}). The Raman maps of Figure 3 were collected using the WITec system. The overview map (2.38 cm × 1.3 mm) was collected with 12 μm step size, the detailed map with 1 μm step size (160 μm × 120 μm). In both cases, spectra were collected using 200 ms exposure time.

Before measurement, we optimized spectral acquisition parameters (especially laser power). Due to the large mapping areas (up to 2 cm × 2 cm), we decided to use the highest laser power while reducing exposure time as much as possible to achieve a high mapping speed. We started with point measurements on all minerals (acquisition time: 0.2 s) using increasing laser power to check for thermal decomposition or oxidation. This way, the maximum laser power for each mineral can be determined. In our samples, sulfides were found to be the most sensitive to laser damage resulting in significant visible and spectroscopical changes. The most sensitive minerals determined the laser power that was used. We did not observe any changes in the spectra of magnetite and hematite. In comparison, the literature reporting on hematite transformation [25], used exposure times from at least several seconds up to several minutes (10 s–5 min). Samples were checked for oxidation effects after measurements with optical reflected light microscopy.

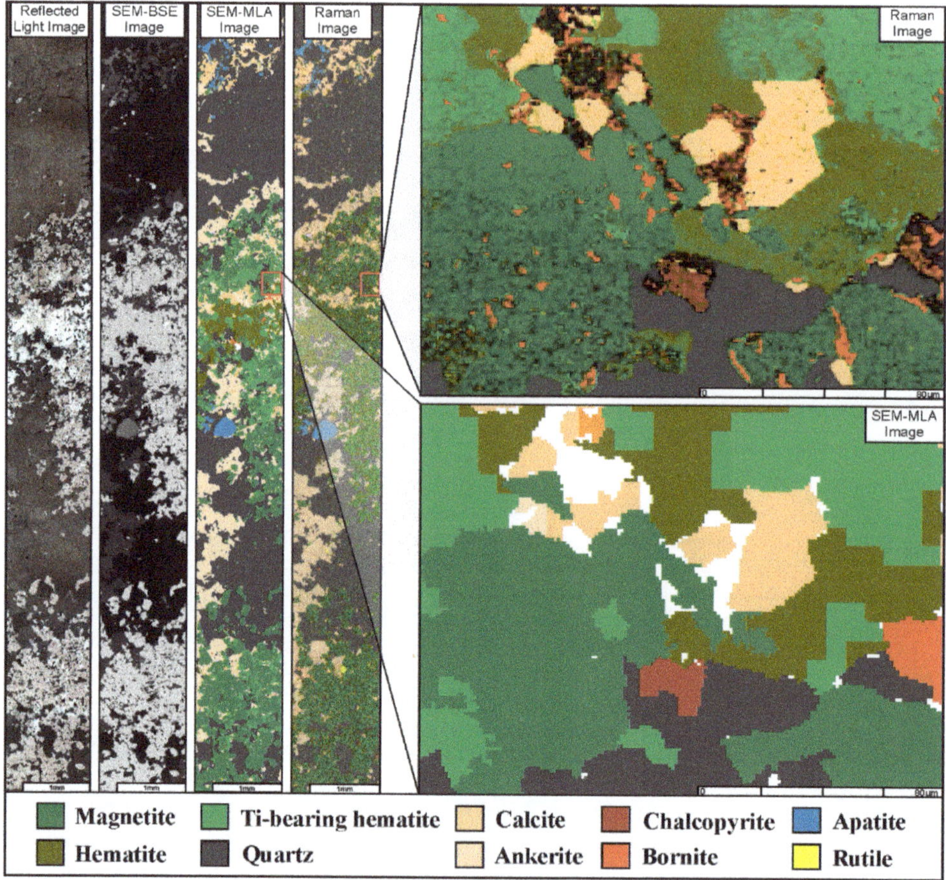

Figure 3. Reflected light, backscattered electron (BSE), false-colored MLA, and Raman images from the same area of sample HA-81155-2. Detailed comparison of magnetite, Ti-bearing hematite, and Ti-poor hematite is shown in zoomed-in maps.

3.3.2. Mineral Reference List

Reference spectra were collected either by spot measurements, isolated by non-negative matrix factorization (NMF), or extracted by averaging mineral-specific areas of Raman maps based on the EDS maps obtained for MLA. Before linear combination, the spectra were background-corrected and normalized. All reference spectra can be found in the Supporting Information (Figure S1).

3.3.3. Offline Data Processing-Data Preprocessing

Spectral datasets were imported into the in-house SpectralImaging software (TU Dresden, Dresden, Germany) on the basis of Matlab provided and developed by Matthias Finger (spectralimaging@outlook.com). First, cosmic ray removal was performed to eliminate spikes. Spectral datasets were typically cut to a wavelength of 100–1900 cm^{-1} where significant spectral differences were found. Baseline correction was performed by asymmetric least squares correction (smoothness 5, asymmetry 2). If necessary, smoothing (Savitzky-Golay, seven points, degree 2) was applied.

3.3.4. Offline Data Processing-Multivariate Analysis

Multivariate analysis was used to identify the number of principal components and to image their distribution. We used two data analyses methods, principal component analysis (PCA) and non-negative matrix factorization (NMF). Both techniques were conducted using the above-mentioned in-house SpectralImaging software based on Matlab.

PCA and NMF are both unsupervised statistical methods to reduce the dimensionality of large datasets. Using PCA, original variables of the dataset are summarized in new variables, the so-called principal components. These components are calculated as the solution of an eigenvalue problem responsible for the whole variance of the data set. As a result, PCA weights a set of principal components with positive and negative values to represent the full spectral information.

NMF is similar to PCA but the weights and factors are constrained to be positive. Therefore, it is well-suited to represent data with non-negative features. We found that the resulting spectral factors are easier to inspect. That is why, if necessary, non-negative matrix factorization (NMF) was used to extract spectral signatures of individual minerals from the total set. To validate these unsupervised methods, a supervised linear combination approach was used to map the distribution of selected components. All maps shown were prepared using linear combination of a set of measured reference spectra (see above).

4. Results

The following results are presented as comparison and verification of SEM-MLA and Raman mapping. Emphasis is given on imaging, discrimination of magnetite and hematite and modal mineralogy in area % with respect to the time demands for measurements and processing.

4.1. Classification and Imaging

The mineral distribution maps of both classifications (SEM-MLA and Raman) correspond well with the thin section of the reflected light photograph (Figure 2). Both methods recognized the present minerals. The classification with MLA was manually adjusted with a BSE grey level range from 150 to 190 for hematite and 191 to 250 for magnetite. Mineral boundaries are pre-defined based on the BSE grey values with the SEM-MLA. This function is not yet available for Raman imaging resulting in a rasterized resolution. General features, textures of the sample material and the abundance and location of important minerals are visible with both methods.

4.2. Discrimination of Magnetite and Hematite

In the reflected light image of the polished sample of apatite iron ore, magnetite, hematite, and gangue minerals (mostly apatite, carbonates, and silicates) can be seen (Figure 3). In SEM-based backscattered electron images, iron oxides are hardly distinguishable from each other due to similar BSE grey values. The processed false-colored mineral maps from MLA and Raman show all present phases. However, the precise discrimination of magnetite and hematite varies with both techniques. A smaller area (see the red square in Figure 3) was selected for high-resolution Raman mapping. Six minerals were spectrally identified within this smaller area, corresponding to magnetite, Ti-free hematite, Ti-bearing hematite, quartz, ankerite, bornite, and chalcopyrite. The different Raman spectra for hematite were assigned to Ti-bearing hematite by EDS data.

4.3. Modal Mineralogy

The modal mineralogy can be extracted in the Dataview software for SEM-MLA measurements. Raman does not commonly offer this data analysis feature as an in-built tool. However, quantification of phase abundances is possible based on the number of pixels identified for each phase in relation to the total number of pixels in the scanned area.

The modal mineralogy of the major and minor components obtained for the area displayed in Figure 2 shows similar results between the slow and the fast SEM-MLA and Raman measurements (Table 2). However, magnetite occurs in higher modal abundance in the measurement with lower step size (35.54 area %) compared to the fast measurement (33.62 area %) and Raman (34.9 area %). Contrary, the modal mineralogy of hematite and Ti-bearing hematite is elevated in the fast GXMAP but lower in Raman (Table 2). Quartz, apatite, and ankerite show slightly lower values compared to SEM-MLA. The modal mineralogy of calcite is in the range of ± 0.2 area % for all measurements.

Table 2. Modal mineralogy (area %) of major and minor components obtained by SEM + MLA and Raman measurements on sample HA-81155-2. Fe-oxides are highlighted for better comparison.

Mineral	MLA GXMAP(s)	MLA GXMAP(f)	Raman
Magnetite	**35.54**	**33.62**	**34.9**
Hematite_Ti	**14.62**	**15.59**	**11.6**
Quartz	22.02	21.60	19.5
Apatite	9.78	9.38	9.0
Ankerite	8.61	8.85	7.4
Calcite	7.68	7.70	7.6
Hematite	**0.72**	**2.21**	**0.7**

4.4. Time Constraints

For this study, SEM-based MLA measurements were performed with two different resolutions and step sizes on the same sample in addition to Raman mapping.

4.4.1. Measurement Time

The measurement time for fast GXMAP mapping with a resolution of 500 × 500 pixels and a step size of 36 μm was 6 h. Mapping with a step size of 12 μm and the same resolution took 12 h. The identical area of the same sample measured with Raman was the most time-intensive with 42 h using a step size of 30 μm. The measurement time of the detailed Raman map (Figure 3) took 1 h using a step size of 1 μm.

4.4.2. Data Processing Time

The processing time is dependent on many variables, including the operator in charge, technical equipment, e.g., computational power, software requirements, and functions. In SEM-MLA data processing of faster measurements using larger step sizes resulted in more unknowns and misidentification of phases. In this case, additional processing time was required due to manual correction of the sample set.

Data processing in Raman imaging depends strongly on the signal-to-noise ratio of the acquired spectra. Since processing is done mainly offline after data acquisition, it is crucial to ensure the best acquisition parameters (e.g., laser power, exposure time) before measurement. To discriminate similar minerals (e.g., Ti-bearing and Ti-poor hematite) longer acquisition times are recommended.

5. Discussion

The presented results and images of SEM-MLA and Raman measurements of magnetite and hematite-bearing ores from apatite iron oxide deposits in the Kiruna area depend, in both cases, mainly on the correct classification of the present minerals. For SEM-MLA, a comprehensive mineral database containing all present minerals that are correctly distinguished by their chemical information is inevitable. Raman imaging does not necessarily require any preexisting knowledge about the existing minerals. Therefore, online data processing is not necessary and data analysis happens solely offline after data acquisition, e.g., using unsupervised multivariate data analysis to identify different minerals.

When it comes to mineral classification, Raman shows some major advantages, especially in the discrimination of magnetite and hematite due to their distinct Raman spectral signatures [25–30]. A diagnostic Raman peak for magnetite is 663 cm^{-1} (T_{2g}), whereas hematite is dominated by strong peaks at 293 (E_g), 410 (E_g), 615 (E_g) and 1320 cm^{-1} (Figure 4). In comparison, Figueroa et al. [13] showed that using the SEM-based MLA method discrimination of magnetite and hematite is possible but needs significant adjustments prior to and during measurements. It was stated, that measurement times did not increase significantly but the process of time-consuming offline modification and the effect of fluctuations in beam current was neglected. The ability of Raman spectroscopy to intrinsically discriminate clearly between magnetite and hematite shows great potential for the characterization of iron oxide deposits. However, it needs to be noted that Fe(II)-containing minerals can be easily converted into Fe(III)-oxides like hematite by high laser powers [25,29,30]. In order to avoid conversion, it is crucial to optimize spectral acquisition parameters prior to mapping (see experimental details).

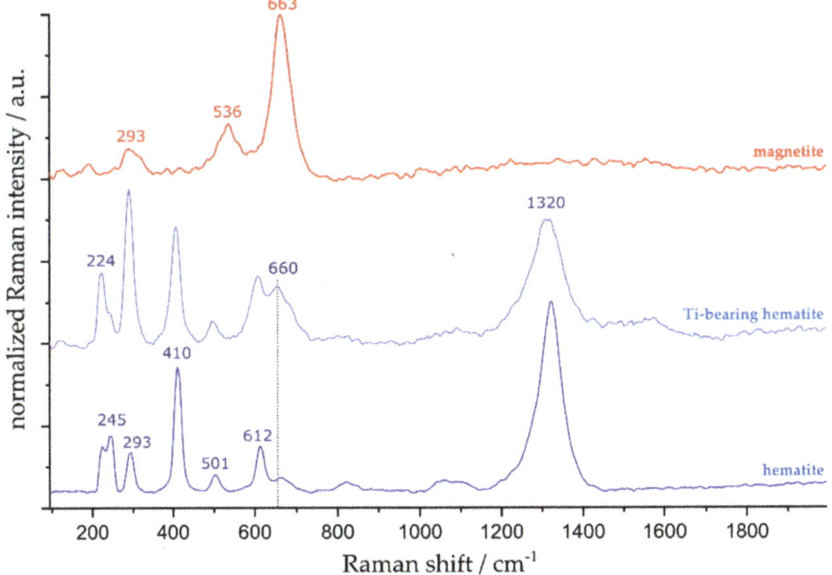

Figure 4. Characteristic Raman spectra of magnetite, hematite, and Ti-bearing hematite. Note that the peak width and intensity for both hematites vary according to elemental changes.

Another great advantage of Raman spectroscopy is its ability to not only identify, but also quantify incorporated foreign elements. Substitution of iron with aluminum [31–33], manganese/chromium [34,35], or titanium [36] results in a shift of Raman peak positions due to differences in mass. In a similar manner, carbonates show changes in peak location and shape depending on the amount of magnesium, iron, and manganese replacing calcium [37–40]. Ti-bearing hematite was classified by both methods because of the incorporation of ca. 3 wt % Ti. In Raman, we observed a change in the peak intensity at 295 cm^{-1} and 660 cm^{-1} (LO E_u) and an increased peak width, especially at 660 cm^{-1} and 1320 cm^{-1}. A shift to higher wavenumbers, as described by Varshney et al. [36] was not observed most likely due to lack of spectral resolution. The IR-active E_u (LO) at 660 cm^{-1} is theoretically not allowed (Raman-inactive). The intensity increase at about 660 cm^{-1} is a common indication for the incorporation of foreign elements. It becomes Raman-active due to structural disorder induced by surface defects or stress [34–36].

Please note that Raman imaging can be compromised by fluorescence excited by the laser source. The use of different laser wavelengths, bleaching, or fluorescence correction algorithms can efficiently counter the possible fluorescence of geological samples. If necessary, fluorescent areas can also be visualized using the Raman system.

Raman spectroscopy offers a variety of additional possibilities to characterize inorganic materials in terms of polymorphism, crystal orientation, crystallinity, phase, stress, and strain [41,42]. The combination of the strong and distinct Raman signatures with its nondestructive nature makes Raman spectroscopy a powerful tool for fine-scale identification and characterization of iron oxides.

A summary of the pros and cons of SEM-MLA and Raman imaging is given in Table 3.

Table 3. Pros and cons of MLA and Raman imaging for characterization of iron ores.

MLA	Raman Imaging
	+ higher resolution
	+ easy discrimination of minerals of identical composition
+ ast acquisition of large areas	+ identification and quantification of incorporated foreign elements/trace minerals
+ quantitative mineralogical data (abundance, associations, distributions)	+ polymorphism, crystal orientation, crystallinity, phase, stress, strain
+ no influence of fluorescence	+ no sample preparation necessary
	+ direct field measurements possible with handheld Raman
	+ confocal measurements possible
+/− resolution	+/− fluorescence imaging possible but impairing Raman signal detection
+/− discrimination of Fe-oxides possible	
− discrimination of minerals with identical composition not possible	− fixed mapping grid
− ample preparation and carbon coating necessary	− slow acquisition of large areas
− no confocal measurements	

Quantification of the mineral abundance is well established using the MLA software (FEI, Brisbane, Australia) [12]. For Raman imaging, quantification using a similar algorithm as described by Fandrich et al. [12] and FEI [21], needs to be implemented for geological applications. For this reason, we decided to compare area % in Table 2, which are easily accessible by both methods. For mineral quantification of the Raman map, we relied solely on the results of linear combination of reference spectra, not taking into account any microscopic data. The results also rely on color thresholds set in the SpectralImaging software. Therefore, it should be noted that the error of the Raman quantification method described here is larger than that of MLA quantification. Nevertheless, both methods reveal similar quantitative analyses of mineral abundance in the range of 1.5% with the exception of Ti-bearing hematite. The discrepancy of >3% is assumed to be caused by the fact that Raman relies on a fixed mapping grid neglecting mineral boundaries visible in the BSE image. Therefore, in some areas the step size of 30 μm is too big to resolve fine structures smaller than 30 μm size. Furthermore, the increased Raman intensity at 660 cm^{-1} for Ti-bearing hematite could, at least partially, result in a misassignment of magnetite using the linear combination approach.

Another important criterion for institutions and companies working on iron ores is the cost of analysis when making feasible assumptions according to improved characterization and extraction of the ore mineral. In purchase and maintenance, the financial advantage is on the Raman side, as the acquisition of an SEM-MLA (~800.000€) is roughly three times higher than a Raman imaging device including software (~250.000€). Furthermore, operating an SEM requires high vacuum pumps and nitrogen supply. Both techniques require a skilled operator to efficiently adjust measurements and they should be operated in an air-conditioned laboratory. In this study, the analyses were conducted on prepared thin sections for both methods. However, the advantage of Raman on the cost side is further supported by the fact that no sample preparation is needed, whereas SEM-MLA requires preparation and carbon coating of the samples.

Taking into account the limits and potentials of both methods, it is essential to define the right questions to a problem in order to find the most suitable analytical solution. Samples that contain very fine-grained aggregates or minerals that have fine intergrowths may need to be resolved in higher resolution. These detailed maps with a spot size in nm range (up to around half the laser wavelength) are one of the major advantages of the Raman technique, whereas the SEM is limited to μm range (~1 μm) even with the best operating conditions. It should be noted that, with higher resolution and smaller spot, size measurements become more time-consuming. In this study, Raman measurements on the same sample area with the same spot size took at least 10 times longer than MLA mapping, thus they are not nearly as time-efficient. Although Ramanaidou and Wells [43], suggested Raman spectroscopy as a potential method for large volume or bulk analysis, this seems not applicable with stationary Raman spectroscopes at this stage. Transportable Raman devices may offer fast field analysis by point measurements, but the loss of spectral information due to the downsized equipment needs to be compensated. Future instrumental research and development may lead to Raman spectroscopy and imaging as the first order application in process mineralogical analysis of iron ore. However, solely imaging will not solve the need for precise characterization of these complex ores, thus parameters, such as mineral liberation, association or locking, especially from processed ore, need to be extracted. Raman manufacturers start to offer software solutions for particle analysis, e.g., ParticleScout (WITec GmbH, Ulm, Germany), applicable to large sample areas typical for geosciences. The authors of this study are currently working on the development of these functions too, to further enhance the application of Raman imaging as a modern tool for analysis, especially for iron ore.

6. Conclusions

This study has demonstrated the potential of both SEM- and Raman-based mineralogical imaging for the characterization of iron oxide apatite deposits (IOA). In polished sections of IOA, magnetite, hematite, apatite, carbonates, and silicates were easily detected and mapped. The discrimination of magnetite and hematite, however, is strongly dependent on operating conditions in SEM-MLA. Raman imaging shows a major advantage due to distinct Raman spectra that enable in situ identification. Raman spectroscopy allows the determination of mineral composition and can detect contaminants in iron oxides, further advancing iron ore characterization. However, the main "struggles" of Raman are the time demands, especially when it comes to large volume or bulk analysis of iron ores. Further enhancement of the Raman technique is pursued for better characterization of iron ores.

Supplementary Materials: The following are available online at http://www.mdpi.com/2075-163X/9/9/544/s1, Figure S1: Reference Raman spectra. Background-corrected and normalized Raman reference spectra used for linear combination.

Author Contributions: Conceptualization, P.K. and A.J.; software and methodology, S.G. and P.K. (MLA) and A.J. (Raman); investigation, P.K.; resources, P.K.; data curation, P.K.; writing—original draft preparation, P.K. and A.J.; writing—review and editing, all authors.; visualization, P.K. (MLA) and A.J. (Raman); supervision and project administration, T.S. and K.N.; funding acquisition, P.K.

Funding: This research is part of the author's doctoral study and fully funded by Luossavaara Kiirunavaara AB (publ.).

Acknowledgments: We thank Richard J. Best and Igor Zlotnikov (BCUBE Dresden) for the Renishaw Raman measurements. The kind help of Enrico Pigorsch (PTS Heidenau) with the WITEC Raman system is highly appreciated. Matthias Finger provided the SpectralImaging software and was helpful in analyzing the Raman data. Patrick Krolop gratefully acknowledges the funding and support of LKAB and involved personnel. Michael Stoll and Roland Würkert are thanked for sample preparation.

Conflicts of Interest: The authors declare no conflict of interest. The funders had no role in the design of the study; in the collection, analyses, or interpretation of data, nor in the writing of the manuscript.

References

1. Andersen, J.C.Ø.; Rollinson, G.K.; Snook, B.; Herrington, R.; Fairhurst, R.J. Use of QEMSCAN® for the characterization of Ni-rich and Ni-poor goethite in laterite ores. *Miner. Eng.* **2009**, *22*, 1119–1129. [CrossRef]
2. Rollinson, G.K.; Andersen, J.C.Ø.; Stickland, R.J.; Boni, M.; Fairhurst, R. Characterisation of non-sulphide zinc deposits using QEMSCAN®. *Miner. Eng.* **2011**, *24*, 778–787. [CrossRef]
3. Hunt, J.; Berry, R.; Bradshaw, D. Characterising chalcopyrite liberation and flotation potential. Examples from an IOCG deposit. *Miner. Eng.* **2011**, *24*, 1271–1276. [CrossRef]
4. Sandmann, D.; Haser, S.; Gutzmer, J. Characterisation of graphite by automated mineral liberation analysis. *Miner. Process. Extr. Metall.* **2014**, *123*, 184–189. [CrossRef]
5. Lund, C.; Lamberg, P.; Lindberg, T. Development of a geometallurgical framework to quantify mineral textures for process prediction. *Miner. Eng.* **2015**, *82*, 61–77. [CrossRef]
6. Niiranen, K. Characterization of the Kiirunavaara Iron Ore Deposit for Mineral Processing with a Focus on the High Silica Ore Type B2. Ph.D. Thesis, Montanuniversität Leoben, Leoben, Austria, 2015.
7. Lund, C. Mineralogical, Chemical and Textural Characterisation of the Malmberget Iron Ore Deposit for a Geometallurgical Model. Ph.D. Thesis, Luleå tekniska universitet, Luleå, Sweden, 2013.
8. Gilligan, M.; Costanzo, A.; Feely, M.; Rollinson, G.K.; Timmins, E.; Henry, T.; Morrison, L. Mapping arsenopyrite alteration in a quartz vein-hosted gold deposit using microbeam analytical techniques. *Mineral. Mag.* **2016**, *80*, 739–748. [CrossRef]
9. Escolme, A.; Cooke, D.; Hunt, J.; Berry, R.; Fisher, L.; Potma, W. Ore characterisation and geometallurgy modelling. Productora Cu-Au-Mo deposit, Chile. In Proceedings of the Second International Seminar on Geometallurgy, Santiago, Chile, 1–3 December 2014; p. 1.
10. Kern, M.; Möckel, R.; Krause, J.; Teichmann, J.; Gutzmer, J. Calculating the deportment of a fine-grained and compositionally complex Sn skarn with a modified approach for automated mineralogy. *Miner. Eng.* **2018**, *116*, 213–225. [CrossRef]
11. Warlo, M.; Wanhainen, C.; Bark, G.; Butcher, A.R.; McElroy, I.; Brising, D.; Rollinson, G.K. Automated Quantitative Mineralogy Optimized for Simultaneous Detection of (Precious/Critical) Rare Metals and Base Metals in A Production-Focused Environment. *Minerals* **2019**, *9*, 440. [CrossRef]
12. Fandrich, R.; Gu, Y.; Burrows, D.; Moeller, K. Modern SEM-based mineral liberation analysis. *Int. J. Miner. Process.* **2007**, *84*, 310–320. [CrossRef]
13. Figueroa, G.; Moeller, K.; Buhot, M.; Gloy, G.; Haberla, D. Advanced discrimination of hematite and magnetite by automated mineralogy. In Proceedings of the 10th International Congress for Applied Mineralogy (ICAM), Trondheim, Norway, 1–5 August 2012; Springer: Berlin, Germany; pp. 197–204.
14. Weihed, P.; Bergman, J.; Bergström, U. Metallogeny and tectonic evolution of the Early Proterozoic Skellefte district, northern Sweden. *Precambrian Res.* **1992**, *58*, 143–167. [CrossRef]
15. Lahtinen, R.; Nironen, M.; Korja, A. Palaeoproterozoic orogenic evolution of the Fennoscandian Shield at 1.92–1.77 Ga with notes on the metallogeny of FeOx–Cu–Au, VMS, and orogenic gold deposits. In Proceedings of the Seventh Biennial SGA Conference "Mineral Exploration and Sustainable Development", Athens, Greece, 24–28 August 2003; pp. 1057–1060.
16. Lahtinen, R.; Korja, A.; Nironen, M. Paleoproterozoic tectonic evolution. In *Developments in Precambrian Geology*; Elsevier: Amsterdam, The Netherlands, 2005; Volume 14, pp. 481–531.
17. Lahtinen, R.; Hölttä, P.; Kontinen, A.; Niiranen, T.; Nironen, M.; Saalmann, K.; Sorjonen-Ward, P. Tectonic and metallogenic evolution of the Fennoscandian Shield. Key questions with emphasis on Finland. *Geol. Surv. Finl. Spec. Pap.* **2011**, *49*, 23–33.
18. Martinsson, O. Genesis of the Per Geijer apatite iron ores, Kiruna area, northern Sweden. In Proceedings of the Abstract Volume, SGA Biennal Meeting, Nancy, France, 24–27 August 2015; pp. 23–27.

19. Martinsson, O. Geology and Metallogeny of the Northern Norrbotten Fe-Cu-Au Province. In *Svecofennian Ore-Forming Environments Field Trip Volcanic-Associated Zn-Cu-Au-Ag and Magnetite-Apatite, Sediment-Hosted Pb-Zn, and Intrusion-Associated Cu-Au Deposits in Northern Sweden*; Rodney, A., Olof, M., Pär, W., Eds.; Society of Economic Geologists Guidebook Series; Society of Economic Geologists: Littleton, CO, USA, 2004; Volume 33, pp. 131–148.

20. Krolop, P.; Niiranen, K.; Gilbricht, S.; Seifert, T. Ore type characterization of the Per Geijer iron ore deposits in Kiruna, Northern Sweden. In Proceedings of the Iron Ore 2019 Conference, Perth, Australia, 22–24 July 2019; The Australasian Institute of Mining and Metallurgy: Perth, Australia, 2019; pp. 343–353.

21. FEI Company. MLA System User Training Course. 2011; 380 p, unpublished work.

22. Sandmann, D. Method Development in Automated Mineralogy. Ph.D. Thesis, TU Bergakademie Freiberg, Freiberg, Germany, 2015.

23. Dobbe, R.; Gottlieb, P.; Gu, Y.; Butcher, A.R.; Fandrich, R.; Lemmens, H. Scanning Electron Beambased Automated Mineralogy-Outline of Technology and Selected Applications in the Natural Resources Industry. In *European Workshop on Modern developments and Applications in Microbeam Analysis: The Book of Tutorials and abstracts of EMAS 2009 11th European Workshop on Modern Developments and Applications in Microbeam Analysis, Gdynia, Poland, 10–14 May 2009*; Gdynia/Rumia: Gdansk, Poland, 2009; Volume 10, pp. 169–189.

24. Gu, Y. Automated scanning electron microscope based mineral liberation analysis an introduction to JKMRC/FEI mineral liberation analyser. *J. Miner. Mater. Charact. Eng.* **2003**, *2*, 33. [CrossRef]

25. De Faria, D.L.A.; Venâncio, S.S.; de Oliveira, M.T. Raman microspectroscopy of some iron oxides and oxyhydroxides. *J. Raman Spectrosc.* **1997**, *28*, 873–878. [CrossRef]

26. Bersani, D.; Lottici, P.P.; Montenero, A. Micro-Raman investigation of iron oxide films and powders produced by sol–gel syntheses. *J. Raman Spectrosc.* **1999**, *30*, 355–360. [CrossRef]

27. Chamritski, I.; Burns, G. Infrared-and Raman-active phonons of magnetite, maghemite, and hematite. A computer simulation and spectroscopic study. *J. Phys. Chem.* **2005**, *109*, 4965–4968. [CrossRef] [PubMed]

28. Ramanaidou, E.; Wells, M.; Lau, I.; Laukamp, C. Characterization of iron ore by visible and infrared reflectance and, Raman spectroscopies. In *Iron ore*; Woodhead Publishing: Cambridge, UK, 2015; pp. 191–228.

29. Colomban, P. Potential and drawbacks of Raman (micro)spectrometry for the understanding of iron and steel corrosion. In *New Trends and Developments in Automotive System Engineering*; Chiaberge, M., Ed.; InTech: London, UK, 2011; pp. 567–584.

30. Shebanova, O.N.; Lazor, P. Raman spectroscopic study of magnetite (FeFe$_2$O$_4$): A new assignment for the vibrational spectrum. *J. Solid State Chem.* **2003**, *174*, 424–430. [CrossRef]

31. Zoppi, A.; Lofrumento, C.; Castellucci, E.; Dejoie, C.; Sciau, P. Micro-Raman study of aluminium-bearing hematite from the slip of Gaul *sigillata* wares. *J. Raman Spectrosc.* **2006**, *37*, 1131–1138. [CrossRef]

32. Zoppi, A.; Lofrumento, C.; Castellucci, E.; Sciau, P. Al-for-Fe substitution in hematite: The effect of low Al concentrations in the Raman spectrum of Fe$_2$O$_3$. *J. Raman Spectrosc.* **2008**, *39*, 40–46. [CrossRef]

33. Liu, H.; Chen, T.; Zou, X.; Qing, C.; Frost, R. Effect of Al content on the structure of Al-substituted goethite: A micro-Raman spectroscopic study. *J. Raman Spectrosc.* **2013**, *44*, 1609–1614. [CrossRef]

34. Varnshey, D.; Yogi, A. Structural and Electrical conductivity of Mn doped Hematite (α-Fe$_2$O$_3$) phase. *J. Mol. Struct.* **2011**, *995*, 157–162.

35. Varnshey, D.; Yogi, A. Influence of Cr and Mn substitution on the structural and spectroscopic properties of doped haematite: α-Fe$_2$-xMxO$_3$ ($0.0 \leq x <0.50$). *J. Mol. Struct.* **2013**, *1052*, 105–111.

36. Varshney, D.; Yogi, A. Structural, vibrational and magnetic properties of Ti substituted bulk hematite. A-Fe$_2$-xTixO$_3$. *J. Adv. Ceram.* **2014**, *3*, 269–277. [CrossRef]

37. Rividi, N.; van Zuilen, M.; Philippot, P.; Ménez, B.; Godard, G.; Poidatz, E. Calibration of Carbonate Composition Using Micro-Raman Analysis: Application to Planetary Surface Exploration. *Astrobiology* **2010**, *3*, 293–309. [CrossRef] [PubMed]

38. Bischoff, W.; Sharma, S.; Mackenzie, F. Carbonate ion disorder in synthetic and biogenic magnesian calcites: A Raman spectral study. *Am. Mineral.* **1985**, *70*, 581–589.

39. Perrin, J.; Vielzeuf, D.; Laporte, D.; Ricolleau, A.; Rossman, G.R.; Floquet, N. Raman characterization of synthetic magnesian calcites. *Am. Mineral.* **2016**, *101*, 2525–2538. [CrossRef]

40. Borromeo, L.; Zimmermann, U.; Andò, S.; Coletti, G.; Bersani, D.; Basso, D.; Gentile, P.; Schulz, B.; Garzanti, E. Raman spectroscopy as a tool for magnesium estimation in Mg-calcite. *J. Raman Spectrosc.* **2017**, *48*, 983–992. [CrossRef]

41. Schmid, T.; Schäfer, N.; Levcenko, S.; Rissom, T.; Abou-Ras, D. Orientation-distribution mapping of polycrystalline materials by Raman microspectroscopy. *Sci. Rep.* **2015**, *5*, 18410. [CrossRef] [PubMed]
42. Schmid, T.; Schäfer, N.; Abou-Ras, D. Raman microspectroscopy provides access to compositional and microstructural details of polycrystalline materials. *Spectrosc. Eur.* **2016**, *28*, 16–20.
43. Ramanaidou, E.R.; Wells, M.A. Raman I Do–Raman spectroscopy for the mineralogical characterisation of banded iron formation and iron ore. In Proceedings of the Iron Ore 2011 Conference, The Australasian Institute of Mining and Metallurgy, Perth, Australia, 11–13 July 2011; pp. 11–13.

Article

Exploration Potential of Fine-Fraction Heavy Mineral Concentrates from Till Using Automated Mineralogy: A Case Study from the Izok Lake Cu–Zn–Pb–Ag VMS Deposit, Nunavut, Canada

H. Donald Lougheed [1,*], **M. Beth McClenaghan** [2], **Dan Layton-Matthews** [1] and **Matthew Leybourne** [1,3]

1   Queen's Facility for Isotope Research, Department of Geological Sciences and Geological Engineering, Queen's University, 36 Union St., Kingston, ON K7L 3N6, Canada; dlayton@queensu.ca (D.L.-M.); m.leybourne@queensu.ca (M.L.)
2   Geological Survey of Canada, 601 Booth St., Ottawa, ON K1A 0E8, Canada; beth.mcclenaghan@canada.ca
3   McDonald Institute, Canadian Particle Astrophysics Research Centre, Department of Physics, Engineering Physics & Astronomy, Queen's University, 64 Bader Lane, Kingston, ON K7L 3N6, Canada
*   Correspondence: 5hdl@queensu.ca

Received: 15 February 2020; Accepted: 26 March 2020; Published: 30 March 2020

Abstract: Exploration under thick glacial sediment cover is an important facet of modern mineral exploration in Canada and northern Europe. Till heavy mineral concentrate (HMC) indicator mineral methods are well established in exploration for diamonds, gold, and base metals in glaciated terrain. Traditional methods rely on visual examination of >250 μm HMC material, however this study applies modern automated mineralogical methods (mineral liberation analysis (MLA)) to investigate the finer (<250 μm) fraction of till HMC. Automated mineralogy of finer material allows for rapid collection of precise compositional and morphological data from a large number (10,000–100,000) of heavy mineral grains in a single sample. The Izok Lake volcanogenic massive sulfide (VMS) deposit, one of the largest undeveloped Zn–Cu resources in North America, has a well-documented fan-shaped indicator mineral dispersal train and was used as a test site for this study. Axinite, a VMS indicator mineral difficult to identify optically in HMC, is identified in till samples up to 8 km down ice. Epidote and Fe-oxide minerals are identified, with concentrations peaking proximal to mineralization. Corundum and gahnite are intergrown in till samples immediately down ice of mineralization. Till samples also contain chalcopyrite and galena up to 8 km down ice of mineralization, an increase from 1.3 km for sulfide minerals in till previously reported for coarse HMC fractions. Some of these sulfide grains occur as inclusions within chemically and physically robust mineral grains and would not be identified visually in the coarse HMC visual counts. Best practices for epoxy mineral grain mounting and abundance reporting are presented along with the automated mineralogy of till samples down ice of the deposit.

Keywords: indicator minerals; heavy mineral concentrates; automated mineralogy; till sampling; VMS; MLA; Izok Lake

---

1. Introduction

Indicator mineral methods applied to sediment samples are important exploration tools in glaciated terrain for diamonds [1] and gold [2–5]. More recently, their potential to aid porphyry Cu [6–8], magmatic Ni–Cu-PGE [5,9], carbonate-hosted Pb–Zn [10] and volcanogenic massive sulfide (VMS) exploration [11] has been demonstrated.

Using current sampling protocols, a large till or stream sediment sample (10–20 kg) is necessary to recover detectable and meaningful numbers of indicator mineral grains in a single sample [12]. Indicator minerals are recovered from these large samples at specialized commercial laboratories using a combination of sizing, magnetic and density concentration methods to reduce the volume of material into a non-ferromagnetic heavy mineral concentrate (HMC). The coarse fraction (>250 μm, medium to coarse sand size) of the HMC is subsequently visually examined to identify and count indicator minerals using a binocular microscope [13]. These methods focus on the recovery of the coarser (>250 μm, medium to very coarse sand) mineral fraction of sediments because it is cost effective and relatively easy to recover and visually examine. The resulting HMC is composed of dense mineral grains (specific gravity (SG) >3.2) displaying varying levels of physical and chemical weathering, and the degree of abrasion and wear can be used to infer the transport distance of certain minerals [3]. Mineral associations can be observed in composite grains, while the degree of liberation of interlocked minerals can also serve as an indicator of mechanical weathering during transport.

Developments in rapid scanning electron microscopy (SEM) such as mineral liberation analysis (MLA, Hillsboro, OR, USA) or Quantitative Evaluation of Minerals by Scanning Electron Microscope (QEMSCAN™, Hillsboro, OR, USA) over the past 10 years make it possible to also examine and analyze the finer (<250 μm) heavy mineral fraction of sediment samples using automated technologies [14]. Automated mineralogy provides the potential for the identification and utilization of additional indicator minerals that traditional visual examination of the >250 μm mineral fraction does not allow.

Volcanogenic massive sulfide (VMS) deposits are an important source of Cu, Pb and Zn in Canada [15,16]. Thus, indicator mineral methods can be an important exploration method in Canada because much of the landscape is covered by glacial deposits. Metamorphism of massive or disseminated sulfide ore bodies results in coarsening of sulfide and alteration mineral grains as a result of recrystallization and grain boundary reduction. Indicator minerals commonly found in glacial dispersal trains extending from metamorphosed VMS deposits can include chalcopyrite, barite, gahnite, spinel-staurolite-sapphirine, kyanite-sillimanite, anthophyllite-orthopyroxene, spessartine, red epidote (Mn), red rutile (Cr), and loellingite [3,17]. Some of these minerals are soft and/or brittle, thus they do not survive glacial transport and deposition or subsequent post-glacial weathering as readily as more resistant phases. Chalcopyrite in glacial sediments is more resistant to weathering than other sulfide minerals and commonly forms dispersal patterns larger than those of other sulfides present in a deposit in greater abundance [3]. Oxide and silicate minerals are typically more physically and chemically robust than sulfide minerals and thus better survive glacial transport, deposition and subsequent postglacial weathering, and their dispersal fans can extend to hundreds of kilometers in length.

To investigate the potential application of MLA to indicator minerals of highly metamorphosed VMS deposits, the heavy mineral concentrates from till samples previously collected around the Izok Lake Zn–Cu–Pb–Ag VMS deposit by the Geological Survey of Canada (GSC), Ottawa, ON, Canada [11,18] were examined using MLA. Four till samples were selected for this study and all had been previously processed at the commercial laboratory Overburden Drilling Management Limited, Ottawa, ON, Canada, to prepare heavy HMC of the <0.25 mm, 0.25–0.5 mm, 0.5–1.0 mm and 1.0–2.0 mm fractions. Hicken et al. [18] and McClenaghan et al. [11] reported on indicator minerals that were visually identified in the 0.25–2.0 mm HMC of the till samples. This study tested and evaluated the efficacy and efficiency of MLA to determine all minerals present in the <250 μm HMC fraction, as well as the key indicator minerals of VMS mineralization.

1.1. Bedrock Geology of the Izok Lake Area

The Izok Lake VMS deposit is located in the central Slave Province, Canada, a granitic-greenstone terrane comprising deformed and metamorphosed Archean rocks hosting the Yellowknife Supergroup, a package of 2.67–2.70 Ga metasedimentary and metavolcanic rocks [19]. The Yellowknife Supergroup is divided into the lower Point Lake Formation (a suite of mafic tholeiitic to felsic calc-alkaline metavolcanic rocks and derived metasedimentary rocks, part of the Banting Group) and the upper Contwoyto

Formation (a series of iron formation-bearing greywacke turbidites) (Figure 1). These formations host numerous granitic plutons along with N–NW trending regional diabase dykes of the Helikian Mackenzie Swarm [20,21].

The deposit is hosted within the Izok Lake volcanic belt of the Point Lake Formation [22,23] that forms an arcuate belt approximately 18 km long and 1 to 5 km wide [24]. The deposit is located near the stratigraphic top of this belt. To the south of Izok Lake, the belt strikes southwest and dips steeply to the southeast, whereas to the north it makes an abrupt shift to striking northwest and dipping steeply northeast. This abrupt shift in strike is the result of regional deformation, referred to as the Izok Lake antiform.

Rocks to the south of the deposit have been interpreted to be greenschist-facies conditions with a mineral assemblage of albite–epidote–chlorite, whereas rocks to the north are interpreted as upper amphibolite–sillimanite grade with a mineral assemblage of hornblende–cordierite–sillimanite [25]. The difference in metamorphic grade to the north and south was interpreted by Thomas [25] to be the result of two regional metamorphic facies, but this idea was challenged first by Morrison [22] and later by Nowak [26]. They suggested that the rocks in the region were affected by a single craton-wide, high-temperature, low-pressure event. This conclusion is supported by similarities between P–T estimates from mineral assemblages at Izok Lake and other rocks from the Slave Craton [22], whereas the local variation in metamorphic mineral assemblages identified is attributed to variations in pre-metamorphic bulk rock composition brought about by varying degrees of hydrothermal alteration [26].

Nowak [26] related the variations in pre-metamorphic bulk rock composition to the different types and intensities of hydrothermal alteration associated with mineralization. His work calculated bulk rock compositions using detailed geochemical whole-rock analysis, and also by combining detailed modal mineral abundance data gathered by automated mineralogy (QEMSCAN) with compositional data gathered by electron microprobe analysis (EMPA). Nowak [26] used discrimination plots to compare data from Izok Lake to several unmetamorphosed to low-grade metamorphosed VMS deposits, and the bulk rock compositions of the mineral assemblages identified are consistent with those of the associated alteration haloes from the other sites.

Mineralized boulders containing upwards of 30% Zn were discovered along the west and south shores of Izok Lake by Money and Heslop [27], and this initiated more detailed local exploration and the discovery of the Izok Lake deposit. The deposit was explored by several companies, including Minnova Corp., Inmet Mining, Wolfden Resources Corp., and the current property owner MMG [28]. The deposit consists mainly of galena, sphalerite, and chalcopyrite, with a variety of other less abundant ore minerals including covellite, digenite, electrum, and native silver [22,23,27].

Izok Lake is a significant mineral resource, with total indicated and inferred resources of 14.8 Mt grading 2.5% Cu, 12.8% Zn, 1.3% Pb, and 71 g/t Ag within a group of five near-surface sulfide lenses (North, Northwest, Central West, Central East, and Inukshuk) [29], making it one of the largest undeveloped Zn–Cu resources in North America [22]. The three westernmost of these lenses subcrop under Izok Lake. The five lenses are arranged with two northern lenses, two central lenses, and one peripheral lens (The Inukshuk lens). The two north lenses are lower grade than the two central lenses and both subcrop under Izok Lake. The North lens is small and located near the surface and is considered to be the remnant trough of a sulfide lens that was eroded by glacial activity. The two central lenses represent the majority of the deposit's metal endowment, being larger and richer in Cu than the two northern lenses, and of the two the central west lens subcrops under Izok Lake. The Inukshuk lens is the least defined of the five sulfide lenses, and does not appear to subcrop in the region [22].

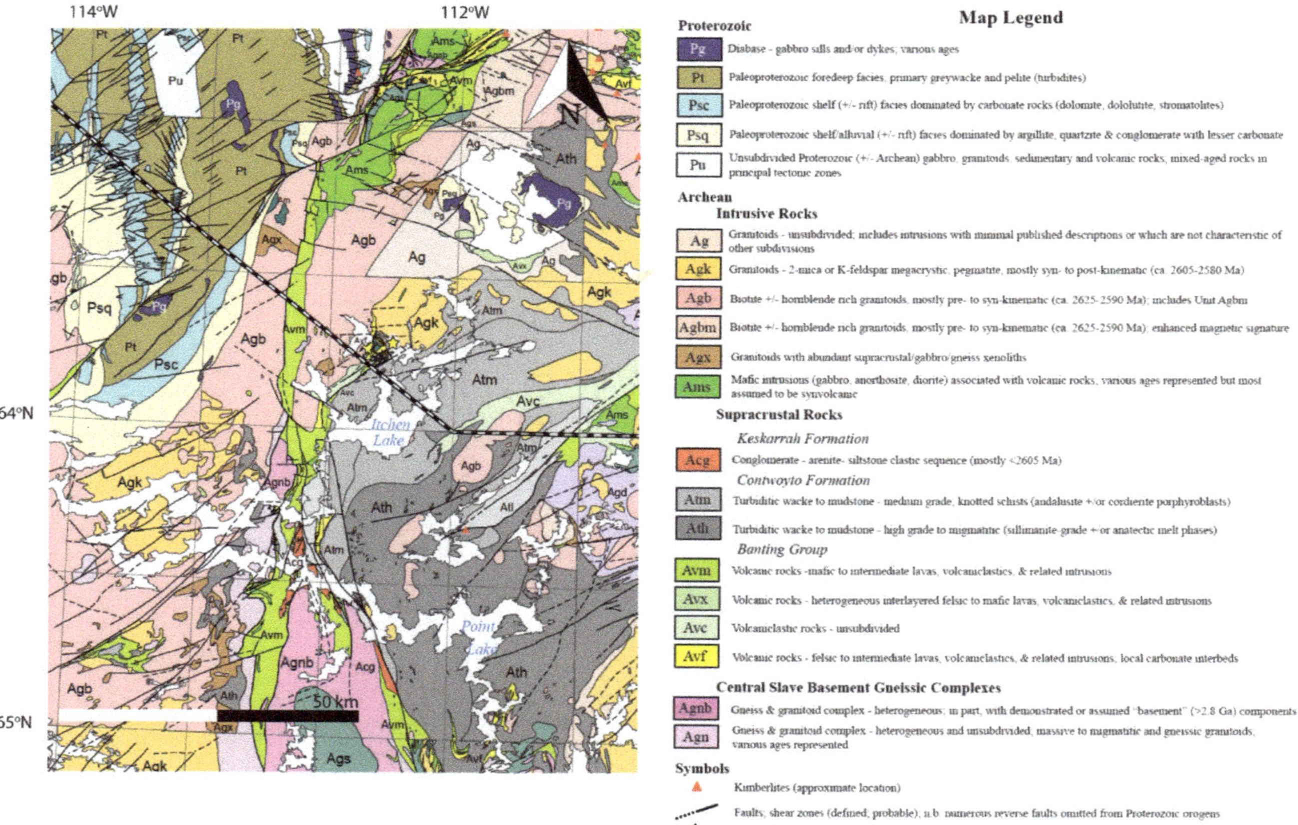

Figure 1. Geological map of the Point Lake belt, northwest Slave Province, Nunavut, Canada. The location of the Izok Lake deposit within the belt is indicated by the yellow star. Heavy black and white dashed line indicates the NWT-Nunavut border. Compiled using map data from Stubley and Irwin [30].

Within the region surrounding the Izok Lake deposit, rocks in the footwall of the deposit and to the north have been interpreted as a rhyolite protolith metamorphosed to upper amphibolite–sillimanite grade with characteristic mineral assemblages of hornblende, cordierite, and sillimanite. Hanging wall rocks are interpreted as meta-andesite (cordierite–biotite–epidote–garnet), aphyric meta-rhyolite (quartz–muscovite–sillimanite–garnet), and metagabbro (amphibole–plagioclase–biotite–garnet–magnetite). Gahnite (Zn-spinel) is a metamorphic mineral occurring in stringer-sulfide mineralized zones proximal to one of the deposit's five sulfide lenses. Gahnite was identified by Hicken et al. [18] and McClenaghan et al. [11] as a useful indicator mineral of mineralization at Izok Lake, present in a dispersal fan down ice of the deposit. In bedrock, gahnite is found in close association with quartz, feldspar, biotite, muscovite, sphalerite, chalcopyrite, and pyrrhotite [22,26,31].

1.2. Surficial Geology

The Izok Lake region was most recently glaciated during the Wisconsinan [32]. Regional surficial mapping in the Izok Lake region was carried out by Stea et al. [33], with large-scale ice flow reconstructions compiled by Dyke and Prest [34] and Dyke [35]. Recent ice flow indicator mapping in the region by Paulen et al. [36] built on the previous work to elucidate four ice flow phases that affected the region. Of these, an older southwest ice flow and younger northwest ice flow were determined to be responsible for the formation of most local glacial landforms and the palimpsest glacial transport of mineralized debris down ice [11,18]. These ice flows deposited a till cover that is relatively thin (<3 m thick) and consists mainly of silty sand till. The Izok Lake region is a permafrost terrain and, therefore, it is not practical to dig to appropriate till horizons by hand. All samples were collected at the surface from frost boils, small circular periglacial features formed by cryoturbation that bring till material from depth to surface [37].

The GSC carried out a reconnaissance-scale till geochemical survey in the Point Lake region, including the Izok Lake area, but geochemical analysis of the <0.002 mm and <0.063 mm till fractions did not indicate the presence of the mineralization at Izok Lake [38]. In 2009, a till geochemical survey was carried out around the Izok Lake deposit [17] along with a detailed bedrock and till indicator mineral survey. The <63 µm fraction of till, analyzed by aqua regia/inductively coupled plasma mass spectrometry (ICP-MS), yielded elevated values for Zn (<346 ppm), Cu (<322 ppm), Pb (<392 ppm), Fe (<1880 ppm), Ag (<1411 ppb), Cd (<1.36 ppm), Bi (<5.12 ppm), Hg (<247 ppb), Se (<0.5 ppm), In (<0.28 ppm), and Tl (<0.20 ppm) up to 6 km down ice to the northwest [11]. Also in 2009, a gossan containing sphalerite and referred to as the West Iznogoudh (WIZ) showing (Figure 2; [27]) was sampled. Oviatt [39] reported elevated contents of Zn, Ag, Cu, Hg, and Bi in the <63 µm fraction of till collected in the vicinity of the showing.

In addition to the till geochemical samples, bulk (10–15 kg) till and bedrock (1 kg) samples were collected around the Izok Lake deposit, concentrated to produce HMC and the >250 µm size fraction was examined for indicator minerals. That study identified gahnite, staurolite, chalcopyrite, sphalerite, pyrrhotite, and pyrite as indicator minerals. Gahnite dispersal was identified to be the broadest and farthest down ice of all of the indicator minerals identified, extending more than 40 km to the northwest with an older component to the southwest [11,18]. Significant abundances of sulfide minerals (chalcopyrite, galena, sphalerite, pyrite) were identified in the coarse fraction of only the most proximal down ice till sample to mineralization. Axinite, a potential indicator of hydrothermal alteration, was identified in bedrock thin sections and HMC, but was not identified in till HMC due to its lack of distinguishing visual characteristics [18,31].

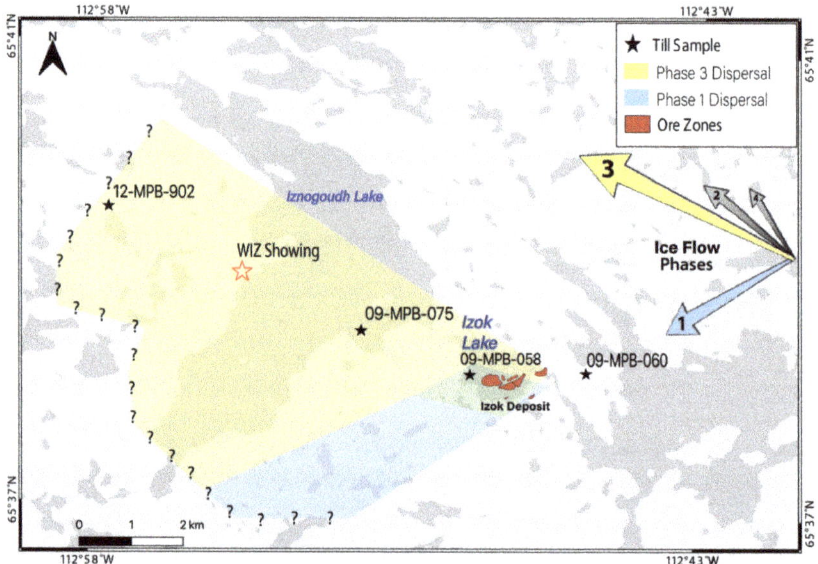

Figure 2. Location of the four till samples used in this study. Arrows indicate relative ice flow chronology (1 = oldest) and vigor (arrow width) of flow events. Colored polygons depict the glacial dispersal fan for gahnite in the non-ferromagnetic 250–500 μm fraction of till heavy mineral concentrate (HMC), as interpreted by McClenaghan et al. [1]. Yellow polygon represents dispersal by the NW ice flow; blue represents dispersal by the older SW ice flow, and the question-marked border represents the terminal sampling distance beyond which data is not available.

2. Materials and Methods

Four till samples collected by Hicken et al. [18,31] were chosen for use in this study. They were selected based on their locations relative to mineralization and the indicator minerals identified in their coarse fraction of HMC, in particular gahnite. The four sample locations are shown in Figure 2, along with the gahnite dispersal fan identified by McClenaghan et al. [11]. Sample 09-MPB-060, located 1 km up ice of mineralization, was the farthest-up ice sample available, and was chosen to represent background values compared to down ice sample locations. However, this up ice sample site is still within the hydrothermal alteration halo surrounding the Izok Lake deposit [22], and may not represent true regional background for certain alteration indicator minerals. Sample 09-MPB-058, located 500 m down ice (west) of mineralization, was the only sample with a significant abundance of sulfide indicator minerals identified in the coarse fraction [18,31]. It is the most proximal down ice sample chosen for this study. Sample 09-MPB-075, located 3 km down ice of mineralization, represents the intermediate down ice sample chosen for this study. Previous work identified gahnite in the coarse fraction HMC of these till samples. Sample 12-MPB-902, located 8 km down ice of mineralization, represents the most distal down ice sample chosen for this study.

2.1. Preparation of Heavy Mineral Concentrate (HMC) in Previous Geological Survey of Canada (GSC) Study

In 2009, the four GSC till samples were processed to recover >3.2 SG HMC at Overburden Drilling Management Limited (ODM), in Ottawa, ON, Canada. Density concentration was carried out using a combination of wet screening and shaking table, followed by heavy liquid separation at 3.2 SG, acid-washing, and ferro-magnetic separation to produce coarse (>250 μm) non-ferromagnetic HMC (>3.2 SG) fraction. This fraction was visually examined, and indicator minerals were counted with results reported by Hicken [18] and McClenaghan et al. [11]. The archived byproduct of the sample processing was the non-ferromagnetic <250 µm fraction. It is this fraction that forms this basis of the study reported here (Table 1).

Table 1. Sample processing weights indicating the original mass of bulk till sample prior to heavy mineral concentrate (HMC) production, the individual mass of each size fraction following sieving of the <250 µm HMC, and the mass of <250 µm HMC lost during sieving.

Sample	Original Mass (g)	Fraction (µm)	Mass (g)	Mass Weighed (g)	Total Loss (g)
09-MPB-060	16.637	185–250	2.668	16.544	0.093
		125–185	5.145		
		64–125	6.624		
		<64	2.107		
09-MPB-058	29.223	185–250	3.426	29.124	0.099
		125–185	7.060		
		64–125	12.064		
		<64	6.574		
09-MPB-075	22.933	185–250	3.132	22.848	0.085
		125–185	5.959		
		64–125	9.584		
		<64	4.173		
12-MPB-902	22.769	185–250	2.949	22.679	0.090
		125–185	4.814		
		64–125	9.716		
		<64	5.200		

2.2. Sieving Methods

The <250 µm HMC fraction of the four till samples was dry-sieved at the Queen's University Facility for Isotope Research (QFIR, Kingston, ON, Canada) into four smaller size fractions: (1) 185–250 µm, (2) 125–185 µm, (3) 64–125 µm, and (4) <64 µm. The sieving was carried out using single-use, nylon-screened sieves following the procedures developed by Lougheed et al. [40]. Table 1 contains the masses of <250 µm HMC prior to sieving, the masses of each resultant size fraction, and the mass loss of HMC material as a result of sieving.

2.3. Epoxy Mounting of Mineral Grains

The grain mounting method used in this study was modified after Blaskovich [41]. In our study, the entire mass of each of the 16 subsamples was mixed with vacuum-evacuated epoxy and poured into a 2.54 cm diameter plastic ring mold. Each ring mold was immediately placed into a vacuum chamber under full vacuum for 5 min. Vacuum impregnation eliminated air bubbles from liquid epoxy and drew epoxy into void spaces in mineral grains, maximizing the adhesion between epoxy and grains [42]. Following removal from the vacuum chamber, the mounts were allowed to cure for 24 h. The resulting grain mount was trimmed and vertically quartered using a Raytech Jemsaw 45 (Raytech Industries, Middletown, CT, USA) fitted with a diamond-impregnated blade. One quarter of each mount was archived. The other three quarter-pieces were remounted into a single plastic ring mold and set with epoxy (Figure 3) and cured for 12 h. Thus, one new epoxy mount containing three quarter slabs was prepared for each of the 16 subsamples. After polishing to 1 µm, these three slabs in the secondary mount were used for MLA analysis and examination of any settling effects or gradients that developed during mixing with epoxy and curing of the primary mount. Following polishing, the mounts were carbon coated using a Denton Vacuum Desk V carbon-coater (Denton Vacuum LLC, Moorestown, NJ, USA).

Grain mount preparation took ~30 min, followed by 12 h of curing time. Several samples were prepared simultaneously, with the limiting factor being the number of mounts that can be placed in the vacuum chamber.

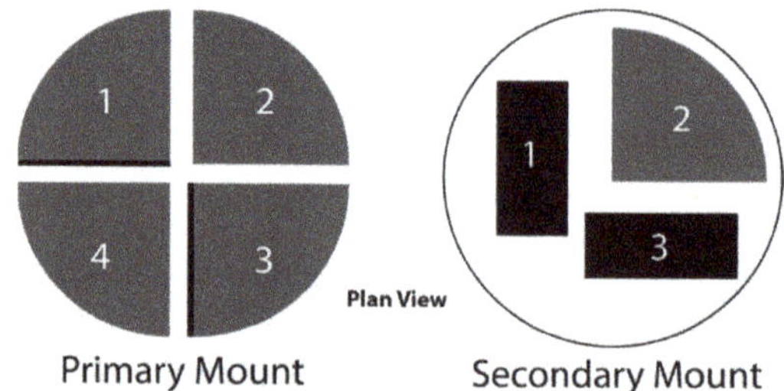

Figure 3. Mounting schematic displaying the basal surface of the two mounting stages. The primary grain mount was quartered, and three quarters were reoriented and made into a second mount to display one basal surface and two cross-sectional surfaces for analysis. Cross-sectional surfaces are indicated by the black grey bar in left pane of figure.

2.4. Automated Mineralogy

MLA is an automated mineralogy software package used in tandem with a field-emission gun scanning electron microscope equipped with electron dispersive spectrometer (EDS) detectors (FEG-SEM-EDS). The software uses high-resolution backscatter emission (BSE) images, image analysis, and elemental chemistry from EDS to create a mosaic image of an epoxy grain mount. The MLA software (version 3.1.4.686, FEI, Hillsboro, OR, USA) delineates grain boundaries based on BSE brightness contrast and subsequently collects a full X-ray spectrum (EDS) at the geometric center of each grain, comparing this to an established EDS mineral library assembled by the user, in order to identify the mineral [43]. The classified grains are subsequently combined into a false-color mineral image, and the program can quickly calculate for each grain, its size, mineral associations (occurrence and interlocking), particle properties (roundness, area, shape), and mineral liberation [44].

Carbon-coated grain mounts were analyzed using XBSE mode on a FEI Quanta 650 FEG SEM (Thermo Fisher Scientific, Waltham, MA, USA) operating under high vacuum at QFIR. The XBSE measurement mode collects a BSE image of each frame and uses variances in grey-level to define phase boundaries on a sample surface. This is followed by collection of a single EDS spot analysis at the calculated centroid of each identified phase. Operating conditions included a beam current of 10 nA, an accelerating voltage of 25 kV, and a spot size of 6. Backscattered electron image brightness and contrast were standardized to a gold (Au) imaging standard. Minimum feature size for image detection and minimum grain size for EDS analysis were both set to 4 μm. Magnification was set to 250× with a resolution of 1000, resulting in a pixel size of 2.14 μm and a horizontal field width of 2.14 mm. Each basal quarter contained between 75 and 210 frames for analysis, whereas each vertical section contained between 100 and 200 frames. Analysis of a single grain mount took between 35 min and 2.5 h, depending on the dimensions of the vertical slabs and the number of particles to analyze (Figure 3, slabs 1 and 3). Basal sections (Figure 3, slab 2) contain between 5000 and 50,000 particles.

Post-processing was performed using MLA Image Processing and Dataview, product version 3.1.4.686. Areas affected by charging effects were removed from the false-color map of each grain mount and each grain was classified using a mineral reference library constructed from spectra collected from the FEI reference library, augmented by spectra gathered manually over several years at QFIR.

Quality control evaluation of mineral identification, grain abundance analysis, and MLA error was carried out on one secondary mount of the 185–250 μm fraction and one of the <64 μm fraction. Both mounts were evaluated by performing a duplicate scan on a sub-section of each mount. Two scans were performed so that the difference in generated modal mineralogy data could be evaluated between scans of the same area under identical operating conditions, within the same scanning routine. After this scanning routine was completed, the two mounts were left in the machine and it underwent routine in-house calibration. Keeping the mounts inside the machine ensured that their orientation would not change. Following calibration, a second scanning routine was performed using the same settings as the first routine. This second scanning routine allowed for the evaluation of differences in generated modal mineralogy data prior to, and immediately following, routine calibration.

The time needed to complete an MLA scan varied depending on the operating parameters and the number of particles being analyzed. In general, decreasing particle size (and subsequently increasing particle abundance) increased the analysis time. Preparation of the mineral reference library was tailored to the needs of the study and the minerals/elements of interest, whereas the post-processing of data and generation of false-color grain maps and mineral abundance tables is a relatively routine process.

3. Results

3.1. Grain Mount Density Gradients

Grains on each epoxy grain mount were classified into three false colors according to their density; red (>4.25 SG) for the heaviest 1/3 of minerals in the mineral reference library, and green (3.24–4.25 SG) and black (<3.24 SG) representing intermediate- and light-density minerals, respectively (Figure 4). This classification was done to determine if any settling had affected the grains during preparation of the epoxy mounts. Heavy minerals (red) were more abundant in the bottom of the vertical slabs 1 and 3 from the primary mounts. The number of grains was also highest in the bottom 1/3 of slabs 1 and 3. In contrast, the basal surface of the primary mount (slab 2) contained a higher relative abundance of heavy minerals to lighter minerals, as well as a higher overall abundance of grains when compared to the vertical slab surfaces. This contrast between slab 1 and 3 versus slab 2 was observed in all secondary mounts for all till samples, for all size fractions examined. The differences between slabs 1 and 3 versus slab 2 were interpreted to be a settling gradient from top to bottom that formed after the grains were stirred into the epoxy during preparation of the primary mount.

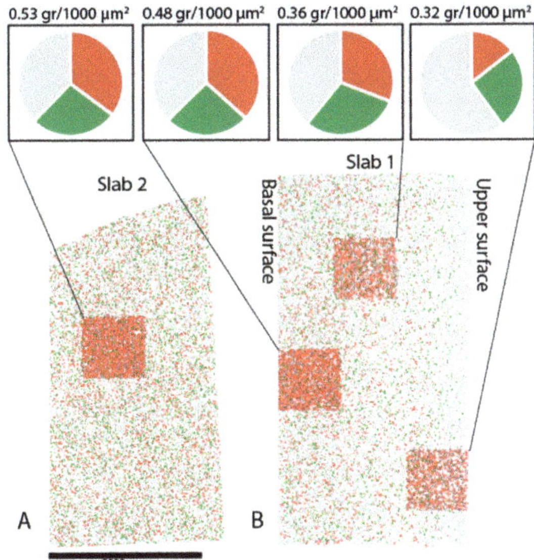

Figure 4. False-color grain maps of: (**A**) a portion of the slab 2 basal surface of a HMC epoxy grain mount, and (**B**) the entire slab 1 vertical cross-sectional surface of a HMC epoxy grain mount. Particle abundance per unit area and the relative abundance of heavy (red), intermediate (green) and light (grey) are depicted in the pie charts above for each red-shaded area. Three sections were analyzed on the vertical slabs and they display variations in particle abundance with depth through the slab.

3.2. Mineral Liberation Analysis (MLA) Error Estimation

Modal mineralogy tables were generated for a portion of basal slab 2 from two different secondary mounts (one of the 185–250 μm fraction and one of the <64 μm fraction) for the duplicate MLA runs (before and after calibration). The absolute difference between the values generated by each scanning routine (in area percentage) was calculated for each mineral identified. The absolute difference was divided by the mean of the two area percentage values to prevent bias between minerals found in very low or very high abundances, and then divided by 2 to give an approximation of the percentage error above or below that mean value.

Values were calculated for successive scans of the same area, within the same scanning routine ("run"). This produced four "in-run" error comparison values, one for each of the two size fractions examined in the run before and the run following gun alignment calibration. In addition to the four in-run values, 8 "out-run" values were computed by comparing all possible permutations of before- and after-calibration scans for each of the two size fractions examined. Table 2 details the resulting in-run error values (before and after calibration) and the out-run error values for several indicator minerals.

Table 2. Calculated error between two identical MLA sample runs, before and after routine gun alignment calibration, for two size fractions of polished HMC grain mount. Out-run error is calculated using all runs, regardless of pre- or post-calibration.

Mineral	Pre-Calibration Scan 1 (area%)	Scan 2 (area%)	In-Run Error	Out-Run Error (avg.)
Axinite	0.0101	0.0073	0.1598	0.6806
Epidote	3.9529	3.8394	0.0146	0.0073
Corundum	0.0250	0.0251	0.0016	0.0042
Hematite	1.8558	1.8207	0.0095	0.0048
Gahnite	0.1110	0.1286	0.0735	0.0420
Staurolite	1.0410	1.0180	0.0112	0.0056

Mineral	Post-Calibration Scan 1 (area%)	Scan 2 (area%)	In-Run Error	
Axinite	0.0026	0.0008	0.5246	
Epidote	3.9400	3.8814	0.0075	
Corundum	0.0248	0.0249	0.0023	
Hematite	1.8324	1.8428	0.0028	
Gahnite	0.1100	0.1096	0.0016	
Staurolite	1.0273	1.0276	0.0002	

The identification of minerals using EDS requires the identification of peaks in EDS spectra, with the x-axis location of a peak representing one of the energy emissions of the element of interest and the area of the peak (PB-ZAF corrected) reflecting the relative abundance of that element. The method is therefore semi-quantitative, and a fully quantitative technique like wavelength-dispersive spectroscopy (WDS) is required to obtain precise measures of elemental abundance in a mineral. The semi-quantitative nature of automated mineralogy (EDS) analysis makes it difficult to discern between minerals with similar composition or highly variable stoichiometry. Conversely, automated mineralogy systems will excel at identifying minerals containing rare elements like U or Zr, or minerals with distinct, simple spectra like many sulfide minerals [45]. Certain minerals were found consistently to represent outlier values. These minerals include garnets, pyroxenes, and other complex silicates, including axinite.

3.3. Modal Mineralogy

The basal slab 2 surface had the highest number of grains per unit area and highest proportion of the heaviest minerals for each sample. We deemed the basal slab 2 the most appropriate for use in the rest of the study and, therefore, mineralogy data are reported only for the slab 2 portion of each of the 16 secondary mounts.

Several reporting metrics can be generated by automated mineralogical software packages, including volume percentage (vol. %), weight percentage (wt. %), grain count, and area percentage. The calculation of 3D vol. % values from a 2D surface requires the generalization of grain volume as a sphere or ellipse using the length of the long axis present on the polished grain surface [43]. This is a powerful tool for many applications of automated mineralogy as it allows a user to establish a "grade" for a sample, when mineral chemistry data is combined with the resultant volume from MLA.

Simandl et al. [46] found a high coefficient of determination ($R^2 = 0.90$) between the wt. % estimation of Nb-bearing minerals using QEMSCAN™ and Nb content determined by X-ray fluorescence (XRF), and a similarly high R^2 value for the abundance of monazite with (Σ La, Ce, Pr, Nd) values obtained from borate fusion/nitric acid digestion/ICP-MS. However, the elements observed are not commonly present in a wide range of minerals. The method of determining wt. % within automated mineralogy software packages assumes either perfect mineral stoichiometry or requires mineral chemistry data (e.g., by electron microprobe) and, therefore, wt. % values are not appropriate for reporting the abundance of most till minerals, which have variable sources and stoichiometry, or where the elements of interest are commonly found in multiple mineral phases. Likewise, volume percentages do not enrich the interpretation of mineral abundance used in indicator mineral exploration and, therefore, area percentage (which both volume percentage and weight percentage are derived from) is the most appropriate reporting metric for till indicator mineral studies.

Although grain counts are an intuitive metric for understanding indicator mineral abundance and are more easily related to traditional visual grain counts for the >250 μm HMC, there is still a small amount of uncertainty introduced by the MLA software when it defines and separates grain boundaries. Area percentage is calculated by dividing the number of pixels assigned to a particular mineral value by the total number of pixels measured. This calculation supplies a metric that, although less intuitive than grain counts, represents a more precise quantitative measure.

Area percentage can be used in tandem with the number of grains to discern information about the occurrence of indicator minerals that would not be observed from either metric alone. Low grain counts with correspondingly high relative area percentage values suggest that a mineral is present as discrete whole grains (Figure 5). High grain counts combined with low area percentage values suggest that a mineral is present as high numbers of smaller grains or disseminated inclusions in other mineral grains. These conclusions must be confirmed by observing the mineral in a false-color MLA grain map. When using grain counts it is important to normalize values to allow for comparison between samples containing varying numbers of grains. Hulkki et al. [47] used two methods for normalizing grain count data, with their disaggregated bedrock samples being normalized using the equation:

(Number of a specific indicator mineral grain in a sample/ Total number of minerals grains in a sample) × 1000

This ratio accommodates the varying number of grains analyzed between subsamples. The stream sediment grain counts of Hulkki et al. [47] were normalized to a 1 kg processed sample weight, although the authors noted that given the extremely low abundance of the Cu minerals of interest, the normalized values were very close to the numbers of observed mineral grains.

Values in this study were normalized using the same method as the bedrock samples described above. Normalizing to total processed sample weight does not consider the varying numbers of grains analyzed between the different size fractions studied, which can range between tens of thousands of grains on 185–250 μm sample mounts to hundreds of thousands on <64 μm sample mounts. Normalizing to indicator grains/1000 grains allows for comparison of abundance values between samples, and between different size fractions of the same sample.

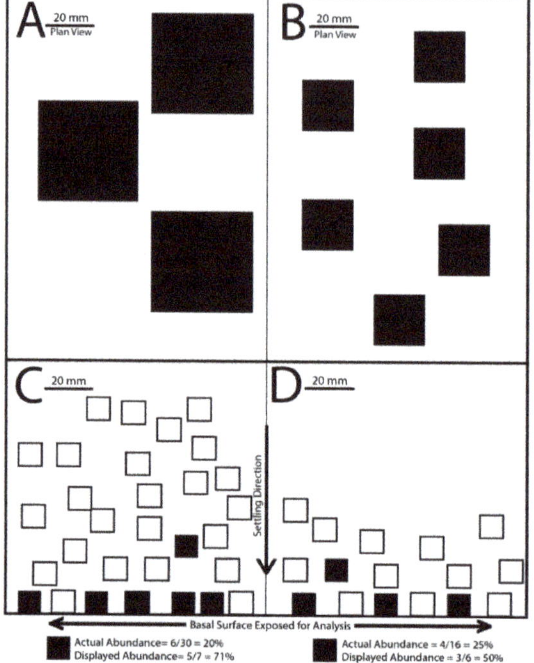

Figure 5. A,B conceptual view of the basal surface of epoxy grain mounts: (**A**) three large heavy (black) mineral grains; (**B**) six small heavy (black) mineral grains. The area (%) values for heavy grains on mounts A and B are identical even though mount A contains half as many grains as mount B. C,D conceptual cross section view through epoxy grain mounts: (**C**) grain mount containing a greater total number of grains (30) than mount D, with four black grains out of a total of seven grains exposed on the basal surface; (**D**) grain mount containing a smaller total number of grains (16) than mount C with three black grains out of a total of six grains exposed on the basal surface The actual abundance of heavy (black) grains per total grains mounted in epoxy is lower in mount C (5 out of 30 grains) than in mount D, but due to the preferential settling of heavier minerals through the curing epoxy, the displayed abundance on the polished basal surface is higher in mount C (71%) than D (50%).

The indicator mineral abundances for each indicator mineral for each slab 2 are reported in two different ways (Tables 3 and 4): (1) number of indicator grains normalized to 1000 grains, and (2) area percentage.

Minerals in the four size fractions of the four till samples are dominated by calcic amphibole, ilmenite, and almandine. Minerals present in lesser abundance include epidote, chlorite, biotite, titanite, rutile, and assorted Fe-oxide minerals (Figure 6). Quartz and albite, both less dense than the 3.2 SG heavy liquid used to process the original till samples, were found in high abundances in all samples, primarily as parts of composite particles containing other, denser minerals. Abundances increased with decreasing particle size for the following minerals: ilmenite, almandine, epidote, rutile, titanite, apatite, zircon, and monazite. Other minerals decreased in abundance with decreasing particle size: quartz, chlorite, and biotite.

Table 3. Abundance of ore indicator minerals in four till sample up- and down ice of the Izok Lake volcanogenic massive sulfide (VMS) deposit. Mineral abundance data are reported in two ways: (1) number of indicator mineral grains per total grains, normalized to 1000 grains; (2) minerals by 2-dimensional percentage area they take up on the polished mount surface. Values for 0.25–0.50 μm material are from the heavy mineral fraction of the same sample obtained by visual grain picking that have been normalized to a 10 kg of sample mass (data from Hicken et al. [2]).

Sample	Location Relative to Mineralization	Size Fraction (mm)	Chalcopyrite		Galena		Pyrite		Sphalerite	
			Grains/Total **	Area (%)	Grains/Total **	Area (%)	Grains/Total **	Area (%)	Grains/Total **	Area (%)
09-MPB-060	1 km up ice	0.185–0.250	0.00	0.0000	0.07	0.0001	0.51	0.0001	0.00	0.0000
		0.125–0.185	0.00	0.0000	0.00	0.0000	0.24	0.0009	0.00	0.0000
		0.064–0.125	0.00	0.0000	0.00	0.0000	0.40	0.0015	0.00	0.0000
		<0.064	0.02	0.0001	0.00	0.0000	0.06	0.0014	0.00	0.0000
		0.250–0.500 *	0.00	ND	0.00	ND	0.00	ND	0.00	ND
09-MPB-058	0.5 km down ice	0.185–0.250	0.40	0.0006	0.40	0.0002	2.50	0.2117	8.09	0.8045
		0.125–0.185	0.27	0.0177	0.09	0.0001	2.41	0.1592	1.50	0.4314
		0.064–0.125	0.21	0.0010	0.03	0.0001	1.11	0.0966	1.14	0.1523
		<0.064	0.07	0.0044	0.00	0.0000	0.25	0.0151	0.60	0.0710
		0.250–0.500 *	9.00	ND	0.00	ND	339.00	ND	1271.00	ND
09-MPB-075	2.5 km down ice	0.185–0.250	0.05	0.0002	0.29	0.0004	0.49	0.0019	0.00	0.0000
		0.125–0.185	0.15	0.0004	0.00	0.0000	0.19	0.0012	0.00	0.0000
		0.064–0.125	0.00	0.0000	0.09	0.0001	0.12	0.0011	0.00	0.0000
		<0.064	0.06	0.0001	0.00	0.0000	0.17	0.0017	0.00	0.0000
		0.250–0.500 *	0.00	ND	0.00	ND	0.00	ND	0.00	ND
12-MPB-902	8 km down ice	0.185–0.250	0.09	0.0001	0.22	0.0006	0.26	0.0007	0.00	0.0000
		0.125–0.185	0.12	0.0001	0.00	0.0000	0.40	0.0013	0.08	0.0001
		0.064–0.125	0.05	0.0000	0.03	0.0004	0.08	0.0004	0.00	0.0000
		<0.064	0.03	0.0000	0.00	0.0000	0.07	0.0004	0.03	0.0001
		0.250–0.500 *	0.00	ND	0	ND	0	ND	0	ND

* Grain Count, data from Hicken et al. [2] ** Normalized to 1000 grains; ND = not determined.

Table 4. Abundance of alteration indicator minerals in four till sample up- and down ice of the Izok Lake VMS deposit. Mineral abundance data are reported in two different ways: (1) number of indicator mineral grains per 1000 grains; (2) minerals by 2-dimensional percentage area they take up on the polished mount surface. Grey-shaded values are grain counts from the coarse (250 to 500 μm) heavy mineral fraction of the same sample obtained by visual grain picking that have been normalized to a 10 kg of sample mass (data from Hicken et al. [2]).

Sample	Location Relative to Mineralization	Size Fraction (mm)	Corundum		Epidote		Staurolite		Gahnite	
			Grains/Total **	Area (%)	Grains/Total **	Area (%)	Grains/Total **	Area (%)	Grains/Total **	Area (%)
09-MPB-060	1 km up ice	0.185–0.250	0.14	0.0024	21.23	5.4712	13.57	3.1242	0.00	0.0000
		0.125–0.185	0.69	0.0335	29.28	6.0791	11.41	2.6086	0.08	0.0283
		0.064–0.125	0.61	0.0553	48.06	8.9677	12.73	2.2661	0.03	0.0000
		<0.064	0.81	0.0366	47.26	7.2992	13.99	1.9695	0.10	0.0064
		0.250–0.500 *		ND	590 ***	ND	3061.00	ND	0.00	ND
09-MPB-058	0.5 km down ice	0.185–0.250	3.59	0.177	26.65	6.7352	12.68	2.0104	2.30	0.3724
		0.125–0.185	1	0.0582	39.24	8.6277	10.57	2.4334	1.55	0.0743
		0.064–0.125	1.07	0.1177	59.23	9.2708	13.24	1.9613	0.90	0.1059
		<0.064	0.87	0.0383	61.71	8.6900	16.21	1.9937	1.14	0.1309
		0.250–0.500 *		ND	1880 ***	ND	2542.00	ND	77.00	ND
09-MPB-075	2.5 km down ice	0.185–0.250	0.24	0.0353	22.09	7.2143	6.56	1.6389	1.03	0.1854
		0.125–0.185	0.41	0.005	27.03	6.4038	5.62	1.6028	0.34	0.0444
		0.064–0.125	0.68	0.0399	48.04	9.3994	8.50	1.4065	1.05	0.1107
		<0.064	0.27	0.0269	42.82	6.7296	8.95	1.208	1.11	0.1458
		0.250–0.500 *		ND	3160 ***	ND	971.00	ND	64.00	ND
12-MPB-902	8 km down ice	0.185–0.250	0.3	0.0065	14.40	4.1317	5.18	1.83	0.35	0.0462
		0.125–0.185	0.32	0.0094	22.55	5.3092	5.83	1.3816	0.83	0.2069
		0.064–0.125	0.32	0.0225	32.15	6.1219	7.35	1.335	0.56	0.0831
		<0.064	0.3	0.0205	33.24	4.1088	7.83	0.8659	0.85	0.1128
		0.250–0.500 *		ND	620 ***	ND	1634	ND	24	ND

* Grain Count, data from Hicken et al. [2]; ** Normalized to 1000 grains; *** Estimated grain count totals (Geological Survey of Canada, unpublished data); ND = not determined.

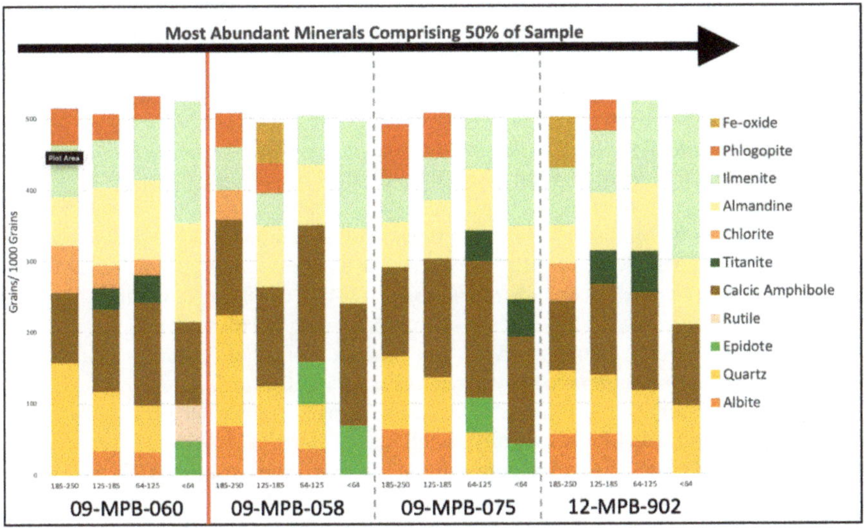

Figure 6. Mineral abundance (normalized to 1000 grains) for the most abundant minerals comprising 50% of <250 μm till heavy mineral concentrates, for four size fractions of four till samples. Red line denotes the relative position of massive sulfide mineralization, and the black arrow denotes the direction of ice flow.

3.4. Alteration Minerals and Metamorphic Equivalents

Gahnite was present in all four till samples, including the sample up ice, with values ranging from 0.03 to 2.30 grains per 1000 grains. Sample 09-MPB-058, immediately down ice of mineralization, contained the most gahnite of the four samples (0.90 to 2.30). Gahnite was three times more abundant in the coarsest (185–250 μm) fraction of this sample as compared to the finer three fractions. Up ice, the <64 μm fraction had the highest abundance of gahnite (0.08), but significantly lower than all down ice samples.

Corundum was identified by Averill [3] as an indicator mineral of metamorphosed massive sulfide deposits and is found intergrown with gahnite grains in till samples down ice of the Izok Lake deposit. Corundum is present in all till samples, with values ranging from 0.14 to 3.59 per 1000 grains. The highest abundances by far (0.87 to 3.59) were found in the four fractions of sample 09-MPB-058, immediately down ice of mineralization. Of these four fractions for this sample, corundum is most abundant in the coarsest fraction. The fewest grains of corundum (0.14) were detected in the coarsest fraction of sample 09-MPB-060, 1 km up ice of the deposit. Corundum grains are intergrown with other minerals, or present as discrete grains.

Corundum grains intergrown with gahnite were identified in all samples down ice of mineralization. These intergrown grains were found in all four size fractions of sample 09-MPB-058, immediately down ice of mineralization, in the 64–125 μm fraction of intermediate distance sample 09-MPB-075 (3 km down ice), and in the 125–185 and <64 μm fractions of distal sample 12-MPB-902 (8 km down ice). Corundum was intergrown with other minerals, including hercynite (Fe-spinel), chlorite, staurolite, Fe-oxide minerals, hornblende and almandine. Corundum, both discrete grains and grains intergrown with gahnite, from proximal down ice sample 09-MPB-058, contained inclusions of the sulfide ore minerals galena and chalcopyrite.

Staurolite abundance ranged from 5.18 to 16.21 grains per 1000 grains. It was present in all four till samples, in all four size fractions. The highest abundance of staurolite (16.21) was detected in sample

09-MPB-058, just down ice of mineralization. The finest fraction of all four till samples consistently had the highest values of staurolite. In general, content in till decreased as distance down ice increased.

Axinite abundance ranged from 0.03 to 0.33 grains per 1000 grains and was identified in all four fractions of all four till samples. The 64–125 µm fraction of sample 09-MPB-075, 3 km down ice of mineralization, contained the most (0.33) axinite. The least amount of axinite was detected in the 125–185 µm fraction of sample 12-MPB-902, 8 km down ice. There were no consistent patterns of abundance related to size fractions for the four till samples.

Epidote abundance ranged from 14.4 to 61.71 grains per 1000 grains and was identified in all four size fractions of all four till samples. The <64 µm fraction of till sample 09-MPB-058, 0.5 km down ice of mineralization, contained the most (61.71) epidote. The least amount was detected in the coarsest (185–250 µm) fraction of distal down ice till sample 12-MPB-902 (8 km down ice). The greatest abundance of epidote grains is in the 64–125 or <64 µm fraction of all till samples.

Fe-oxide minerals were identified in all size fractions of all till samples from Izok Lake, with a notable increase in abundance in the coarsest (185–250 µm) fraction in the till samples down ice of mineralization. Fe-oxide mineral abundance ranged from 15.65 to 72.16 grains per 1000 grains and was identified in all four size fractions of all four till samples. The 185–250 µm fraction of till sample 12-MPB-902, 8 km down ice of mineralization, contained the most (72.16) Fe-oxide. The least amount was detected in the 125–185 µm fraction of till sample 09-MPB-075, located 3 km down ice of mineralization.

3.5. Ore Minerals

Chalcopyrite abundance ranges from 0.02 to 0.40 grains per 1000 grains and was identified in all four till samples. It is most abundant (0.40) in sample 09-MPB-058, just down ice of mineralization, in which it is present in all four size fractions. In this sample, it is most abundant in the coarsest fraction and least abundant in the finest fraction. In fact, chalcopyrite was found in all four fractions (except for the 64–125 µm of one sample) of the three till samples down ice of mineralization. In sample 09-MPB-060, 1 km up ice of the deposit, chalcopyrite was only identified in very low abundance (0.02) in the finest fraction (<64 µm).

Sphalerite is the least abundant of the four ore minerals reported here. It was present only in two till samples, with values ranging from 0.03 to 8.09 grains per 1000 grains. It was found in all four size fractions of sample 09-MPB-058 and in two size fractions (125–185 µm and <64 µm) of sample 12-MPB-902, 8 km down ice. In sample 09-MPB-058, it is most abundant in the coarsest fraction and least abundant in the finest fraction. No sphalerite was detected in the till sample up ice.

The abundance of galena in till samples varies from 0.03 to 0.40 grains per 1000 grains. It is most abundant in sample 09-MPB-058, with the highest abundance in the coarsest (185–250 µm) fraction. For each till sample, it is most abundant in the coarsest size fraction. It was also detected in minor amounts in the coarsest fraction of up ice sample 09-MPB-060.

Pyrrhotite abundance ranges from 0.07 to 1.00 grains per 1000 grains. It is most abundant in sample 09-MPB-058, with the highest abundance in the 125–185 µm fraction. It is identified in all size fractions of all till samples, with the highest abundances in the coarsest size fractions observed (185–250 and 125–185 µm).

Pyrite was found in all four size fractions of all four samples. Values vary between 0.06 and 2.50 grains per 1000 grains. Abundance was highest in sample 09-MPB-058, with pyrite most abundant (2.50) in the coarsest size fraction and least abundant (0.25) in the finest size fraction. Pyrite abundance in till samples 09-MPB-075 and 12-MPB-902, 3 and 8 km down ice of mineralization respectively, is lower than in till sample 09-MPB-060, 1 km up ice of mineralization.

Sulfide ore minerals are present in the sample as both discrete, liberated grains and as components in other composite mineral particles. The proportion (area%) of each ore mineral contained as liberated grains or as part of composite particles is contained in Table 5.

Table 5. The proportion (area %) of each ore mineral present as either liberated grains or as a component in a composite particle.

Sample	Location	Size Fraction (mm)	Chalcopyrite		Galena		Pyrite		Pyrrhotite		Sphalerite	
			Liberated (area%)	Composite (area%)	Liberated (area%)	Composite (area%)	Liberated (area%)	Composite (area%)	Liberated (area%)	Composite (area%)	Liberated (area%)	Composite (area%)
09-MPB-060	1 km up ice	0.185–0.250	NA	NA	0.00	100.00	0.00	100.00	0.00	100.00	NA	NA
		0.125–0.185	NA	NA	NA	NA	0.00	100.00	0.00	100.00	NA	NA
		0.064–0.125	NA	NA	NA	NA	0.00	100.00	0.00	100.00	NA	NA
		<0.064	NA	NA	NA	NA	0.00	100.00	0.00	100.00	NA	NA
09-MPB-058	0.5 km down ice	0.185–0.250	55.86	44.14	0.00	100.00	0.00	100.00	0.00	100.00	82.72	17.28
		0.125–0.185	96.43	3.57	0.00	100.00	2.86	97.14	0.00	100.00	66.91	33.09
		0.064–0.125	0.00	100.00	0.00	100.00	14.29	85.71	0.00	100.00	37.58	62.42
		<0.064	93.68	6.32	NA	NA	15.01	84.99	86.56	13.44	96.71	3.29
09-MPB-075	2.5 km down ice	0.185–0.250	0.00	100.00	45.05	54.95	0.00	100.00	0.00	100.00	NA	NA
		0.125–0.185	0.00	100.00	NA	NA	0.00	100.00	0.00	100.00	NA	NA
		0.064–0.125	NA	NA	0.00	100.00	0.00	100.00	0.00	100.00	NA	NA
		<0.064	0.00	100.00	NA	NA	0.00	100.00	0.00	100.00	NA	NA
12-MPB-902	8 km down ice	0.185–0.250	0.00	100.00	35.09	64.91	0.00	100.00	0.00	100.00	NA	NA
		0.125–0.185	0.00	100.00	NA	NA	0.00	100.00	0.00	100.00	0.00	100.00
		0.064–0.125	0.00	100.00	100.00	0.00	0.00	100.00	0.00	100.00	NA	NA
		<0.064	0.00	100.00	NA	NA	NA	NA	0.00	100.00	0.00	100.00

NA = Not identified in sample.

4. Discussion

4.1. Consideration for Working with Ultrafine-grained Heavy Minerals

The size of mineral grains in till reflects the original grain size in bedrock, as well as the effects of glacial erosion, transportation, comminution and deposition on the grains. Dreimanis and Vagners [48] described a bimodal size distribution for specific minerals in till, and the "terminal" grade at which mineral grains become resistant to further comminution. The terminal grade size for most minerals is <250 μm, including garnets and other heavy minerals and, therefore, certain indicator minerals may be enriched in a specific size fraction that is <250 μm. The transport distance and distribution of this fine, terminal grade material in till is not currently known.

Previous work by Pickett [49] analyzing the fine (<63 μm) fraction HMC of till samples identified sample cross-contamination as a significant concern when dealing with fine-grained HMCs. Processing finer-grained sample material increases the difficulty in cleaning sieve surfaces as grains are more easily entrained in void spaces, adhered by electrostatic forces, or lost to aerosolization. Trapped grains may not be readily visible without magnification, making thorough sieve and related sieve equipment cleaning difficult and time consuming. Because low concentrations of indicator minerals (a couple of grains in a 10 kg sample) can constitute a significant 'anomaly', the potential for false-positive results stemming from sample cross-contamination, or false-negative results stemming from lost indicator grains, requires meticulous sample handling and sieving measures that mitigate mineral grain loss or cross contamination [3]. To address this need, this study utilized single-use nylon-screened sieves following the methods outlined in Lougheed et al. [50]. Single-use sieves eliminate the potential for sample cross-contamination and the need for time-consuming sieve cleaning.

4.2. Mineral Liberation Analysis (MLA)

The option to gather information only for grains above a specific brightness threshold is available in MLA. This option was successfully utilized by Hulkki et al. [47], using the grey level of hematite as a lower limit, to increase the detection of Cu-bearing mineral phases in stream sediment HMC samples. These techniques have the added benefit of decreasing the number of EDS analysis necessary during scanning routines, which decreases the overall analysis time for each sample. However, our study analyzed all grains on each polished grain mount.

The parameters that can be generated and queried by MLA (modal mineralogy, mineral associations, grain size, and grain angularity) would be impractical to measure manually for every grain in a sample, and thus automated SEM-based techniques offer the ability to capture both data that are indiscernible to the eye (spectral EDS imagery) and morphological data that would be impossible to measure for every grain with any consistency. These data are collected simultaneously during a single scan, and once collected can be manipulated by a user to examine individual parameters (e.g., prevalence of a single mineral phase) or to examine the relationships between them (e.g., grain size distributions for individual minerals, mineral association data). Because the number of grains available for analysis increases with decreasing size of the mineral grains being examined, the use of automated mineralogy on <250 μm HMC allows for rapid, efficient and cost-effective collection of large amounts of accurate, relatable quantitative data for thousands of grains.

The <250 μm fractions of processed HMC produced in mineral exploration and government surveys are commonly archived immediately after sample processing that recovers the coarse (>250 μm) HMC fraction. The <250 μm of HMC may sometimes be pulverized and analyzed geochemically [51,52] but most commonly it is set aside and never examined. Utilizing this previously unused finer size fraction presents an opportunity to gain new insights without the added cost of revisiting the sampling areas or processing additional bulk samples. The false-color mineralogy maps generated by MLA can be used to identify rapidly and accurately indicator minerals that are already known to be in the samples, identify additional indicator minerals that were not previously reported, and identify new indicator minerals that could assist in characterizing the local bedrock geology. These minerals can then be examined further using more precise elemental evaluation techniques like laser ablation inductively-coupled plasma mass spectrometry (LA-ICP-MS) or EMPA.

4.3. Density Gradient and Grain Mounting

The vertical cross-section slabs 1 and 3 of the primary mounts were prepared in order to evaluate the methods described by Blaskovich [41]. The two vertical slabs for each sub-sample displays a visible trend towards higher particle density (grains per unit area) at the bottom (basal surface) of each mount (Figure 4). They also display a greater proportion of the heaviest (red) particles towards the basal surface of the mount. These slabs demonstrate that a vertical density settling gradient exists through the vertical extent of cured epoxy grain mounts and that the number of grains per unit area is greatest at the base of the vertical slabs. The basal surface of a primary grain mount (Slab 2) consistently contains more heavy mineral grains per unit surface area than throughout the vertical slabs.

The basal surface area of a grain mount that is available for grains to be exposed and imaged is constant (5.07×10^8 μm^2). However, if variable masses of minerals (e.g., 0.2 vs. 0.3 vs. 0.5 g) are mounted, the preferential settling of the heaviest minerals in the aliquot to be mounted will lead to over representation of the heaviest minerals on the basal surface. This concept is visualized in Figure 5. Therefore, it is important that grain mounts within an individual sampling program be prepared from a consistent mass of sample to ensure accurate comparison between normalized indicator abundances.

Lastra and Petruk [53] used 0.4 g of material to prepare epoxy grain mounts for automated mineralogy. In our study, 0.3 g was determined to be a sufficient mass to mount that allowed the most grains to settle on the basal surface of a circular 25 mm diameter epoxy mount for polishing. This determination was made by experimenting with Almonte till blank heavy mineral concentrate. Grains were added to a ring mount until the base was thinly covered and the mass was recorded. The mass of sample necessary to cover the base of a mount varied depending on the density of the medium being sampled as well as the size fraction being mounted. It is important for each sampling program to determine the appropriate mass of sample to prepare before mounting, and to ensure that this mass is used for all samples.

Automated mineralogical analysis of fine (<250 μm) HMC permits the counting of very small grains quickly and accurately. The data produced can be used for robust statistical analysis due to the large number of grains sampled and the consistency in the automated EDS collection method. Formerly qualitative assessments of grain morphology or mineral association can now be examined quantitatively using the grain association and angularity statistics calculated within the software. Reducing grain size and subsequently increasing the grains on a mount surface could result in a reduction in bulk sample size if future work finds that indicator mineral concentrations remain representative in a lower bulk sample volume.

It should be noted that the primary mount slabs examined in this study (Slab 2) were all prepared by quartering so there is the potential for biased representation of heavier minerals on the basal surface of the primary mount. The slab 2 surface analyzed only represents $\frac{1}{4}$ of the total basal surface that would routinely be analyzed, thus the total number of grains counted and analyzed are $\frac{1}{4}$ of what would be expected for analysis of the entire basal surface of a 25 mm epoxy grain mount using the recommended 0.3 g near-monolayer method described above.

4.4. MLA Error Estimation

The error in indicator mineral abundance (area%) between MLA runs was approximated using the difference between the two values collected from scans of the same section of basal slab 2 on an epoxy grain mount. This is modeled after the work of Voordouw et al. [54], who performed similar repeat scans on thin sections of platinum group mineral (PGM)-bearing mineralized bedrock. Their findings indicate variation of ±7 wt. % for individual base metal sulfide minerals, and ±5 wt. % for individual silicate minerals. When using the MLA software, a section to be scanned in a sample run can be copied and re-applied to the exact same section, ensuring that all frames are analyzed in the same location, with the same operating parameters. This overcomes the potential for nugget effects skewing results as described in Hulkki et al. [47]. Their estimation of error had the caveat that their repeat scan could not be performed on precisely the same area, leaving the possibility that grains present along the edges of the scanned area could be missed between consecutive scans. This resulted in differing amounts of Cu minerals detected between their two runs. Their two runs were performed with differing operating parameters (BSE standards) and, therefore, are not directly comparable to the results of this study.

Error evaluation indicates that MLA reliably identifies and counts most indicator minerals, with negligible variation observed in consecutive runs within and outside of the same scanning routine (Table 2). Axinite-(Mn) ($Ca_2Mn^{2+}Al_2(BO_3)Si_4O_{12}(OH))$), an alteration mineral previously identified by Hicken [17] in the coarse (>250 μm) fraction of disaggregated mineralized bedrock from Izok Lake, was identified by MLA in the fine (<250 μm) fraction of till HMC by this study using MLA. However, the evaluation of error determined that the variability in axinite abundance calculated between runs was consistently higher than that of the other minerals examined (0.3–0.5%). The accelerating voltages used by this study are much higher than the optimal overvoltage for light elements like B (~5 keV) and, therefore, the peak to background ratio will be too low for effective detection of B [55]. The other elements comprising axinite are relatively common, and the high error evaluation by this study suggests that the MLA determination of axinite is being confused with other Ca-, Fe-, or Mn-containing aluminosilicates. The previous visual identification and subsequent EMPA confirmation of axinite in the coarse fraction of till from Izok lake indicates that at least some of the axinite identified by MLA is likely correctly attributed, but the variation in abundance between subsequent scans of the same surface is too high to use as an indicator mineral with this method. The difficulty in consistently and accurately distinguishing axinite from other minerals with similar stoichiometry indicates that axinite evaluation with MLA is not an effective tool for exploration.

The error, as defined by this study, is greater in magnitude when analyzing the coarser (185–250 μm) fraction. The difference in errors between the coarse (larger error) and fine size fractions (smaller error) is likely due to the greater number of grains/per unit area analyzed (8000 grains in the <64 μm fraction, compared to 4000 grains in the 185–250 μm fraction) and, therefore, the difference in statistical sample size. The error is smaller for both size fractions studied after the machine underwent routine gun alignment calibration. These values are acceptable for this study, for two reasons: (1) no minerals of interest were identified in one scan but not in another, and (2) the minerals of interest are present in large enough amounts that these minor fluctuations would not impact the identification of an anomalous abundance. Further, this error evaluation emphasizes the need for regular maintenance on FEG-SEM systems, frequent in-house machine calibration and repeated evaluation of error to ensure consistent results.

4.5. Modal Mineralogy

The grain abundance of key indicator minerals at each sample location are displayed in Figure 7 (ore minerals) and Figure 8 (alteration minerals). The X-axis represents horizontal distance down ice (NW) of mineralization (at zero). The Y-axis displays the normalized grain counts for each mineral, and the four till sample locations are joined by a curve to visualize the changes in indicator abundance along the length of the glacial dispersal train. Note that the values in the 250–500 μm fraction are normalized to a 10 kg sample mass and were reported in a previous study [11,18]. Some of these minerals do not have data as they were not counted during that study.

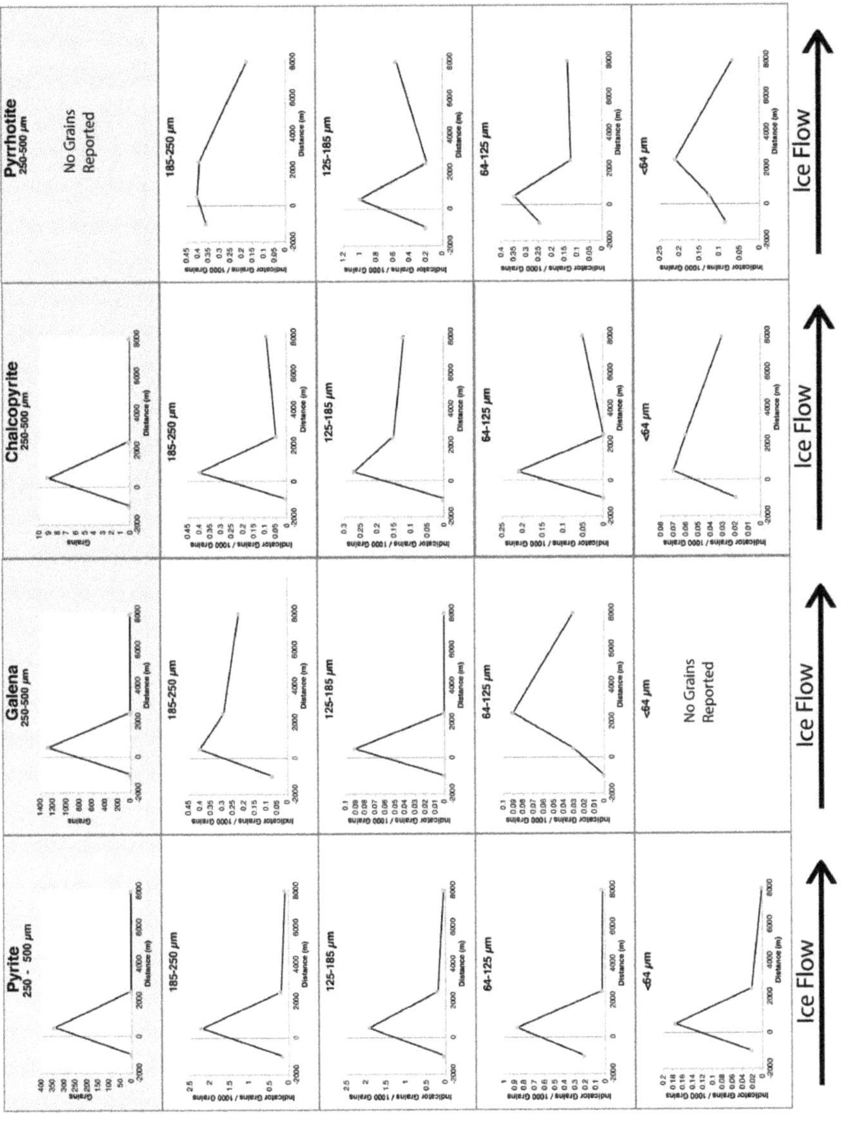

Figure 7. The grain abundance (normalized to 1000 grains) of ore indicator minerals at each sample location, for each size fraction examined. The X-axis of each is a scaled representation in the direction of ice flow (black arrow). The location of mineralization is denoted by the vertical grey line, and the black trend-line depicts the inferred decay in indicator mineral abundance along the dispersal train extending from the deposit.

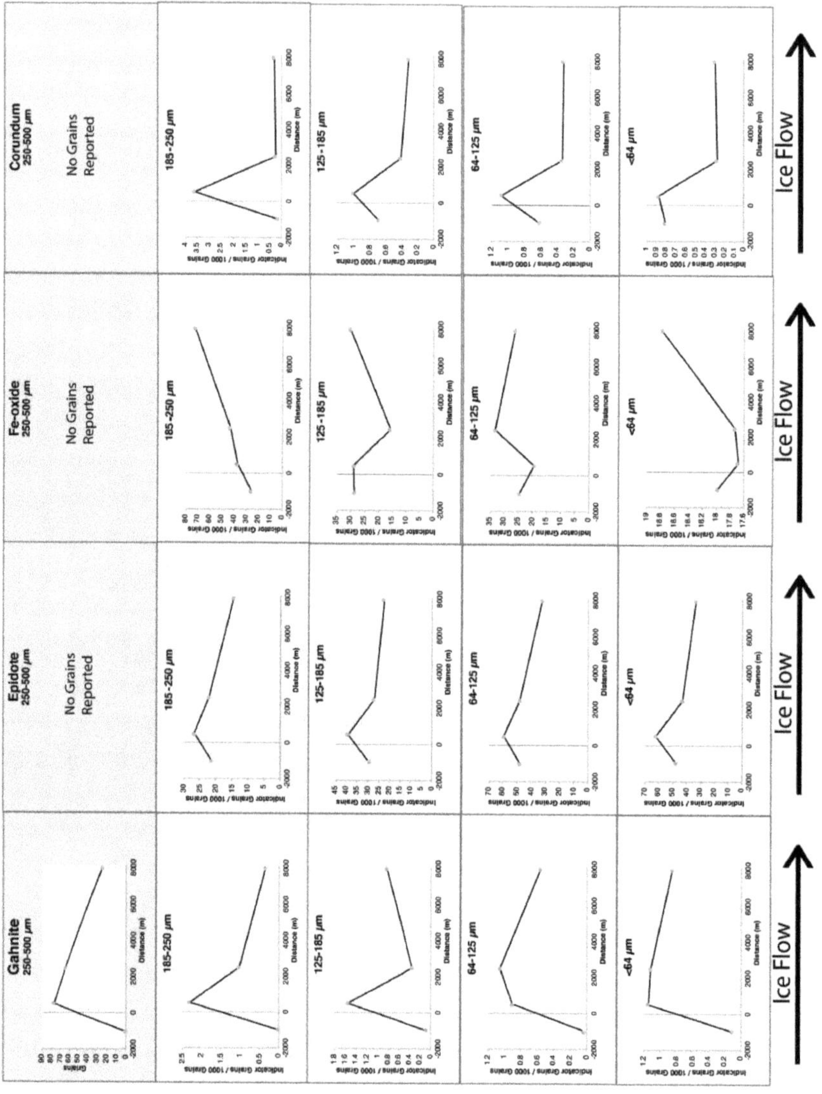

Figure 8. The grain abundance (normalized to 1000 grains) of key alteration indicator minerals at each sample location, for each size fraction examined. The X-axis of each is a scaled representation in the direction of ice flow (black arrow). The location of mineralization is denoted by the vertical grey line, and the black trend-line depicts the inferred decay in indicator mineral abundance along the dispersal train extending from the deposit.

4.5.1. Gahnite

Hicken [17] reported that gahnite in polished thin-section (PTS) ranged between 0.2 and 3.0 mm in size, and disaggregated bedrock HMC of mineralized bedrock ranged between 0.25 and 1.0 mm in size. This range reflects the original size of the gahnite presented to the overriding glacier during erosion of the deposit and generation of the till. As these grains were transported down ice they would have been crushed to smaller sizes with increasing transport distance, producing the pattern observed in our samples, i.e decreasing abundance and size of gahnite in till with increasing distance down ice.

The pattern of gahnite abundance in the <250 μm HMC is similar to that observed by McClenaghan et al. [11] and Hicken et al. [18] for the 250 to 500 μm HMC. Abundance is highest in sample 09-MPB-058, immediately down ice of the deposit, and decreases with increasing distance down ice. Of the four size fractions of the most proximal sample, gahnite is most abundant in the coarsest (185–250 μm) fraction, whereas in the sample farthest down ice, gahnite is most abundant in the finest size fraction. This is consistent with a model of grain comminution by abrasion and/or crushing during transport down ice, where grains are gradually reduced to a terminal grain size from their original size in bedrock [48]. No gahnite was recovered from the coarse (>250 μm) HMC fraction of sample 09-MPB-060 up ice of the deposit [18], however, in our study, a few gahnite grains are found in the sample. Because sample 09-MPB-060 is located only 1 km up ice (east) of massive sulfide mineralization, it is possible that an alteration halo containing fine-grained gahnite extends this far east. Note minor amounts of 0.25–0.5 mm gahnite grains (few grains/10 kg) were previously reported in regional till samples farther up ice of Izok Lake by McClenaghan et al. [56].

4.5.2. Corundum

Morrison [22] identified corundum as a component of the chlorite-biotite-cordierite (CBC) alteration zone, thought to represent the metamorphosed equivalent of the hydrothermally altered feeder-zone for the mineralizing fluids at Izok Lake. Both Hicken [17] and Morrison [22] identify corundum in cordierite-bearing bedrock, whereas Nowak [26] identifies it in the "moderately-altered rhyolite" assemblage defined in his work.

Intergrowths of corundum were previously identified in gahnite grains recovered from till samples at Izok Lake and discrete grains were observed in the polished thin section of one bedrock sample from 2 km northeast of the deposit by Hicken [18]. Only blue and red corundum were visually scanned for and counted in the 250–500 μm fraction of Izok Lake till samples, but no grains were found in the till samples selected for use in this study. All other colours of corundum were not scanned for in the 250–500 μm fraction. In this study, corundum is identified by MLA in all four size fractions of the <250 μm HMC of all four samples, with a pattern of abundance similar to that of gahnite.

Gahnite-corundum intergrowth is observed in the finer size fractions of the down ice till samples. Corundum has been previously noted in association with gahnite at the Geco VMS deposit, Canada, by Spry [57] and is considered part of a mineral assemblage formed following high-grade metamorphism of chloritic precursor rocks [58]. These chloritic precursor rocks are characteristic of the footwall hydrothermal alteration pipes of VMS deposits [15]. Backscatter SEM images of corundum grains from the 185–250 μm fraction reveal that it is intergrown with gahnite, but this relationship is not observed in this coarsest (185–250 μm) fraction at any of the other sample sites located down ice (Figure 9). The presence of coarse-grained (185–250 μm) intergrown corundum only in the most proximal till sample suggests that corundum and gahnite are liberated from one another by crushing and/or abrasion in the subglacial environment, and that their presence as coarse composite grains could serve as an indicator of close proximity (<1 km) to a gahnite-rich mineralized source.

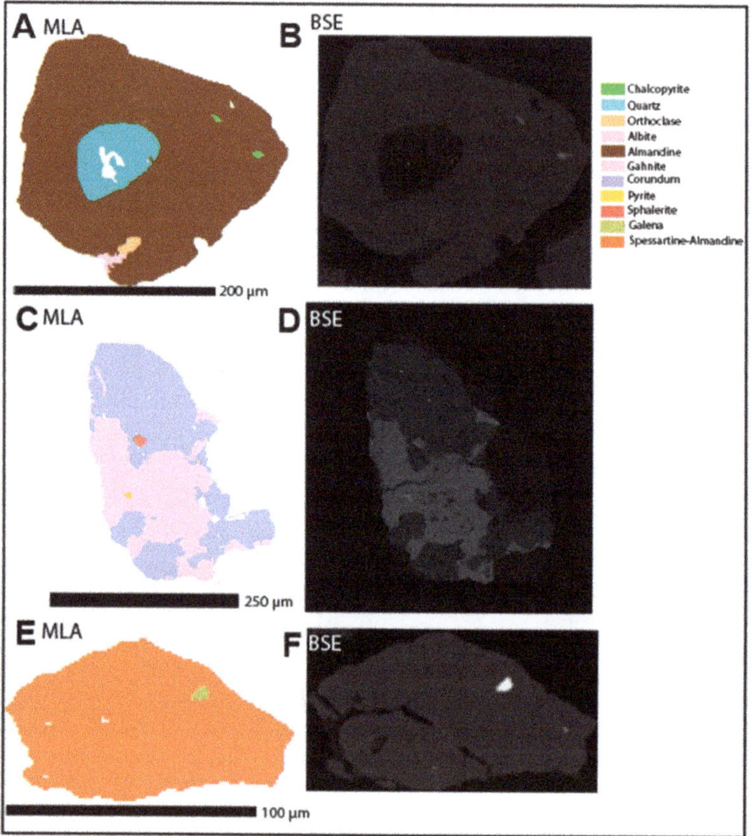

Figure 9. Mineral liberation analysis (MLA) and backscatter emission (BSE) images of mineral grains from till: **(A,B)** chalcopyrite inclusions (green) in almandine; **(C,D)** intergrown gahnite (pink) and corundum (blue); **(E,F)** galena inclusion (brown) in garnet. Note: scale bars below images vary.

4.5.3. Epidote

Epidote was previously reported by Hicken [17], Nowak [26], and Hicken et al. [31] in bedrock PTS and till HMC. Nowak [26] and Hicken [17] identified epidote grains in bedrock. Nowak [26] divided bedrock into domains based on the degree of hydrothermal alteration of the protolith, and epidote was reported in metamorphosed bedrock corresponding to least- and weakly altered protolith, with coarser-grained epidote being found in the weakly altered (0.2–0.7 mm) assemblages than in the least-altered (0.1–0.3 mm) assemblages.

Hicken [17] reported epidote grains in four bedrock PTS samples. Two samples (diabase dyke and iron formation) were collected to the west of Iznogoudh Lake (>5 km down ice of the Izok Lake deposit), and two samples (schist and gneiss) were collected from metamorphic rocks <1.5 km from the Izok Lake deposit. Epidote in the two proximal bedrock samples was coarser-grained than in the two more distal samples. Green epidote abundance was estimated by ODM for the 0.250–0.500 mm fraction of till HMC for the four samples examined by this study (Geological Survey of Canada, unpublished data). The counts were only accurate to 10–100 grains, depending on the total number of epidote grains present, but indicate an increase in abundance overlying mineralization, peaking 3 km down ice at sample site 09-MPB-075 and diminishing at sample site 12-MPB-902, 8 km down ice (Table 4).

Epidote identified by this study increases in abundance in the most proximal till sample down ice of mineralization (09-MPB-058), and till samples 09-MPB-060 (1 km up ice) and 09-MPB-075 (3 km down ice) have relatively equal epidote abundance in all size fractions. Till sample 12-MPB-902, 8 km down ice of mineralization, has the lowest abundance of epidote in all size fractions examined. This suggests that till sample 09-MPB-060 is within the hydrothermal alteration halo of the Izok Lake deposit, which is supported by the extent of hydrothermal alteration described in Morrison [22], who describes a late-stage, widespread zone of calcic alteration containing epidote surrounding the deposit. Epidote has been previously identified as an important indicator in porphyry terranes [8], where epidote abundance and trace element composition can be used in tandem to identify hydrothermal carbonate and propylitic alteration halos and assess ore fertility in porphyry systems. However, the chemical changes associated with carbonate and propylitic alteration lead to bulk-rock compositions similar to those of calc-silicate rocks of sedimentary or metasomatic origin [59] and, therefore, epidote abundance will need to be identified along with other indicator minerals to serve as a vector to VMS mineralization. Future work is needed to investigate the utility of epidote as an indicator mineral for VMS deposits, with careful attention paid to collecting true regional background samples outside the potentially wide zone of propylitic alteration.

4.5.4. Staurolite

Staurolite is a common metamorphic mineral in upper amphibolite facies terranes [60]. Its presence in all four till samples is expected as the deposit is in a high metamorphic grade terrane [3]. Staurolite with a significant Zn content (>5 wt. %) was identified as a useful marker when exploring for metamorphosed massive sulfide deposits, specifically as an intermediate mineral to form gahnite through the reaction of Zn-bearing biotite and staurolite during metamorphism [61–64]. However, EMPA analysis of several hundred staurolite grains collected from the >250 μm fraction of till HMC at Izok Lake by Hicken [17] identified very few grains containing >5 wt. % Zn. Spectra collected from these grains were added to our study's mineral reference library and no significant abundance of Zn-rich staurolite was observed in our samples.

4.5.5. Fe-Oxide Minerals

Fe-oxide minerals hematite, goethite, and magnetite have been used as indicator minerals of porphyry Cu (e.g., Kelley et al. [7]) and VMS mineralization (e.g., Makvandi et al. [65] and McClenaghan et al. [66]). These minerals can be derived from fresh bedrock, sulfide gossans that were glacially eroded and incorporated into glacial sediments, or from postglacial weathering of sulfide grains in till. Automated mineralogical platforms such as MLA cannot distinguish the valence state of individual elements and also has difficulty discerning grain boundaries between minerals of similar atomic (Z) number due to similarities in the grey level in BSE. This means that Fe-oxide minerals (goethite, limonite, hematite, magnetite) are difficult to separate and distinguish using EDS or BSE [45]. The MLA mineral reference library for this project has one entry for Fe-oxide and while it is possible to further differentiate minerals using the data gathered by the MLA based on the amplitude of individual elemental peaks this would have required more time than was permitted during our routine MLA analysis.

The increase in Fe-oxide abundance down ice of the Izok Lake deposit likely represents the incorporation of iron formation, weathered gossanous material from the Izok Lake deposit, the weathering of sulfide grains within till following deposition, or some combination of all three. Mineralized iron formation makes up a portion of the hanging wall immediately east of the deposit [22]. Fe-oxide abundance in the 185–250 μm fraction increases in the till progressively farther down ice, with the highest value (72.16 grains per 1000 grains) observed in the 185–250 μm fraction of till sample 12-MPB-902, 8 km down ice of mineralization. It is normal for indicator mineral abundance in till dispersal trains to peak down ice of the weathered source as source material is incorporated relative to the reduction in material transported from up ice [67] and likely accounts for the lower abundance of

Fe-oxide identified in till sample 09-MPB-058 (0.5 km down ice) compared to till sample 09-MPB-075 (3 km down ice). The further increase in Fe-oxide abundance in distal till sample 12-MPB-902 may reflect the influx of debris from the WIZ showing, 2.5 km up ice of till sample 12-MPB-902.

4.5.6. Sulfide Minerals

Sulfide mineral instability in oxidizing surface environments combined with their low resistance to physical abrasion and crushing result in poor preservation in till [3]. Sulfide grains (chalcopyrite, pyrite, sphalerite) were only detected in the coarse (250 μm) fraction of till HMC by Hicken et al. [3] and only in sample 09-MPB-058, immediately down ice of mineralization. Galena was not identified in the >250 μm fraction of any till samples.

This study identifies several sulfide minerals in the <250 μm till HMC, including chalcopyrite, galena, pyrite, sphalerite, and pyrrhotite. The highest abundances of all sulfides are in sample 09-MPB-058, immediately down ice of mineralization, and this establishes values for metal-rich surface till proximal to a VMS deposit in permafrost terrain. In fact, this till sample contains the only significant abundance of sphalerite identified in this study. The relative abundance of sulfide minerals in this proximal metal-rich till sample is sphalerite > pyrite > pyrrhotite > chalcopyrite > galena.

Coarse sphalerite was only visually observed in two of 53 till samples in the GSC study, 09-MPB-058 and 09-MPB-052, both down ice of mineralization between Izok Lake and Iznogoudh Lake. Our study detected sphalerite as discrete grains that occasionally contain inclusions of chalcopyrite or pyrite. The low sphalerite abundance in till in all four size fractions is likely the result of some combination of the following: rapid physical abrasion during glacial transport due to its low hardness (Hardness 3.5–4), chemical weathering of sphalerite postglacially, and perhaps few sphalerite-bearing zones of the deposit being directly exposed to glacial erosion. The low preservation of sphalerite grains in till in all size fractions make it a poor candidate for a till HMC indicator mineral.

Pyrite is a common mineral in mineralized and unmineralized rocks, so its usefulness as an indicator mineral of sulfide mineralization is less obvious. McClenaghan et al. [56] recovered coarse (>250 μm) pyrite in five till samples: 09-MPB-058, -016, -081, -052 and -030, with the highest abundance in samples 09-MPB-081 (56 grains) just up ice of mineralization, and 09-MPB-058 (338 grains) and -016 (82 grains) 500 m down ice of mineralization. Lower abundances were recovered from samples 09-MPB-052 (24 grains) and -030 (13 grains) 1 km and 5 km down ice of mineralization, respectively. No pyrite grains were recovered from till sample 09-MPB-060 up ice of the deposit.

Pyrite abundance in <250 till HMC at Izok Lake is different. Abundance is highest immediately down ice of mineralization but reaches background levels (0.2–0.4 grains per 1000 total grains) in most size fractions at sample site 09-MPB-075, 3 km down ice of mineralization. In the <64 μm fraction, pyrite values remain elevated above background (0.06 grains per 1000 total grains) up to sample site 12-MPB-902, 10 km down ice of mineralization. This pattern suggests that till sample site 09-MPB-060 (1 km up ice) may not represent the background for most size fractions, or that pyrite in the coarse fraction is rapidly comminuted to the finer fraction by abrasion and crushing during glacial transport. The high concentrations of pyrite overlying and just down ice of base metal mineralization indicates that it can be a useful indicator in combination with other sulfide minerals, and the dispersal distance may be greater when examining the finest (<64 μm) fraction. Pyrite occurs in both discrete grains and as inclusions in other grains (Table 5), and some grains containing pyrite inclusions also contain inclusions of chalcopyrite (Figure 9) and pyrrhotite. Rapidly identifying these fine pyrite inclusions can generate targets for more precise analytical tools (EMPA, LA-ICP-MS) as pyrite geochemistry has been used in exploration for lode Au [68] and Sedimentary Exhalative (SEDEX) base metal [69] deposits.

Chalcopyrite has been noted as more resistant to weathering in oxidized till by Averill [3]. Chalcopyrite was reported in disaggregated bedrock HMC and bedrock PTS by Hicken [17] and Hicken et al. [31]. Chalcopyrite in bedrock HMC ranged in size between 0.015 and 1.0 mm, and in bedrock PTS ranged in size between 0.1 and 5 mm. McClenaghan et al. [56] reported chalcopyrite in the 250–500 μm fraction of till HMC from Izok Lake, primarily in samples located within 1 km down ice of the deposit.

In our study, chalcopyrite is present in the <250 μm fraction of the three till samples overlying and down ice of mineralization, along with a small number of grains in the finest fraction of sample 09-MPB-060, directly up ice of the deposit. The presence of chalcopyrite in the fine fraction of till is significant because it represents an increase in the dispersal distance down ice from 1 km for the >250 μm HMC fraction to at least 8 km for the <250 μm HMC fraction. Chalcopyrite is predominantly present as inclusions in other, more robust grains (Figure 9) such as epidote, hornblende, garnet, and Fe-oxide.

Chalcopyrite and pyrite are commonly present in small amounts in mineralized and barren metamorphosed mafic rocks, and thus the presence of chalcopyrite or pyrite inclusions in other minerals alone, although most likely related to mineralization at this specific site, can be an unreliable indication of mineralization in a regional exploration context [47]. The preservation of chalcopyrite and pyrite as inclusions in more robust grains, combined with the common occurrence of both minerals in mineralized and unmineralized bedrock, means that establishing accurate background values for both minerals is important when using automated mineralogical methods. The presence of whole grains of chalcopyrite and pyrite in till HMC (Table 5) remains a valuable indicator of proximity to mineralization given the relatively rapid physical and chemical weathering of sulfide grains following glacial transport and deposition. The identification of decreasing abundance in dispersal trains at a more detailed survey level can still be useful when vectoring to targets, and the ability of automated mineralogical systems to detect small inclusions of sulfide minerals on polished surfaces preserved as inclusions in other minerals makes them a better tool for this task than optical identification methods.

Hicken [17] reported the presence of galena in the 250–500 μm HMC fraction of only one till sample, collected 1 km down ice of mineralization. The abundance of galena (area percentage) in the samples down ice of the deposit is minor, but the grain abundance is anomalously elevated in the coarsest (180–250 μm) of the four fractions of each sample (Tables 3 and 4), highlighting the utility of reporting both area percentage and grain abundance data for indicator minerals. Area percentage values alone would not qualify this sample as anomalous but combined with grain abundance information indicates that galena is present as multiple very small grains.

The anomalous grain counts are due to the presence of small (10–15 μm) galena inclusions in other minerals (garnet, hornblende, chlorite) with most inclusions in the coarsest fraction, extending up to 8 km down ice of mineralization. Galena is not identified in the finest (<64 μm) HMC fraction of the four till samples. This distribution represents a significant increase of galena dispersal distance from a single sample 1 km down ice to 8 km down ice. This suggests that very fine galena inclusions in coarser grains, previously not identifiable with optical methods, are an important indicator of this style of mineralization, significantly extending the dispersal train of an ore mineral that is poorly preserved as whole grains in till.

No pyrrhotite was recovered from the coarse non-ferromagnetic HMC of till samples in the previous GSC study because pyrrhotite would have been removed during ferromagnetic separation. Pyrrhotite detected in the <250 μm fraction of non-ferromagnetic till HMC by this study is present as inclusions in composite grains where the pyrrhotite content is so minor that the magnetic forces of attraction could not overcome the mass of the grain. The abundance of pyrrhotite in the <250 μm HMC fraction of sample 09-MPB-060, 1 km up ice of mineralization, is comparable to the amount identified in till sample 09-MPB-075, located 3 km down ice (west) of mineralization. This distribution suggests that either (1) the abundance of pyrrhotite in till has diminished to regional background levels within 3 km of down ice, or (2) that up ice sample 09-MPB-060 (1 km up ice) does not represent the regional background for pyrrhotite.

4.5.7. Silver-Bearing Minerals

The Izok Lake deposit is a significant Ag resource [70], and the previous study of local tills around the deposit by [17,18] and McClenaghan et al. [11] identified Ag as a pathfinder element in the <63 μm matrix fraction of till [71]. Chalcopyrite in the Izok Lake deposit is known to be Ag-rich, containing up to 200–550 ppm Ag [70]. Because the lower detection limits for most elements using EDS are ~1000 ppm, the concentrations of Ag in chalcopyrite grains identified in our study will only be detectable in concentrations >1000 ppm. None were found to contain such high amounts of Ag which explains the lack of Ag-bearing minerals identified in this study.

4.5.8. Pathfinder Minerals

Green epidote and Fe-oxides are common minerals in many rock types and deposits and, therefore, are typically found in much higher abundance in till than traditional indicator minerals (S.A. Averill, Pers. Comm), thus their abundance are not routinely determined in the >250 μm fraction. MLA is ideally suited to quantifying their abundances as XBSE scans count every mineral present on a sample surface. Once regional background abundance values are established, anomalous elevated abundances, similar to those observed for epidote and Fe-oxide in our study, can be recognized. Epiote and Fe-oxide minerals are common and the presence in till would not be of note on their own; however, coincident increases in their abundance over the same spatial area highlight a potential area of interest for follow-up mineral exploration, especially when combined with the presence of sulfide minerals. The relationship between these minerals and traditional ore indicator minerals is similar to that of "pathfinder elements" and "indicator elements" in regional geochemical exploration and underscores the potential for automated mineralogy to be a powerful tool in an exploration context.

5. Conclusions

This study is the first examination of <250 μm heavy mineral fraction of till HMC from around a high grade (upper amphibolite-sillimanite) metamorphosed VMS deposit in permafrost terrain using automated mineralogical techniques. It also presents a comprehensive methodology for handling and preparing the sample material to mitigate both grain loss and sample cross contamination. Data reported here can be compared to future automated mineralogy studies of the fine fraction of till to help to establish background and "anomalous" abundances that could be expected in till close to such a deposit. Major findings of our study include:

1) The basal surface of epoxy grain mounts contains a greater number of heavy mineral grains than cross-sectional surfaces and represents the optimal surface for MLA studies. A mass of 0.3 g of heavy minerals will cover the basal surface of a 25 mm mount, although the mass can be adjusted to suit the density of the minerals to be mounted. Once the appropriate sample mass has been established, all mounts should be prepared using that mass to ensure relatable results between the samples.

2) Error between analytical runs of identical sections can be minimized with regular calibration of FEG-SEM systems. Calibration takes ~20 min to complete and should be performed prior to each batch of samples being run. The average error (as measured by this study) is not significant (<±1% shift/mineral). Increasing the number of grains presented for analysis improves statistical accuracy by decreasing the influence of outliers and nugget effects on indicator mineral counts.

3) Reporting indicator mineral abundance as both a normalized grain count and area percentage allows for inference about mineral occurrence that would not be possible using only one metric. Normalization of grain count values to 1000 grains ensures that abundance data can be compared between samples and projects and are intuitive to use. Interpreted along with area percentage values, these combined datasets can suggest whether a mineral is present as smaller numbers of larger grains or many small grains.

4) Common ore (chalcopyrite, galena, pyrite, sphalerite, pyrrhotite) and alteration (gahnite, axinite, corundum, epidote, Fe-oxide, staurolite) minerals of metamorphosed VMS deposits were detectable in the fine (<250 μm fraction) of till HMC from the Izok Lake area. Elevated

abundances of chalcopyrite and galena were detected up to 8 km down ice, a significant increase over the 1.3 km sulfide dispersal distance reported by McClenaghan et al. [11].

5) Epidote and Fe-oxide form a dispersal train down ice of the Izok Lake deposit. Epidote is a characteristic mineral in carbonate and propylitic alteration halos surrounding hydrothermal deposits, as well as in calc-silicate rocks of sedimentary or metasomatic origin and, therefore, epidote abundance must rely on the presence of other indicator minerals to be of use to exploration efforts. Care must be taken to ensure that accurate regional background abundance is established outside of wide-spread carbonate and propylitic alteration halos. Future work should investigate the use of trace-element compositional analysis of epidote grains to assess terrane fertility, after the work of Plouffe et al. [8] on porphyry systems. Fe-oxide minerals can indicate the incorporation of gossanous material into till [72] or the weathering of sulfide grains during transport or following deposition. Not all gossans will be found with associated mineralized sulfide bodies (i.e., the WIZ showing) but the fact that they indicate that a mineralized system existed at one point in time makes them an important exploration target.

6) Automated mineralogy can identify indicator minerals that are difficult or impossible to distinguish using traditional optical methods, or that are present only as small inclusions (galena) in other, more robust grains. Another advantage of automated mineralogy is that mineral associations within grains can be quickly identified and quantified (corundum/gahnite). Conversely, visual color of grains when using optical methods can be used to rapidly distinguish minor-element enrichment in some minerals (e.g., red for Mn-epidote, red for Cr-rich rutile). MLA can only recognize these grains if corresponding X-ray spectra have already been identified and added to the mineral reference library.

7) Large numbers of galena and chalcopyrite were identified in the coarsest two size fractions (125–185 μm and 185–250 μm) of till, at greater distances (≥8 km) down ice than previously identified in the >250 μm size fraction of the same till samples. Corundum, found in composite grains with gahnite, was found in the coarsest fraction examined in the sample most proximal to the deposit. All other alteration minerals (staurolite, gahnite, epidote, Fe-oxides) were found in these coarsest two fractions. Using the 125–250 μm (fine sand) size fraction will reduce the scanning time necessary for each sample, while still presenting 50,000–100,000 grains for analysis on a polished grain mount surface. We believe that at this location the fine sand (125–250 μm) fraction of till HMC is the most effective fraction to use with automated mineralogy for detection of indicator mineral anomalies.

Author Contributions: Conceptualization, H.D.L. and M.B.M. and D.L.-M.; methodology, H.D.L. and D.L.-M.; software, H.D.L; validation, H.D.L. and M.B.M. and D.L.-M. and M.L.; formal analysis, H.D.L; investigation, H.D.L.; resources, D.L.-M. and M.B.M.; data curation, H.D.L.; writing—original draft preparation, H.D.L.; writing—review and editing, M.B.M. and D.L.-M. and M.L.; visualization, H.D.L.; supervision, M.B.M. and D.L.-M. and M.L.; project administration, H.D.L.; funding acquisition, M.B.M. and D.L.-M. All authors have read and agreed to the published version of the manuscript.

Funding: This research was funded by Natural Resource Canada's Targeted Geoscience Initiative Program (TGI-5) through a Research Affiliate Program (RAP) bursary to the senior author as part of his Ph.D. research.

Acknowledgments: The authors would like to thank the team at Overburden Drilling Management Limited their assistance with all questions relating to sample processing and indicator mineral methods, Agatha Dobosz at Queen's University for invaluable help with instrument operation, and Jan Peter at the Geological Survey of Canada for sharing his knowledge of the mineralogy of the deposit.

Conflicts of Interest: The authors declare no conflict of interest

References

1. McClenaghan, M.B.; Kjarsgaard, B.A. Indicator Mineral and Surficial Geochemical Exploration Methods for Kimberlite in Glaciated Terrain: Examples from Canada. In *Mineral Deposits of Canada: A Synthesis*

      of Major Deposit-Types, District Metallogeny, the Evolution of Geological Provinces, and Exploration Methods; Goodfellow, W., Ed.; Geological Association of Canada: St. John's, NL, Canada, 2007; pp. 983–1006.

2. Averill, S.; Zimmerman, J. The Riddle Resolved—The Discovery of the Partridge Gold Zone Using Sonic Drilling in Glacial Overburden at Waddy Lake, Saskatchewan. In *CIM Bulletin*; Canadian Institute of Mining, Metallurgy, and Petroleum: Montreal, QC, Canada, 1984; Volume 77, p. 88.

3. Averill, S.A. The application of heavy indicator mineralogy in mineral exploration with emphasis on base metal indicators in glaciated metamorphic and plutonic terrains. *Geol. Soc. Spec. Publ.* **2001**, *185*, 69–81. [CrossRef]

4. Averill, S.A. Discovery and Delineation of the Rainy River Gold Deposit Using Glacially Dispersed Gold Grains Sampled by Deep Overburden Drilling: A 20 Year Odyssey. In *New Frontiers for Exploration in Glaciated Terrain*; Paulen, R., McClenaghan, M.B., Eds.; Open File 7374; Geological Survey of Canada: Ottawa, ON, Canada, 2013; pp. 37–46.

5. McClenaghan, M.B.; Cabri, L.J. Review of gold and platinum group element (PGE) indicator minerals methods for surficial sediment sampling. *Geochem. Explor. Environ. Anal.* **2011**, *11*, 251–263. [CrossRef]

6. Hashmi, S.; Ward, B.; Plouffe, A.; Leybourne, M.; Ferbey, T. Geochemical and mineralogical dispersal in till from the Mount Polley Cu–Au porphyry deposit, central British Columbia, Canada. *Geochem. Explor. Environ. Anal.* **2015**, *15*, 234–249. [CrossRef]

7. Kelley, K.D.; Eppinger, R.G.; Lang, J.; Smith, S.M.; Fey, D.L. Porphyry Cu indicator minerals in till as an exploration tool: Example from the giant Pebble porphyry Cu–Au–Mo deposit, Alaska, USA. *Geochem. Explor. Environ. Anal.* **2011**, *11*, 321–334. [CrossRef]

8. Plouffe, A.; Ferbey, T.; Hashmi, S.; Ward, B. Till geochemistry and mineralogy: Vectoring towards Cu porphyry deposits in British Columbia, Canada. *Geochem. Explor. Environ. Anal.* **2016**, *16*, 213–232. [CrossRef]

9. Averill, S.A. Viable indicator minerals in surficial sediments for two major base metal deposit types: Ni–Cu-PGE and porphyry Cu. *Geochem. Explor. Environ. Anal.* **2011**, *11*, 279–291. [CrossRef]

10. McClenaghan, M.B.; Paulen, R. Application of Till Mineralogy and Geochemistry to Mineral Exploration. In *Past Glacial Environments*; Menzies, J., Van der Meer, J., Eds.; Elsevier: Amsterdam, The Netherlands, 2018; pp. 689–751.

11. McClenaghan, M.; Paulen, R.; Layton-Matthews, D.; Hicken, A.; Averill, S. Glacial dispersal of gahnite from the Izok Lake Zn–Cu–Pb–Ag VMS deposit, northern Canada. *Geochem. Explor. Environ. Anal.* **2015**, *15*, 333–349. [CrossRef]

12. Plouffe, A.; McClenaghan, M.; Paulen, R.; McMartin, I.; Campbell, J.; Spirito, W. Processing of glacial sediments for the recovery of indicator minerals: Protocols used at the Geological Survey of Canada. *Geochem. Explor. Environ. Anal.* **2013**, *13*, 303–316. [CrossRef]

13. McClenaghan, M. Overview of common processing methods for recovery of indicator minerals from sediment and bedrock in mineral exploration. *Geochem. Explor. Environ. Anal.* **2011**, *11*, 265–278. [CrossRef]

14. Lehtonen, M.; Lahaye, Y.; O'Brien, H.; Lukkari, S.; Marmo, J.; Sarala, P. *Novel Technologies for Indicator Mineral-Based Exploration*; Geological Survey of Finland: Espoo, Finland, 2015; pp. 23–62.

15. Galley, A.G.; Hannington, M.D.; Jonasson, I. Volcanogenic Massive Sulphide Deposits. In *Mineral Deposits of Canada: A Synthesis of Major Deposit-Types, District Metallogeny, the Evolution of Geological Provinces, and Exploration Methods*; Goodfellow, W., Ed.; Geological Association of Canada: St. John's, NL, Canada, 2007; Volume 5, pp. 141–161.

16. McClenaghan, M.B.; Peter, J.M. Till geochemical signatures of volcanogenic massive sulphide deposits: An overview of Canadian examples. *Geochem. Explor. Environ. Anal.* **2016**, *16*, 27–47. [CrossRef]

17. Hicken, A. Glacial Dispersal of Indicator Minerals from the Izok Lake Zn–Cu–Pb–Ag VMS Deposit, Nunavut, Canada. Master's Thesis, Queen's University, Kingston, ON, Canada, 2012.

18. Hicken, A.; McClenaghan, M.; Layton-Matthews, D.; Paulen, R.; Averill, S.; Crabtree, D. *Indicator Mineral Signatures of the Izok Lake Zn-Cu-Pb-Ag Volcanogenic Massive Sulphide Deposit, Nunavut: Part 2 Till*; Open File 7343; Geological Survey of Canada: Ottawa, ON, Canada, 2013.

19. Bleeker, W.; Hall, B. The Slave Craton: Geological and Metallogenic Evolution. In *Mineral Deposits of Canada: A Synthesis of Major Deposit-Types, District Metallogeny, the Evolution of Geological Provinces, and Exploration Methods*; Goodfellow, W., Ed.; Geological Association of Canada: St. John's, NL, Canada, 2007; Volume 5, pp. 849–879.

20. Bleeker, W.; Ketchum, J.W.; Jackson, V.A.; Villeneuve, M.E. The Central Slave Basement Complex, Part I: Its structural topology and autochthonous cover. *Can. J. Earth Sci.* **1999**, *36*, 1083–1109. [CrossRef]

21. Buchan, K.L.; Ernst, R. *Diabase Dyke Swarms and Related Units in Canada and Adjacent Regions*; Geological Survey of Canada: Ottawa, ON, Canada, 2004.

22. Morrison, I. Geology of the Izok massive sulfide deposit, Nunavut Territory, Canada. *Explor. Min. Geol.* **2004**, *13*, 25–36. [CrossRef]

23. Morrison, I.; Balint, F. Geology of the Izok Lake Massive Sulphide Deposits, Northwest Territories, Canada. In Proceedings of the World Zinc'93 Symposium, Hobart, Australia, 10–13 October 1993; pp. 161–170.

24. Bostock, H.H. *Geology of the Itchen Lake Area, District of Mackenzie*; Geological Survey of Canada: Ottawa, ON, Canada, 1980.

25. Thomas, A. Volcanic Stratigraphy of the Izok Lake Greenstone Belt, District of Mackenzie, NWT. Ph.D. Thesis, University of Western Ontario, London, ON, Canada, 1978.

26. Nowak, R. The Nature and Significance of High-grade Metamorphism and Intense Deformation in the Izok VHMS Alteration Halo and Deposit. Master's Thesis, Colorado School of Mines, Golden, CO, USA, 2012.

27. Money, P.; Heslop, J. Geology of the Izok Lake massive sulphide deposit. *Can. Min. J.* **1976**, *97*, 24–27.

28. MMG Limited Izok Corridor. Available online: https://www.mmg.com/our-business/development-projects/ (accessed on 29 March 2020).

29. Costello, K.; Senkow, M.; Bigio, A.; Budkewitsch, P.; Ham, L.; Mate, D. *Nunavut Mineral Exploration, Mining and Geoscience Overview 2011*; Aboriginal Affairs and Northern Development Canada: Ottawa, ON, Canada, 2012.

30. Stubley, M.; Irwin, D. *Bedrock Geology of the Slave Craton, Northwest Territories and Nunavut*; NWT Open File 2019-01; Northwest Territories Geological Survey: Yellowknife, NT, Canada, 2019.

31. Hicken, A.; McClenaghan, M.; Layton-Matthews, D.; Paulen, R.; Averill, S.; Crabtree, D. *Indicator Mineral Signatures of the Izok Lake Zn–Cu–Pb–Ag Volcanogenic Massive Sulphide Deposit, Nunavut: Part 1 Bedrock Samples*; Open File 7173; Geological Survey of Canada: Ottawa, ON, Canada, 2013.

32. Dredge, L.; Kerr, D.; Ward, B. *Surficial Geology, Point Lake, District of Mackenzie, Northwest Territories*; Geological Survey of Canada: Ottawa, ON, Canada, 1996.

33. Stea, R.; Johnson, M.; Hanchar, D.; Paulen, R.; McMartin, I. The Geometry of Kimberlite Indicator Mineral Dispersal Fans in Nunavut, Canada. In *Application of Till and Stream Sediment Heavy Mineral and Geochemical Methods to Mineral Exploration in Western and Northern Canada, Short Course Notes*; McMartin, I., Paulen, R., Eds.; Geological Association of Canada: St. John's, NL, Canada, 2009; Volume 18, pp. 1–13.

34. Dyke, A.; Prest, V. Late Wisconsinan and Holocene history of the Laurentide ice sheet. *Géographie Physique et Quaternaire* **1987**, *41*, 237–263. [CrossRef]

35. Dyke, A.S. An Outline of North American Deglaciation with Emphasis on Central and Northern Canada. In *Developments in Quaternary Sciences*; Elsevier: Amsterdam, The Netherlands, 2004; Volume 2, pp. 373–424.

36. Paulen, R.C.; McClenaghan, M.B.; Hicken, A.K. Regional and local ice-flow history in the vicinity of the Izok Lake Zn–Cu–Pb–Ag deposit, Nunavut. *Can. J. Earth Sci.* **2013**, *50*, 1209–1222. [CrossRef]

37. Shilts, W.W. Nature and genesis of mudboils, central Keewatin, Canada. *Can. J. Earth Sci.* **1978**, *15*, 1053–1068. [CrossRef]

38. Dredge, L.; Ward, B.C.; Kerr, D.E. *Trace Element Geochemistry and Gold Grain Results from Till Samples, Point Lake, Northwest Territories (86H)*; Open File 3317; Geological Survey of Canada: Ottawa, ON, Canada, 1996.

39. Oviatt, N.M. Till geochemistry of the West Iznogoudh Zn Showing, Point Lake Map Sheet (86H), Nunavut. Bachelor's Thesis, University of Calgary, Calgary, AB, Canada, 2010.

40. Lougheed, H.D.; McClenaghan, M.B.; Layton-Matthews, D.; Leybourne, M. *Evaluation of Single-Use Nylon Screened Sieves for Use with Fine-Grained Sediment Samples*; Open File 8613; Geological Survey of Canada: Ottawa, ON, Canada, 2019; p. 14.

41. Blaskovich, R.J. Characterizing Waste Rock Using Automated Quantitative Electron Microscopy. Master's Thesis, University of British Columbia, Vancouver, BC, Canada, 2013.

42. Kjellsen, K.; Monsøy, A.; Isachsen, K.; Detwiler, R. Preparation of flat-polished specimens for SEM-backscattered electron imaging and X-ray microanalysis—Importance of epoxy impregnation. *Cem. Concr. Res.* **2003**, *33*, 611–616. [CrossRef]

43. Sylvester, P.J. Use of the Mineral Liberation Analyzer (MLA) for Mineralogical Studies of Sediments and Sedimentary Rocks. In *Mineralogical Association of Canada Short Course 42*; Mineralogical Association of Canada: St. John's, NL, Canada, 2012; pp. 1–16.

44. Layton-Matthews, D.; Hamilton, C.; McClenaghan, M. Mineral Chemistry: Modern Techniques and Applications to Exploration. In *Application of Indicator Mineral Methods to Mineral Exploration*; McClenaghan, M., Plouffe, A., Layton-Matthews, D., Eds.; Open File 7553; Geological Survey of Canada: Ottawa, ON, Canada, 2014; pp. 9–18.

45. Sandmann, D. Method Development in Automated Mineralogy. Ph.D. Thesis, Technischen Universität Bergakademie Freiberg, Freiberg, Germany, 2015.

46. Simandl, G.J.; Mackay, D.A.R.; Ma, X.; Luck, P.; Gravel, J.; Grcic, B.; Redfearn, M. Direct and Indirect Indicator Minerals in Exploration for Carbonatite and Related Ore Deposits—An Orientation Survey, British Columbia, Canada. In Proceedings of the 27th International Applied Geochemistry Symposium, Tucson, AZ, USA, 20–24 April 2015; pp. 33–39.

47. Hulkki, H.; Taivalkoski, A.; Lehtonen, M. Signatures of Cu (–Au) mineralisation reflected in inorganic and heavy mineral stream sediments at Vähäkurkkio, north-western Finland. *J. Geochem. Explor.* **2018**, *188*, 156–171. [CrossRef]

48. Dreimanis, A.; Vagners, U. Bimodal Distribution of Rock and Mineral Fragments in Basal Tills. In *Till, a Symposium*; Goldthwaite, R.P., Ed.; Ohio State University Press: Columbus, OH, USA, 1971; pp. 237–250.

49. Pickett, J. Method Development for the Systematic Separation of the <63 μm Heavy Mineral Fraction from Bulk Till Samples. Master's Thesis, Queen's University, Kingston, ON, Canada, 2018.

50. Lougheed, H.D.; McClenaghan, M.B.; Layton-Matthews, D. Mineral Markers of Base Metal Mineralization: Progress Report on the Identification of Indicator Minerals in the Fine Heavy Mineral Fraction. In *Targeted Geoscience Initiative: 2017 Report of Activities*; Rogers, N., Ed.; Open File 8373; Geological Survey of Canada: Ottawa, ON, Canada, 2018; Volume 2, pp. 101–108.

51. Day, S.J.A.; Lariviere, J.M.; McNeil, R.J.; Friske, P.W.B.; Cairns, S.R.; McCurdy, M.W.; Wilson, R.S. *Regional Stream Sediment and Water Geochemical Data, Horn Playeau area, Northwest Territories (Parts of NTS 85E, 85F, 85K, 85L, 95H, 95I and 95J) Including Analytical, Mineralogical, and Kimberlite Indicator Mineral Data*; Open File 5478; Geological Survey of Canada: Ottawa, ON, Canada, 2007.

52. Falck, H.; Day, S.; Pierce, K.L.; Cairns, S.R.; Watson, D. *Geochemical, Mineralogical and Indicator Mineral Data for Stream Silt Sediment, Heavy Mineral Concentrates and Waters, Flat River area, Northwest Territories, (Part of NTS 95E, 105H and 105I)*; NWT Open Report 2015-002; Northwest Territories Geoscience Office: Yellowknife, NT, Canada, 2015.

53. Lastra, R.; Petruk, W. Mineralogical characterization of sieved and un-sieved samples. *J. Miner. Mater. Charact. Eng.* **2014**, *2*, 40–48. [CrossRef]

54. Voordouw, R.J.; Gutzmer, J.; Beukes, N.J. Zoning of platinum group mineral assemblages in the UG2 chromitite determined through in situ SEM-EDS-based image analysis. *Miner. Depos.* **2010**, *45*, 147–159. [CrossRef]

55. Berlin, J. Analysis of boron with energy dispersive X-ray spectrometry. *Imaging Microsc.* **2011**, *13*, 19–21.

56. McClenaghan, M.B.; Hicken, A.; Paulen, R.; Layton-Matthews, D. *Indicator Mineral Counts for Regional till Samples Around the Izok Lake Zn–Cu–Pb–Ag VMS Deposit, Nunavut*; Open File 7029; Geological Survey of Canada: Ottawa, ON, Canada, 2012.

57. Spry, P.G. An unusual gahnite-forming reaction, Geco base-metal deposit, Manitouwadge, Ontario. *Can. Mineral.* **1982**, *20*, 549–553.

58. Theart, H.; Ghavami-Riabi, R.; Mouri, H.; Gräser, P. Applying the box plot to the recognition of footwall alteration zones related to VMS deposits in a high-grade metamorphic terrain, South Africa, a lithogeochemical exploration application. *Geochemistry* **2011**, *71*, 143–154. [CrossRef]

59. Bonnet, A.; Corriveau, L. Alteration Vectors to Metamorphosed Hydrothermal Systems in Gneissic terranes. In *Mineral Deposits of Canada: A Synthesis of Major Deposit-Types, District Metallogeny, the Evolution of Geological Provinces, and Exploration Methods*; Goodfellow, W., Ed.; Geological Association of Canada: St. John's, NL, Canada, 2007; Volume 5, pp. 141–161.

60. Zaleski, E.; Froese, E.; Gordon, T.M. Metamorphic petrology of Fe–Zn–Mg–Al alteration at the Linda volcanogenic massive sulfide deposit, Snow Lake, Manitoba. *Can. Mineral.* **1991**, *29*, 995–1017.

61. Stoddard, E.F. Zinc-rich hercynite in high-grade metamorphic rocks; a product of the dehydration of staurolite. *Am. Mineral.* **1979**, *64*, 736–741.

62. Dietvorst, E.J. Biotite breakdown and the formation of gahnite in metapelitic rocks from Kemiö, Southwest Finland. *Contrib. Mineral. Petrol.* **1981**, *75*, 327–337. [CrossRef]

63. Spry, P.G.; Scott, S.D. The stability of zincian spinels in sulfide systems and their potential as exploration guides for metamorphosed massive sulfide deposits. *Econ. Geol.* **1986**, *81*, 1446–1463. [CrossRef]

64. Ghosh, B.; Praveen, M. Indicator minerals as guides to base metal sulphide mineralisation in Betul Belt, central India. *J. Earth Syst. Sci.* **2008**, *117*, 521–536. [CrossRef]

65. Makvandi, S.; Ghasemzadeh-Barvarz, M.; Beaudoin, G.; Grunsky, E.C.; McClenaghan, M.B.; Duchesne, C. Principal component analysis of magnetite composition from volcanogenic massive sulfide deposits: Case studies from the Izok Lake (Nunavut, Canada) and Halfmile Lake (New Brunswick, Canada) deposits. *Ore Geol. Rev.* **2016**, *72*, 60–85. [CrossRef]

66. McClenaghan, M.B.; Budulan, G.; Parkhill, M.A.; Layton-Matthews, D.; Crabtree, D. *Indicator Mineral Signatures of the Halfmile Lake Zn–Pb–Cu Volcanogenic Massive Sulphide Deposit, Bathurst, New Brunswick: Part 2—Till Data*; Open File 8111; Geological Survey of Canada: Ottawa, ON, Canada, 2016.

67. DiLabio, R. Glacial Dispersal Trains. In *Glacial Indicator Tracing*; Kujansuu, R., Saarnisto, M., Eds.; August Aimé Balkema: Rotterdam, The Netherlands, 1990; pp. 109–122.

68. Gregory, D.D.; Large, R.R.; Bath, A.B.; Steadman, J.A.; Wu, S.; Danyushevsky, L.; Bull, S.W.; Holden, P.; Ireland, T.R. Trace element content of pyrite from the kapai slate, St. Ives Gold District, Western Australia. *Econ. Geol.* **2016**, *111*, 1297–1320. [CrossRef]

69. Mukherjee, I.; Large, R. Application of pyrite trace element chemistry to exploration for SEDEX style Zn–Pb deposits: McArthur Basin, Northern Territory, Australia. *Ore Geol. Rev.* **2017**, *81*, 1249–1270. [CrossRef]

70. Harris, D.C.; Cabri, L.J.; Nobiling, R. Silver-bearing chalcopyrite, a principal source of silver in the Izok Lake massive-sulfide deposit; confirmation by electron-and proton-microprobe analyses. *Can. Mineral.* **1984**, *22*, 493–498.

71. Hicken, A.; McClenaghan, M.; Paulen, R.; Layton-Matthews, D. *Till Geochemical Signatures of the Izok Lake Zn–Cu–Pb–Ag Volcanogenic Massive Sulphide Deposit*; Open File 7046; Geological Survey of Canada: Ottawa, ON, Canada, 2012.

72. McClenaghan, M.; Parkhill, M.; Pronk, A.; Seaman, A.; McCurdy, M.; Leybourne, M. Indicator mineral and geochemical signatures associated with the Sisson W–Mo deposit, New Brunswick, Canada. *Geochem. Explor. Environ. Anal.* **2017**, *17*, 297–313. [CrossRef]

 *minerals*

Article

Automated Quantitative Mineralogy Applied to Metamorphic Rocks

Nynke Keulen [1,*], **Sebastian Næsby Malkki** [1] **and Shaun Graham** [2]

[1] Department of Petrology and Economic Geology, Geological Survey of Denmark and Greenland (GEUS), Øster Voldgade 10, DK-1350 Copenhagen K, Denmark; shb@geus.dk

[2] Carl Zeiss Microscopy GmbH, ZEISS Group, 50 Kaki Bukit Place, Singapore 415926, Singapore; shaun.graham@zeiss.com

* Correspondence: ntk@geus.dk

Received: 10 December 2019; Accepted: 30 December 2019; Published: 3 January 2020

Abstract: The ability to apply automated quantitative mineralogy (AQM) on metamorphic rocks was investigated on samples from the Fiskenæsset complex, Greenland. AQM provides the possibility to visualize and quantify microstructures, minerals, as well as the morphology and chemistry of the investigated samples. Here, we applied the ZEISS Mineralogic software platform as an AQM tool, which has integrated matrix corrections and full quantification of energy dispersive spectrometry data, and therefore is able to give detailed chemical information on each pixel in the AQM mineral maps. This has been applied to create mineral maps, element concentration maps, element ratio maps, mineral association maps, as well as to morphochemically classify individual minerals for their grain shape, size, and orientation. The visualization of metamorphic textures, while at the same time quantifying their textures, is the great strength of AQM and is an ideal tool to lift microscopy from the qualitative to the quantitative level.

Keywords: automated quantitative mineralogy (AQM); scanning electron microscopy; ZEISS Mineralogic; Fiskenæsset complex; Feret angle; element concentration map; visualization; mineral association; bulk composition; grain size

---

1. Introduction

Automated mineralogy, or more correctly automated quantitative mineralogy (AQM) was developed in the 1980s to analyse the mineralogy, chemistry, and microstructures of mineral ores, fly ashes, and sediments with an energy dispersive spectrometry (EDX) detector mounted to a scanning electron microscope (SEM) [1–3]. This developed from a range of automated particle analysis procedures into software platforms, e.g., Qemscan, MLA, ZEISS Mineralogic, AMICS, or TIMA-X, dedicated to multiphase materials in a wide range of research fields including, but not limited to, forensic sciences, archaeometry, oil reservoir geology, urban mining, and material sciences [4–13]. Within geosciences, AQM is most widely used on ore minerals (for modal mineralogy, liberation, association, etc.) [13–15] and oil reservoir rocks (e.g., to describe mineralogy and porosity or provenance) [16–18]. However, other areas within geosciences have so far received less attention, (but see e.g., [19–21]). Apart from AQM with SEM-EDX systems, AQM can also be applied with wavelength dispersive spectrometry (WDX), micro-energy-dispersive X-ray fluorescence, laser-induced breakdown spectroscopy and hyperspectral mineral analyses [22–24].

Many AQM systems, like Qemscan, MLA, or TIMA-X, apply spectrum-matching for the classification of the mineralogy of the samples. This AQM technique is based on generating unquantified EDX spectra in user-defined steps or specific spots or a raster on the sample surface. The EDX spectra are not matrix-corrected or quantified but matched against a library of known referenced EDX spectra (based on analyses of standards, or calculated from mineral formulas) [5,6,25,26]. The development of this library typically requires an extensive workflow on mineral spectrum validation using electron microprobe validation or pre-validated mineral standards.

AQM in geosciences (outside the mining industry) has mainly been used as a qualitative instrument, rather than a laboratory tool for quantitative measurements on mineralogy, chemistry, and morphology. There are several reasons for this: It takes a large effort to obtain reproducible data between AQM systems. This is caused by a lack of precise chemical data for most AQM systems, and for complex and variable mineral systems reduces the reliability and therefore the range of application of these AQM techniques in these research environments. Resulting from this is the fact that each mineral list for each spectrum-matching analysis is as precise as the geological and mineralogical knowledge of the operator. To analyze the same sample in a different AQM system, or under different analytical conditions requires the development of a new mineral library. Furthermore, to produce valuable petrological results, good quality data on textures and on minerals are needed, preferably with the chemical data derived from exactly the minerals that are visualized.

Here, we will present examples where the latest AQM solutions, such as ZEISS Mineralogic can provide new insights into metamorphic textures using advanced visualization and quantification methods. To our knowledge, no study with a metamorphic petrology focus exist based on ZEISS Mineralogic software, and only few studies based on other automated mineralogy platforms exist with metamorphic petrology as a main focus outside a mining and exploration setting [27–29]. AQM serves as an ideal tool to visualize metamorphic textures and to simultaneously quantify the mineralogy, chemistry, and grain properties of these textures. The applied software includes the possibility to obtain precise element chemistry and therefore to analyze minor-element contributions to variations in individual mineral compositions and to measure grain properties within multi-phase composites; both features are of interest for metamorphic petrologists. The ZEISS Mineralogic software also allows to exchange mineral lists between Mineralogic users or samples, or to change acceleration voltages without the need to create new mineral lists. These advanced chemical, mineralogical, and textural properties are applied here to visualize mineralogy and textures in a new way.

The examples used in this study are thin sections derived from the Geological Survey of Denmark and Greenland's (GEUS) sample collection of rocks from southern West-Greenland (Figure 1). This region comprises Archean basement rocks that are part of the North Atlantic Craton. The majority of the rocks are grey and brown tonalite–trondhjemite–granodiorite (TTG) orthogneisses, intruded by granitic and granodioritic bodies, and by TTG-composition as well as mafic sheets. Intercalated in the orthogneisses are enclaves of supracrustal rocks including amphibolites, mafic granulites and mica-schists, but also lenses of ultramafic rocks. The area also includes the well-known Meso-Archean Fiskenæsset complex (a leucogabbro–anorthosite intrusive complex). The rocks of the Fiskenæsset complex intruded 2.97–2.95 Ga in amphibolites that already were deformed and metamorphosed by the time of the intrusion. The formation age of the amphibolites is estimated to be 2.90 Ga. [30–39].

At least three deformation phases affected the rocks in the area, where deformation was accommodated mainly by folding, but also by thrusting at a meter- to kilometer-scale. The region consists of several blocks or terrane that assembled into a larger unit at the latest part of the Meso-Archean age. The first regionally recognized deformation phase in the rocks can be observed in finely foliated isoclinally folded units (mica-schists, amphibolites, leucogabbroic rocks of the Fiskenæsset complex). The main deformation event metamorphosed the rocks in the area at amphibolite facies and granulite facies conditions. The effects of this main folding phase were overprinted by a later folding and thrusting phase at amphibolite facies conditions, before minor retrogressive overprint at greenschist facies conditions localized around late fault and shear zones. Uplift and erosion brought the rocks to the Earth's surface [30–32,34,36]. The tectonometamorphic processes affecting the region developed a range of metamorphic features that are used as examples to highlight the visualization and quantification of metamorphic textures by AQM applying the Mineralogic software.

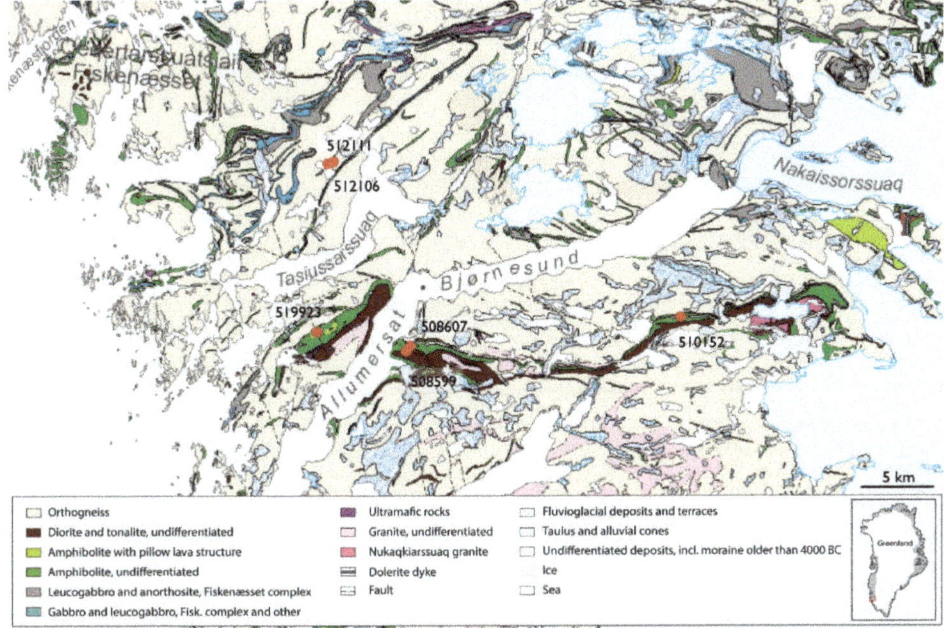

Figure 1. Geological map of southern West-Greenland near the village Fiskenæsset. The localities of the samples used in this study are indicated with red dots. Map modified from GEUS [40].

2. Materials and Methods

2.1. Sample Material

Six polished thin sections from metamorphic rocks from southern West-Greenland were coated with Carbon and investigated with SEM (ZEISS SIGMA 300VP; Cambridge, UK), see below. In the light of this study, no cross-checks were performed on the optical microscope. More details about the geological history of the samples were published previously [37,41]. Table 1 presents the analytical details for the investigated samples and Supplementary 1 gives sample and outcrop images as well as rock and mineral compositions as analysed with Mineralogic.

Table 1. Analytical conditions for Mineral mapping. Mapping step size: three samples were mapped several times at different step sizes, see Section 3 (Examples of Applications) for details.

Sample No.	Acceleration Voltage (kV)	Aperture Size (µm)	Mapping Step Size (µm)	Magnification	Through-Put Rate (kcps)	Dwell Time (s)
508599	20	120	15	160	275	0.006
508607	15	120	5 & 20	235	400	0.008
510152	20	120	20	235	275	0.006
511923	20	120	5–40	235	275	0.004
521106	20	120	5–40	235	275	0.004
521111	15	120	20	235	400	0.007

Samples 508599 and 508607 represent a finely layered volcanoclastic sediment collected from the Bjørnesund Supracrustal Belt (BSB). Hand specimen of sample 508599 is near-black with a slatey cleavage. It consists of quartz, andesine (plagioclase), biotite, amphibole, apatite, and pyrite. The rocks in the BSB were isoclinally folded at amphibolite facies conditions. The sample was collected close to an isoclinal fold core, and biotite in the sample clearly defines the shape of the fold and the foliation.

Sample 508607 was collected from the same volcanoclastic unit in the RSB. However, in this sample contains quartz, anorthite, pale amphibole (anthophyllite based on field observations), biotite, and large garnets with inclusions of quartz, apatite, ilmenite, and rutile were grown over the main foliation.

Samples 521106 and 521111 show the reaction products between an ultramafic rock in contact with anorthosite or leucogabbro and intruded by a tonalite sheet. The anorthosite, leucogabbro and ultramafic rocks are part of the Fiskenæsset complex. The intruding tonalitic sheet causes a desilification reaction resulting in the formation of amphibole (anthophyllite, cummingtonite, gedrite or pargasite), phlogopite, and one or more aluminium-rich phases (cordierite, spinel, sapphirine and corundum (ruby)). Reaction temperatures are estimated to ca. 600 °C, using iso-chemical P–T-sections [36], with a gradient from lower granulite facies conditions in the western part of the Fiskenæsset complex to upper amphibolite facies close to the Inland ice sheet and in the southern part of the complex. The reaction occurred during the main folding phase in the region. Sample 521106 and 521111 are collected from Aappaluttoq, south of the village of Fiskenæsset, and here ruby is the main aluminium-rich phase resulting from the reaction. The sample consists of pale amphibole (gedrite), biotite, corundum, sapphirine, spinel, cordierite, anorthosite, hornblende, epidote, magnesite. Note that the peak mineral assemblage (gedrite, hornblende, ± biotite, cordierite, sapphirine; ca. 630 °C; 7 kbar [38]) does not contain quartz.

Sample 510152 represents a leucogabbro from the Fiskenæsset complex (Figure 1). The sample was collected in the southern part of the complex, where the complex lies in contact with the BSB. It consists of the minerals anorthite, Mg-rich hornblende, chromite and minor rutile. The amphibole in the sample is a result of deformation and metamorphism at amphibolite facies conditions, where it replaces pyroxene.

Sample 519923 is an amphibolite collected from the BSB. The hand specimen is dark green and finely foliated with a slatey cleavage. The minerals in the sample are distributed homogeneously, apart from tiny small faults filled with gouge and secondary minerals. Reactivated shear zones in the area accumulated gold in slightly enhanced concentrations. The sample contains anorthite, hornblende and the minor phases quartz, zoisite, rutile, sphene, apatite and zircon.

2.2. Whole Rock Geochemistry

Whole-rock geochemistry was carried out on a sample to characterize the bulk composition of these rocks. The analyses were carried out by Activation Laboratories (Actlab), Ancaster, ON, Canada, using their Research Lithium Metaborate/Tetraborate Fusion—ICP/MS analytical package. Samples 519923 was investigated, see Table 2. Both samples are described in more detail above.

2.3. Mineralogic Method

The micro-analysis on the selected thin section was performed on a ZEISS SIGMA 300VP SEM equipped with a back-scattered electrons (BSE) detector and two Bruker XFlash 6|30 EDX detectors, with 129 eV energy resolution and with the ZEISS Mineralogic automated quantitative mineralogy software platform located at the Geological Survey of Denmark and Greenland, Copenhagen, Denmark. The description of the Mineralogic method in this section, including screenshots (Figures S1–S7) from the set-up of the Mineralogic recipes can be found in the Supplementary 2. Within each thin section, a region of interest was selected and imaged to provide a high-resolution BSE mosaic of stitched images. Also, on this region of interest, a quantitative mineralogical analysis was carried out using Mineralogic, creating a mineral map with a user defined step size (or pixel size) as well as a list with parameters for grains in the sample. The acceleration voltage of the primary electron beam was set to 15 or 20 kV, to ensure X-ray excitation for all relevant elements (e.g., Fe, Cu, Zn). The 120 μm aperture providing 80 μA beam current used was used to obtain a high input count rate for the EDX detectors. The EDX software is fully integrated with the Mineralogic software, allowing for matrix corrections of each mineral. Therefore, the exact element concentrations can be calculated for each acceleration voltage. A detailed description of the Mineralogic method is given below:

Table 2. Bulk geochemistry of each analysed thin section. For sample 511923, this data is compared to a whole rock geochemistry analysed by Actlab. Other data generated with Mineralogic (EDX analyses) in element wt% and recalculated to wt% oxides (and ppm for sample 511923). LOI: loss on ignition. Numbers in italics are estimates only.

511923	Al%	Ca%	Cl%	Cr%	Cu%	F%	Fe%	K%	Mg%	Mn%	Na%	O%	P%	S%	Si%	Ti%	Zr%	LOI%
MIN: element	8.05	10.2	0.01	0.02	0.03	0.01	10.6	0	5.22	0.01	0.86	38.6	0.03	0	26.1	0.36	0.01	
MIN: oxide	13.9	13					12.4	0	7.91	0.02	1.06		0.06	0.01	51	0.55		
MIN: ppm				295	316												116	
Actlab: oxide	14.9	11	x			x	13.2	0.12	8	0.2	2.01		0.07	x	49	1.02		0.72
Actlab: ppm				250	70												62	
508599																		
MIN: element	7.74	2.25	*0.01*	*0.01*	*0.04*	*0.03*	3.96	1.59	2.17	*0.07*	2.94	41.43	*0.01*	0.06	37.51	*0.05*	*0.00*	
MIN: oxide	12.94	2.79		*0.01*	*0.04*		4.51	1.70	3.19	*0.08*	3.50		*0.02*	0.13	71.02	*0.07*	*0.00*	
508607																		
MIN: element	9.55	5.03	*0.02*	*0.02*	*0.03*	*0.03*	6.45	0.75	2.99	*0.02*	0.30	39.95	*0.01*	0.09	34.25	*0.35*	*0.00*	
MIN: oxide	15.85	6.18		*0.02*	*0.04*		7.29	0.80	4.36	*0.02*	0.35		*0.03*	0.21	64.35	*0.52*	*0.00*	
510152																		
MIN: element	16.88	11.31	*0.01*	2.00	*0.03*	*0.02*	4.42	*0.01*	3.13		1.08	38.28		*0.00*	22.66	*0.02*	*0.00*	
MIN: oxide	27.37	13.59		2.51	*0.03*		4.88	*0.01*	4.46		1.25			*0.01*	41.62	*0.02*	*0.00*	
521106																		
MIN: element	15.28	*0.11*	*0.02*	*0.12*	*0.04*	*0.07*	4.03	0.96	14.92	*0.01*	*0.12*	38.95	*0.00*	0.02	25.08	*0.05*	*0.00*	
MIN: oxide	25.26	*0.14*		*0.16*	*0.04*		4.54	1.01	21.64	*0.02*	*0.14*		*0.00*	0.04	46.95	*0.07*	*0.00*	
521111																		
MIN: element	15.10	0.23	*0.04*	*0.11*	*0.03*	*0.06*	4.67	1.52	14.38	*0.01*	*0.14*	40.58	*0.00*	0.01	22.87	*0.02*	*0.00*	
MIN: oxide	25.96	0.29		*0.15*	*0.04*		5.46	1.66	21.69	*0.02*	*0.17*		*0.00*	0.03	44.51	*0.03*	*0.00*	

2.3.1. Zeiss Mineralogic Mining

The ZEISS Mineralogic software platform has a Mining and a Reservoir rock plug-in, though analytical functionalities between both branches largely overlap. The difference between both lies in how obtained data are visualized, e.g., as target and byproduct (Mining), or integrated with porosity measurements (Reservoir). Here, the Mining part of the software was applied. The software offers a recipe-based solution for all steps in the analysis (SEM parameters, holders and stubs used, calibration for the EDX and BSE detectors, image analysis, morphological analysis, mineralogical analysis, lithological analysis, mining output parameters, and criteria that specify how and when the analytical run is performed and terminated). These recipes can be saved, mixed, and changed individually, thus adapting each analysis to the sample at hand. Recipes can also be in- and exported allowing for an exchange between different internal and external users. Routine analyses can be set up allowing a non-geologist/mineralogist can run samples from a batch of similar samples, while the more experienced user gets the freedom to vary many parameters for each analysis [8]. The description here is based on Mineralogic software version 1.6.

2.3.2. Image Navigation

The Image Navigation tool of the ZEISS SEM applies digital images of thin sections, the entire sample holder, or overview SEM-images to navigate within the sample (Figure S1). The tool allows the user to define three fiducial points on the navigation image and connect those to stage coordinates, afterwards the movements of the stage can be controlled from the navigation image. The tool is especially ideal to set up analyses on holders with several samples, or on fine grained material in a fast manner.

2.3.3. SEM Recipe

In this recipe, the SEM can be operated in a normal mode to set and save the operating conditions for the analysis. The sample can be placed to the correct position, the beam is focused and optimized at the required acceleration voltage (often 15–25 kV) and aperture size (often 60–120 μm). The brightness and contrast of the image are set. The SEM recipe saves the current settings of the SEM (like vacuum settings, beam parameters) and regulates the SEM imaging parameters (dwell time to allow for stage movements, scan speed), see Figure S1.

2.3.4. Holder Recipe

In the Holder recipe the size and shape of both the sample holder and of the individual stubs can be defined and adjusted. First the size and shape of holder are defined (holder set-up), afterward the size and shapes of the individual stubs (stub setup). The recipe lists a number of standard sample holders, but own holders can be defined as well (Figure S2). In the stub details part of the menu, the geometrical parameters for the analysis of each particle can be defined: Z-coordinate of the stage, magnification during the analysis, step size for the mapping of the sample, and the area or areas in the sample that are analysed (Figure S2).

2.3.5. Calibration Recipe

Both the BSE brightness and the EDS peak position can be monitored and corrected during larger mapping sessions. The BSE brightness is regulated using a standard with a bright- and a dark phase next to each other, e.g., a copper–aluminium stub. The brightness of the BSE image is set such that the bright phase in the standard (e.g., copper) has grey-values in the upper grey-level segment, but is not completely white, while the dark phase (e.g., aluminium) has grey values in the dark-grey segment. The interval for monitoring can be selected, e.g., every hour and for every new sample. At each monitoring event, the SEM resets the brightness and contrast values of the SEM to match the predefined values for the bright and dark phase in the standard (Figure S3).

The EDS peak position can be monitored by regularly measuring suitable EDS peak, typically Cu Kα, for the beam energy settings chosen for the analysis.

2.3.6. Image Processing Recipe

Mineralogic analyses are typically performed overlying a BSE image, but other types of images based on electron microscope detector input, like secondary electrons or cathodoluminescence (CL) signal can be applied too, where required. The image can be thresholded to only include certain grey levels for the forthcoming Mineralogic analysis, e.g., with the aim only to investigate the brightest phases in the sample, or to exclude epoxy and porosity from the analysis. The quality of the image can be improved with a large number of image processing techniques like arithmetic functions (e.g., addition, substraction, inversion), logical operations (e.g., image AND image, image OR constant), convolution and filters (e.g., median filter, sharpening, edge detection), histogram and threshold operations, morphological operations (e.g., dilation, opening, hole fill, skeletonize), segmentation and region based operations (e.g., watershed) or geometric and linear transformations (e.g., downsize with a power of two).

The image processing recipe tap can also be applied on pre-set functions like the search for bright phases in the sample (Figure S4), which can be applied to find suitable zircon or monazite minerals for isotope dating or certain ore minerals). Here, the BSE signal is thresholded to only find the brightest particles in a large sample and to store their chemistry and coordinates.

2.3.7. Morphology Recipe

In the Morphology recipe the grain size, grain shape and other physical parameters of the sample that are analysed can be selected. The software offers following possibilities: area, length, breadth, elongation, roughness, Feret maximum, minimum, and mean length, Feret maximum, minimum, and mean angle, porosity, and grey value (Figure S5). These grain morphology parameters are provided as output data for every individual analysed grain and can therefore be correlated to grain mineralogy or chemistry. Morphological analyses are performed on the BSE (or other input) image of the sample, not on the map produced during the analysis.

The morphological parameters can be classified such that grains full-filling certain size or shape parameters are registered separately. The grains fulfilling these parameters can be filtered out after the analysis.

2.3.8. Mineral Recipe

The Mineral recipe is the central part of the Mineralogic Mining plugin. It regulates the EDS and mapping type parameters, holds the list of minerals, and allows for morphochemical classification criteria (Figure S6). The Mineralogic analysis can be performed in five different ways:

- Mapping analysis, where the user defines the step size (pixel size) for the EDS map, and an analysis is performed for every single point (pixel) of the map. Typical pixel sizes are 5–30 µm, but can be larger for coarse grained, homogeneous samples, or as small as 200 nm [20]. This method presents the full detail of the sample but is also the most time-consuming method.
- Spot centroid analysis, where the sample is segmented by BSE value of the grains. The method is typically used on grains in an epoxy matrix. For each grain determined, the geometrical centre of the grain is calculated, and a single EDS analysis on this point is performed. This method is especially suitable for homogeneous grains, as a small variation in the chemistry, like inclusions or zonations, will not be picked up.
- Feature scan analysis, where—like for the spot centroid analysis—the grain boundaries are determined by segmenting the BSE image. In this analysis mode, the beam is rastering within the boundaries of the grain, giving an average chemical composition. In case of zonations, a more correct average grain chemistry will be found; for heavily included samples, a false mineral

classification might be generated. The method can be nearly as fast as a spot centroid analysis, depending on the pre-set amount of counts in the spectrum.

- Fast scan is an intermediate form of Feature scan and spot centroid analysis. It scatters a series of point analyses across a grain and thus arrives at an average composition of the grain. The user gets to determine the density of the spots for the analyses in an area, but each grain is analysed at least once, independent of its size.
- Line scan analysis, where all EDS analyses are made along a line with a pre-defined step-size across the centre of a particle or frame. All variations in composition are accounted for, and a first impression on the grain size and texture can be obtained. The method is very fast compared to a full Mapping analysis.
- Grey level mapping, where the sample is imaged (usually with BSE), without applying EDS measurements. For samples with a simple chemistry, grey values can be used to separate individual minerals. The method is very fast, but less precise than EDS-based investigations.

For all EDS-based methods, the EDS spectrum deconvolution can be specified in the same way as for regular EDS analyses (Figure S6a). Thus, elements can be excluded from quantification (e.g., carbon used for coating is set to deconvolution-only), matrix quantification methods can be chosen (ZAF vs. Phi-Rho-Z), spectra output data can be normalized, and elements can be chosen to be always or never included, if wished for, thus potential sources of peak overlap can be avoided. EDS dwell time, and detector throughput rate can be chosen by the user. Every single generated spectrum during the analysis is fully quantified and the weight percentages of each element in each pixel on the false coloured mineral map is available after the analysis. User can choose to add standards-based quantification of the spectra. These latter two points are unique for the Mineralogic software. The matrix quantification for each spectrum also allows the user to switch between different acceleration voltages in between samples, without the need to specify a new mineral list. Most AQM systems require the user to make separate mineral lists for each acceleration voltage.

The mineral list for the sample analysis is created by the user based on the element wt% of the mappable phases and can be used to produce a variety of visualizable informative image and data outputs. For each mineral the range of tolerated concentrations for each element, as well as for ratios between two elements, can be used to define the mineral phase (Figure S6b). Minerals are placed in a list, which is checked against the chemistry of each analysed point, following a first-match principle and based on the order of minerals in the mineral list, which is defined by the operator. The same principle to build mineral lists can also be used to define element concentration maps or element ratio maps, where individual phases are identified and coloured by the concentration of one or a few elements. Mineral lists can be exported and imported between projects and adjusted to fit the exact mineral specifics for the sample area of interest, allowing for slight differences in mineral chemistry resulting from different whole rock compositions or metamorphic temperatures.

In an addition tab in the recipe, minerals can be clustered into groups (e.g., albite, labradorite and anorthite as plagioclase). Clustered minerals can be integrated as a group when particle data is exported after the analysis.

Additionally, morphological criteria can be used in the classification of minerals. For example, textural properties can be defined based on the chemically mapped grains, where the grain size or shape properties are used to synergize the chemical and textural characteristics. For example, zircons of a size large enough to be dated with a laser connect to a mass spectrometer can be classified from the main zircon population separately. All different ways of describing the sample (e.g., element map, mineral map, morphochemical map) can be calculated offline from the generated data. There is no need to reanalyze the sample after the mineral list was changed.

Many parameters affect the speed and quality of an AQM analysis, and a Mineralogic analysis is no different. Each analysis must be optimized prior to an AQM run. Different types of analyses (see above) will yield a difference in analysis speed and data quality. Acceleration voltage and aperture size both control how much signal reaches the EDX detectors, where a high acceleration voltage and a large aperture size result in a greater signal. These parameters are typically tailored to the nature of the question (i.e., the data required), according to which the resolution and speed can be adapted. Detector throughput rate can be adjusted to optimize the measurement accuracy and precision. The throughput amount (i.e., counts input) impacts the elemental energy peaks full width half maximum (FWHM) and therefore the accuracy and precision. The dwell time (time the beam stays at one spot before moving on) has a major impact on the analysis speed. The lower a dwell time the faster the analysis is, however if the dwell time is too low it will result in insufficient spectrum counts for a good quality analysis. Image capture time, determining the quality of the BSE image, and frame magnification, determining the amount of stage movements, also affect the analysis time, but these effects are minor and have no real influence during a mapping analysis. The last parameter to affect the analysis speed, is the step size (pixel size) in the mineral map, where a smaller step size will increase the analysis time, but at the same time will provide a more detailed in-depth analysis.

2.3.9. Mining Recipe

In this part of the software, particular minerals of interest and elements can be selected to perform advanced textural and chemical quantification. In the "Assay" part of the user interface, the user has the ability to select the elements of interest in this sample (Figure S7). The software will then perform an assay measurement, whereby the chemically quantified pixel data and the specific gravity (data that is added to each mineral classification) are used to calculate a mass. This is done for all pixel/elements selected across the sample and is used to calculate an assay. This can therefore be used as a rough "bulk rock assay" where the analysis is based on a single 2D plane throughout the entire sample.

Additional value can be gathered from this aspect of the software whereby "elemental distribution" data can be gathered. This is where the chemical distribution is quantified. For example, when the chemical distribution of Mn is of interest: the data will give the wt% amount of Mn in all specified minerals, and also a distribution percentage based the total percentage of Mn found in each phase. This data is provided from the directly chemical quantification during the automated analysis and not from idealized or pre-defined concentrations. This provides researchers with a more reliable AQM technique to understand and quantify chemical and mineralogical variations. However, in cases where element concentrations cannot be measured reliably (e.g., for Boron in tourmaline), a concentration can be assigned in the mineral list for that specific element.

The minerals to be analysed for the map can be selected, as Target minerals. Byproducts and Gangue minerals can also be defined. The example shows a sample with a mineral list with a morphochemical classification (Figure S7). The mutual interconnection between minerals can be described by defining the liberation parameters, which can be exported for all minerals after analysis, together with the association and interlocking data.

3. Examples of Applications

3.1. Mapping Speed and Composition

Sample 511923 was analyzed with the purpose of investigating how different types of Mineralogic analyses (line scan and mapping) and a variation in step size affect the analysis speed and results in a fairly homogenous sample. In total six different analyses at 235.2× magnification were run on the same 0.5 × 0.5 cm area of the sample (Figure 2) consisting of 20 frames. Two line-scan analyses with 10 and 20 μm step size, plus four mapping analyses with step sizes ranging from 40 to 5 μm were carried out on the same area of the sample. The parameters for the analyses were the same using 20 kV acceleration voltage, 120 μm aperture, 275 kcps throughput rate for the EDX detector, and 0.004 sec dwell time.

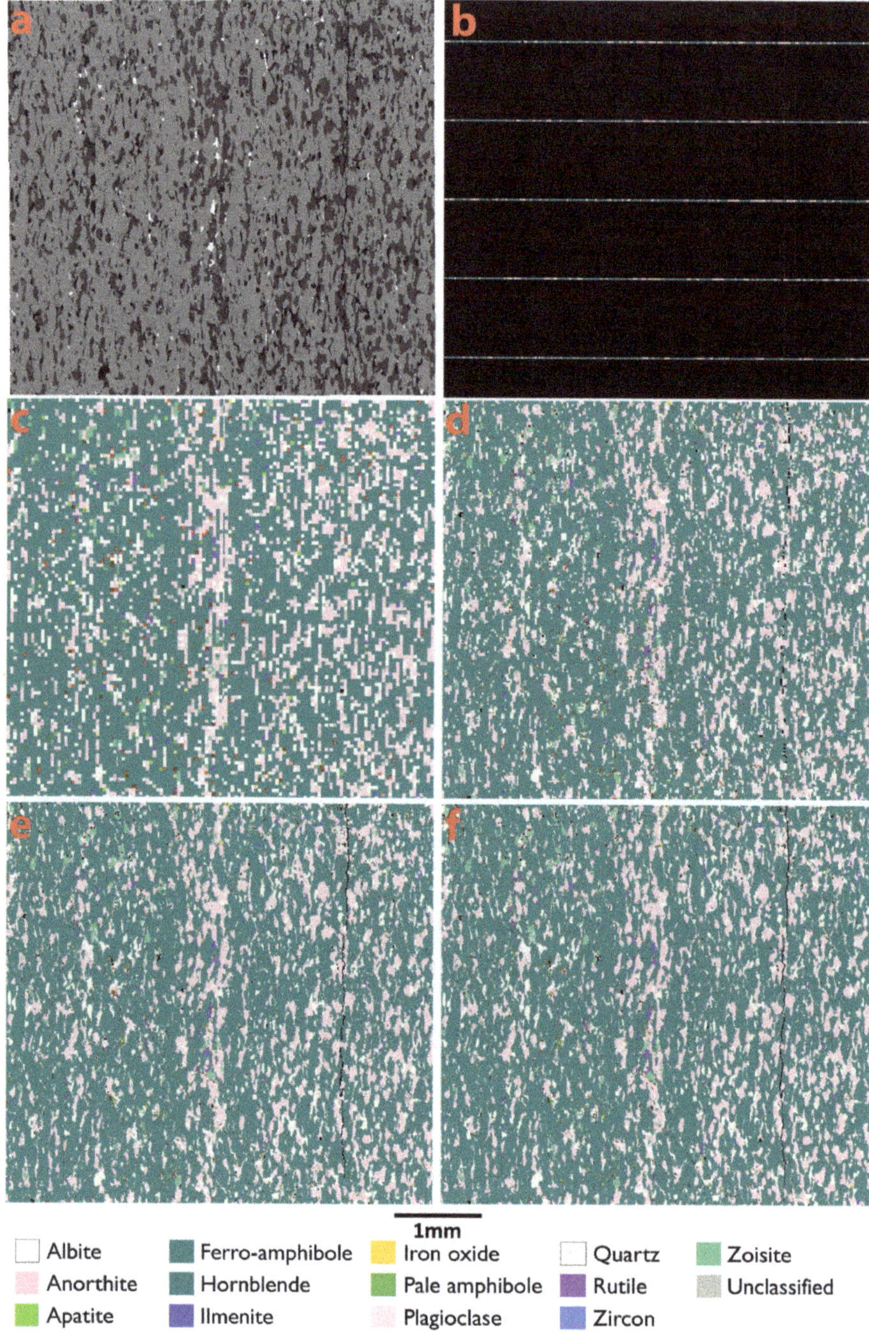

Figure 2. Same area of sample 511923 analyzed with different methods. (**a**) BSE image. (**b**) 20 μm step size line-scan. (**c**) 40 μm step size mapping. (**d**) 20 μm step size mapping. (**e**) 10 μm step size mapping. (**f**) 5 μm step size mapping. For analysis times see Table 3. The 10 μm step size map consists of 244606 analysed pixels.

The line-scan analyses both completed in just around 4 min, and the most time consuming of this type of analysis is to capture the (BSE) image prior to analysis. For the mapping analyses, there is a huge increase in analysis time, when decreasing the step size (see Table 3).

Table 3. Comparison between most abundant and selected other minerals between the different analysis methods in area % for different step sizes in line scanning and mapping. STD = standard deviation. Sample 511923: 24 frames at 235.2× magnification (ca. 0.5 × 0.5 cm). Sample 521106: 220 frames at 235.2× magnification (1.5 × 2 cm).

Sample 519923	10 µm Line Scan	20 µm Line Scan	40 µm Map	20 µm Map	10 µm Map	5 µm Map	STD
Time (minutes)	4	3	8	21	73	282	
FerroAmphibole	46.9	46.8	48.0	47.4	47.6	47.6	0.009
Hornblende	22.1	22.1	20.2	21.0	20.9	20.9	0.03
Anorthite	11.6	11.6	13.2	13.0	12.8	12.8	0.05
Quartz	6.5	6.7	5.7	5.8	5.9	5.9	0.06
Plagioclase	5.5	5.3	5.8	5.9	5.9	5.8	0.04
Zircon	0	0	0.007	0.005	0.004	0.005	0.17
Apatite	0.27	0.31	0.15	0.15	0.13	0.14	0.32
Rutile	0	0	0.04	0.02	0.03	0.03	0.19
Sphene	0.16	0.23	0.20	0.16	0.17	0.16	0.14
Unclassified	2.9	3.2	2.1	2.1	2.1	2.2	0.17
Sample 521106							
Time (minutes)		40	83	298	765	3589	
Pale orthoamphibole		69.0	69.8	68.7	69.6	68.8	0.006
Biotite		9.4	9.7	10.9	9.7	10.9	0.07
Corundum		7.8	7.6	8.0	7.6	8.0	0.02
Cordierite		5.8	6.4	5.6	6.4	5.6	0.06
Sapphirine		2.5	2.6	2.6	2.6	2.6	0.01
Chlorite		1.2	1.3	1.5	1.3	1.5	0.09
Hornblende		0.59	0.49	0.53	0.52	0.53	0.06
Rutile		0.007	0.02	0.02	0.02	0.02	0.29
Zircon		0	0.005	0.003	0.004	0.003	0.24
IronOxide		0.007	0.001	0.001	0.0005	0.0001	1.50
Unclassified		2.6	1.1	1.3	1.3	1.1	0.40

As seen in Figure 3, the distribution of mineral phases between the runs are very similar. Also, when looking on the standard deviation for each mineral phase between the different analysis, there are only small differences (Table 3). However, there are more minor phases picked up by the 5 µm mapping method than the other methods. With the line-scan methods no zircon or rutile were analyzed, however both phases were present in all of the mapping methods. Line scan does not find all minor phases in the sample, and in foliated samples detected small equidimensional phases may get overrepresented when scanning perpendicular to the foliation.

We use the more inhomogeneous sample 521106 to investigate how the different analysis methods apply on a more complex sample. The setup is similar to that used for sample 519923, except for the area analyzed, which is 220 frames at 235.2× magnification (1.5 × 2 cm) (Figure 4), and only one line-scan, at 20 µm, was applied (Figure 5). The results are similar as well, with zircon not identified in the line-scan analysis. Otherwise, the different step sizes in the mapping analysis produce results with minimal variations (Figure 5; Table 3).

The line-scan feature has proven to be very useful as a way of obtaining a quick and precise overview of the geochemistry and major element mineralogy of the sample. Also, a more detailed mineral list can be built from the line scan runs. Where the focus of investigation lies on the chemistry and mineralogy, line scan is a good alternative to running a full map. However, many textural features (e.g., zonations, reaction rims, included minerals) and accessory minerals are not picked up by line-scan analysis.

Mineral mapping at a large step size (here 40 µm in 8 min for 20 frames and 83 min for 220 frames) also is a very fast analysis and gives a good overview of the entire sample (see Figures 2 and 4 to compare image quality). Afterwards ares can be chosen for a more detailed analysis with the mapping feature at a small step size. This two-step procedure is faster than mapping the entire sample at a small step-size. The difference in step sizes generally has little influence on the mineralochemical data (Figures 3 and 5), however details like zonations or reaction rims are often not detected with a larger step size (Figure 4).

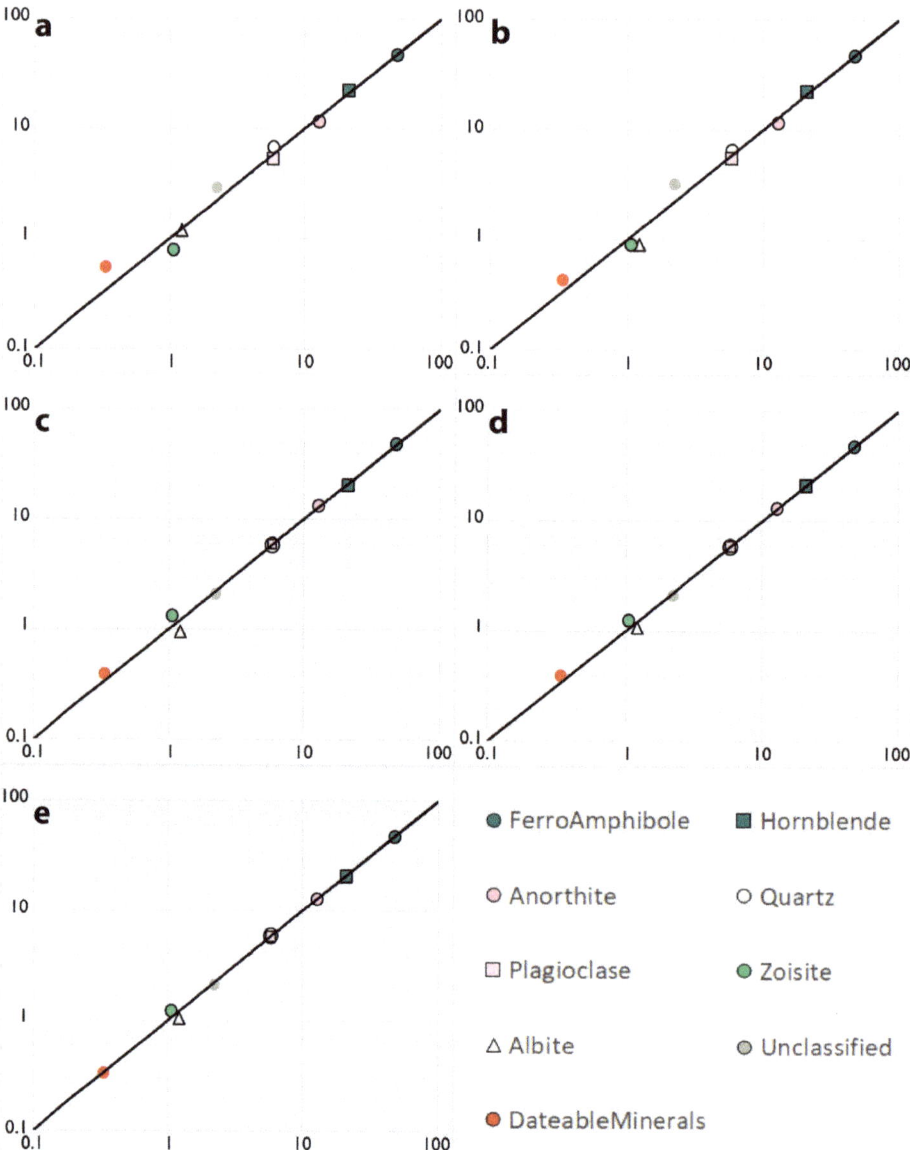

Figure 3. Logarithmic diagrams comparing the different analysis methods (y-axis) against the most time consuming and most detailed 5 μm step size mapping method (x-axis). Data is shown as area %. (**a**) 20 μm step size line-scan. (**b**) 10 μm step size line-scan. (**c**) 40 μm step size mapping. (**d**) 20 μm step size mapping. (**e**) 10 μm step size mapping. Dateable minerals are zircon, apatite, rutile and sphene.

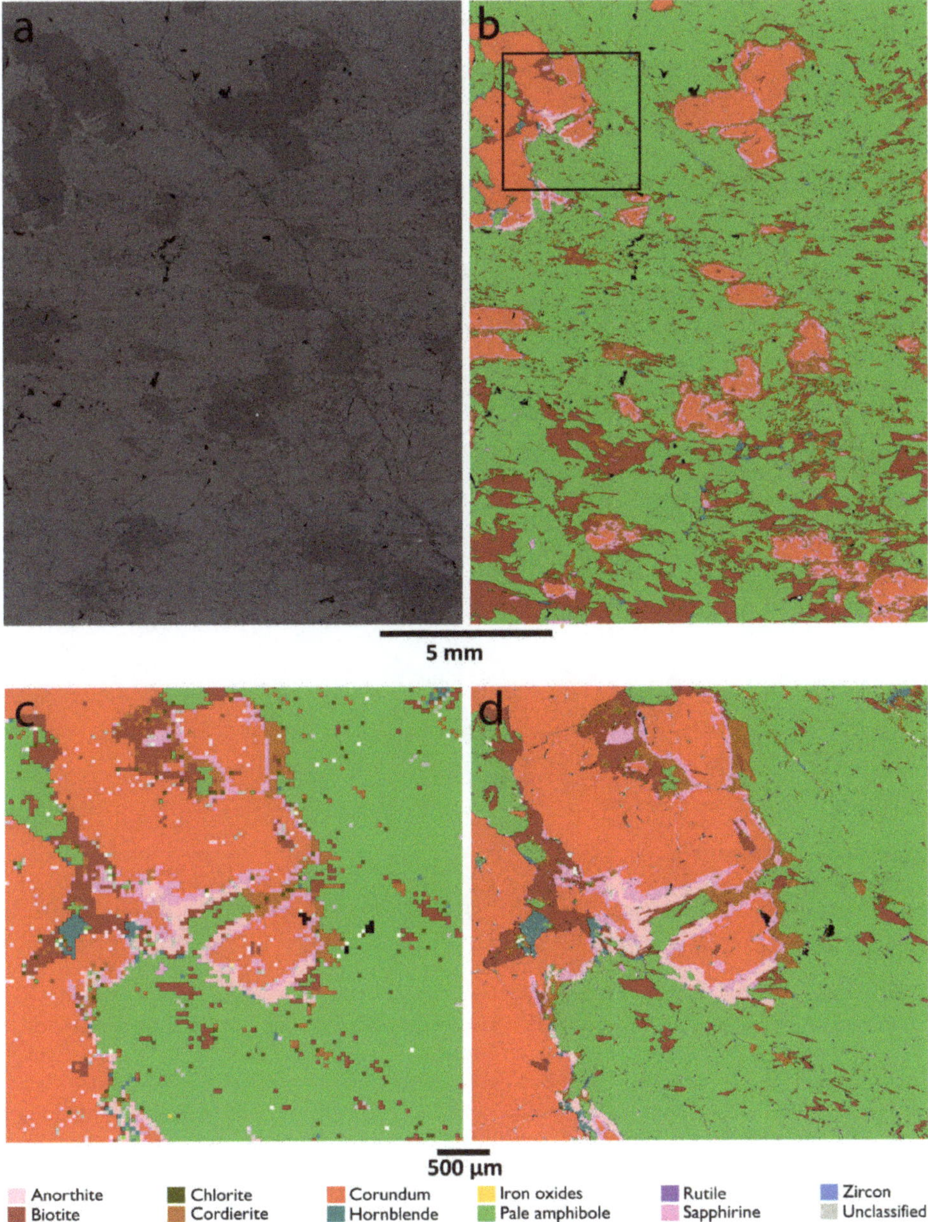

Figure 4. BSE and Mineral maps of sample 521106 and a selected area for comparing the map quality between different step sizes. (**a**) BSE image consisting of 220 frames. (**b**) 5 µm step size mineral map (ca. 9 million pixels). The square indicates the position of details in (**c**,**d**). (**c**) Selected area mapped with 40 µm step size. (**d**) Selected area mapped with 5 µm step size.

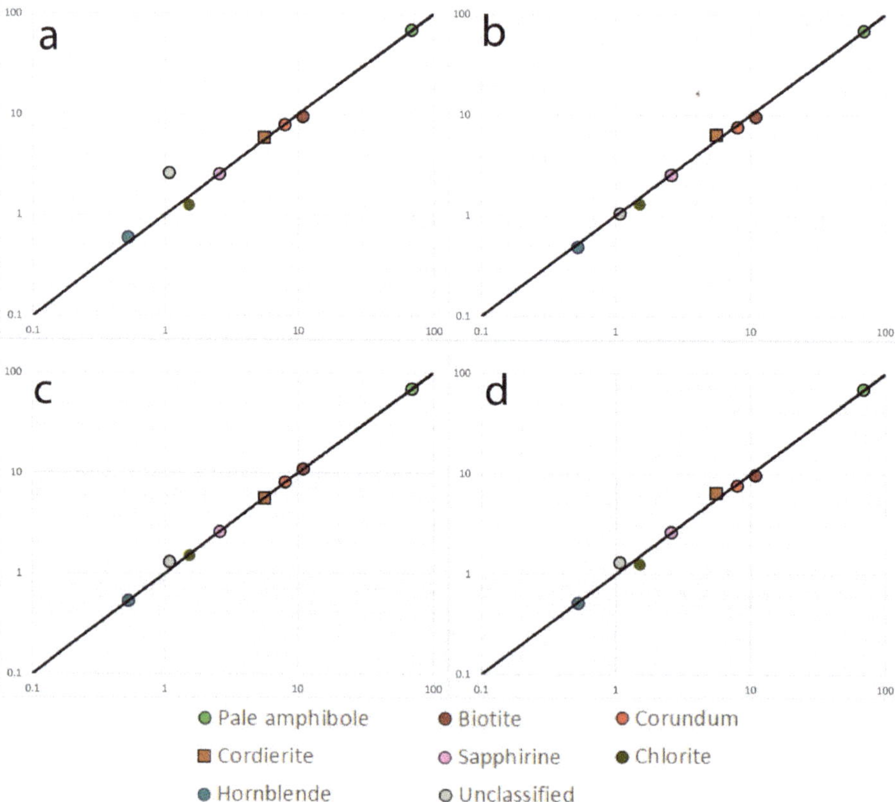

Figure 5. Logarithmic diagrams showing the correlation between different analysis methods (y-axis) and 5 μm step size mapping (x-axis). Data is shown as area %. (**a**) 20 μm step size line-scan. (**b**) 40 μm step size mapping. (**c**) 20 μm step size mapping. (**d**) 10 μm step size mapping.

3.2. Garnet with Quartz Inclusions

In sample 508607 large garnet grains were observed, which were investigated for their chemistry and mineralogy in order to gain more information on the metamorphic history of the volcanoclastic rocks. The sample was collected from the same unit of volcanoclastic sediments as sample 508599 discussed below. However due to lithological variation, its bulk composition is different (less Na, less K, less Si, more Ca, more Fe, more Al, see Table 2), which is expressed in a different mineralogy: quartz, anorthite, biotite, pale amphibole, garnet, apatite and pyrite. In the most biotite-rich amphibole-poor layers, large garnets have grown with the foliation bending around them (Figure 6). In the garnets, inclusions of apatite, rutile, ilmenite and especially quartz are present. The quartz is irregularly shaped with lobes intruding into the garnet. All inclusions are only found in the center of the garnets, but not in the rim. Pyrite is partially oxidized and is found in association with biotite.

The sample was imaged with the BSE and CL detectors, as well as analyzed for its mineralogy and chemistry, all with the Mineralogic software. In the SEM recipe of the Mineralogic software, the detector for imaging can be selected and changed to CL in order to make a stitchable series of images automatically. After creating a Mineral map (Figure 6), the Mineral recipe was applied to generate element concentration maps for the main elements in garnet (in Figure 7 Mg, Fe, and Ca are shown). The element concentrations displayed are true wt% concentrations, not relative intensities.

Furthermore, the quality of the EDS systems was monitored with point analyses on silicate standards and yields an error < 1 wt% compared to electron microprobe analyses, with slightly too low values for light elements and too high on the heavier elements (see Supplementary Table S1).

The mineral map (Figures 6 and 7a) reveal that the inclusions in garnet are mainly observed in the core and inner rim of the garnet. The core of the garnet mainly yields quartz and biotite (plus pyrite) inclusions, while ilmenite, rutile, amphibole and apatite mainly, but not exclusively, are observed in the inner rim. The inner rim does not show biotite inclusions.

CL investigations show that the quartz inclusions in the garnet consist of smaller grains healed into larger inclusions (black arrows in Figure 7b). Anorthite adjacent to garnet reveals growth rims in CL (white arrows in Figure 7b), which are most strongly on the left and right sights of the grains.

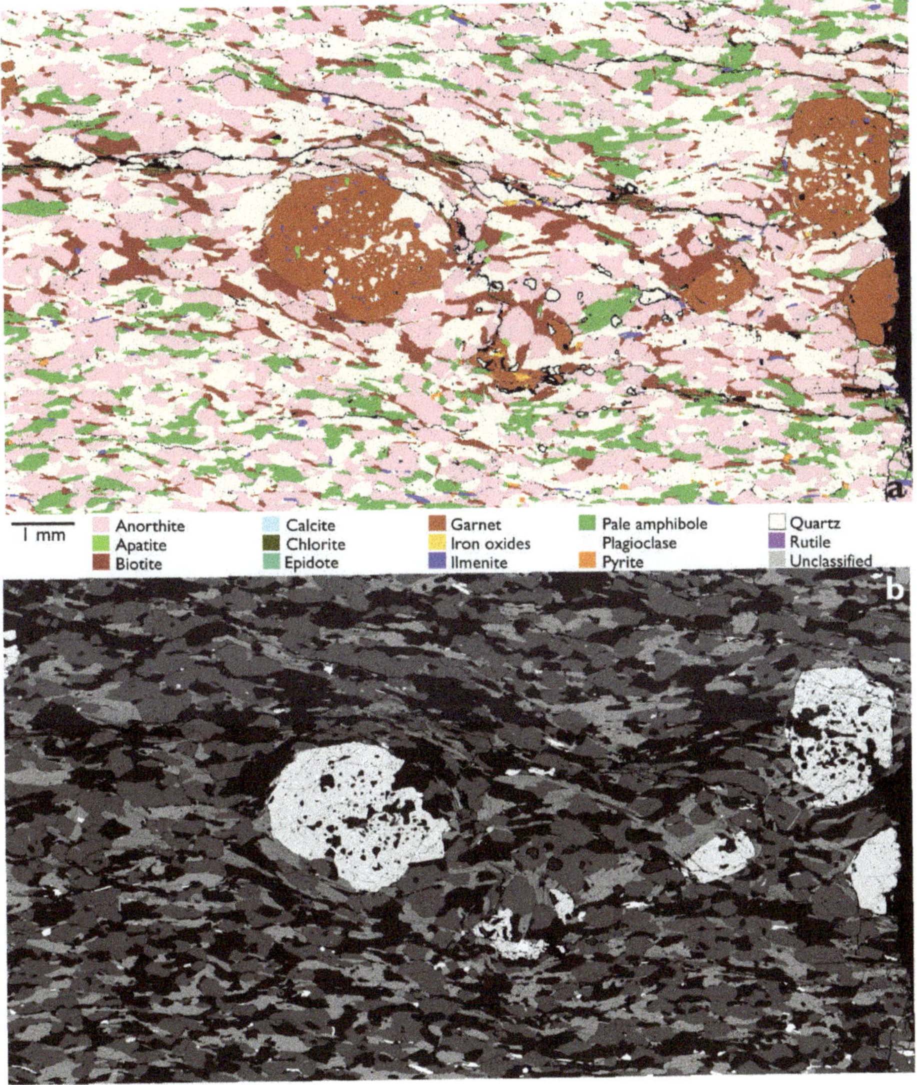

Figure 6. Mineral map (**a**) and BSE micrograph (**b**) for the volcanoclastic sediment with garnet porphyroblasts. The mineral map contains 281714 pixels.

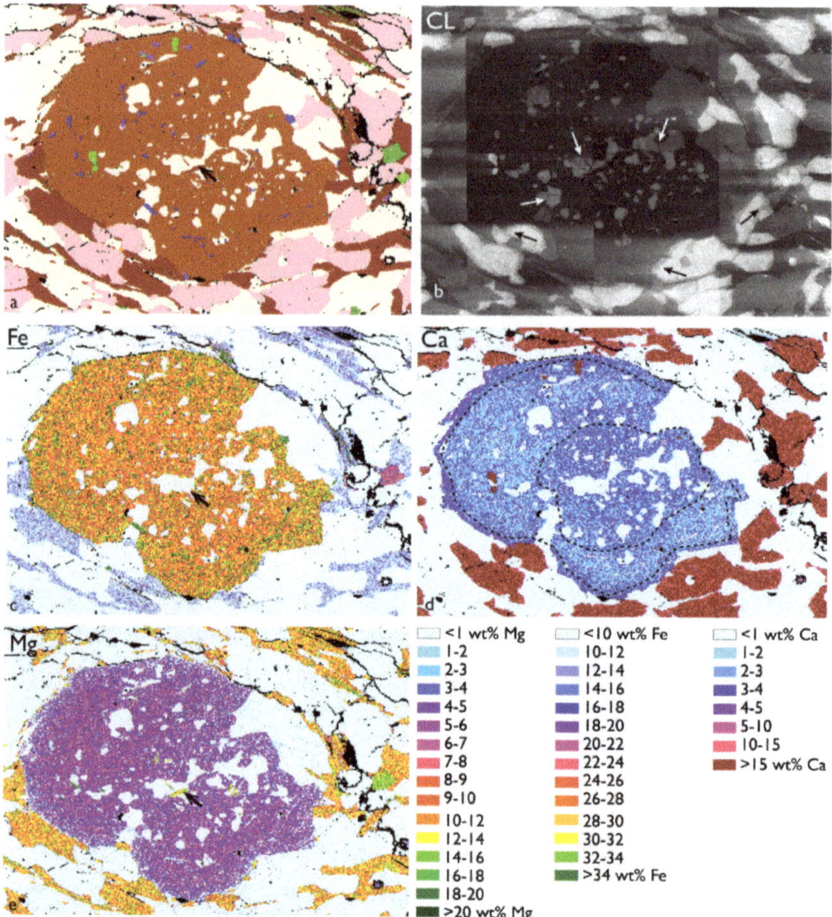

Figure 7. Detailed maps and image of the central garnet in Figure 6. (**a**) Mineral map. See Figure 6 for a legend of the colours applied. (**b**) Cathodoluminescence image mosaic of the same garnet. White and black arrows point to growth features in anorthite and quartz, respectively. (**c**) Mg element concentration map. (**d**) Ca element concentration map with black contours indicating the weak zonation in garnet. (**e**) Fe element concentration map. Arrows in (**c**,**e**) point to an example of an included biotite grain with different Mg–Fe ratios.

EDS analyses of the garnet showed ca. 19% Si, 13% Al, 2% Ca, 5% Mg and ca. 26% Fe. There is some Mn present as well (~0.5 wt%), but no Cr (less than 0.5 wt%, usually not detectable). There is a minor compositional variation between the core and the two rims of the garnet, with no difference in Si and Al (less than 1 wt%), but slightly lower Ca and higher Mg and Fe in the inner rim and slightly higher Ca, lower Mg in the outer rim (the difference in concentrations between the core and rims is always less than 2 wt%). These zonations are just visible in Figure 7c–e. With image manipulation (inversion of Ca-image, followed by adding the images together in Adobe Photoshop©), these differences can be enhanced and mapped. Figure 7d shows the contours for the weak zonation, based on Fe, Mg and Ca concentrations. The small variations in the chemical composition are visible in the garnet ternary diagram (Figure 8), where the XMg–XFeMn–XCa composition of the garnets in Figure 6 are plotted based on pixel-by-pixel data exported from Mineralogic analyses of the garnet. The garnet is an almandine, which are typical for aluminous rocks deformed at amphibolite facies conditions.

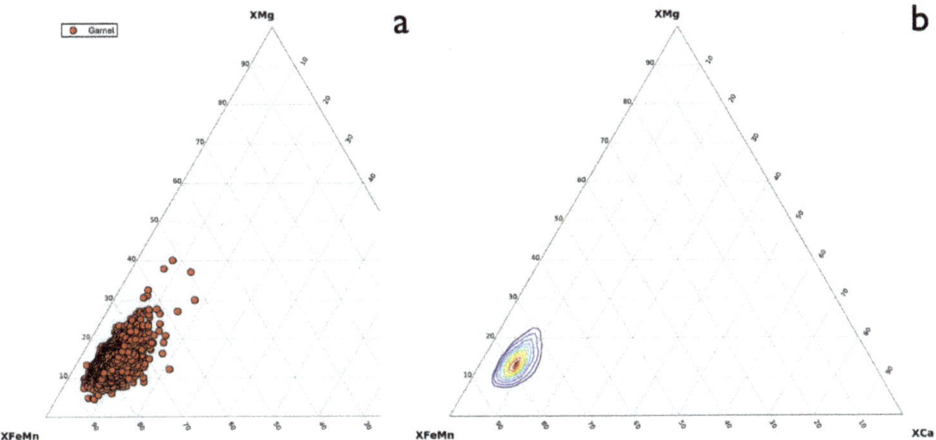

Figure 8. Ternary diagram showing the garnet porphyroblast minor element composition expressed as XMg–XFeMn–XCa. (**a**) Data for 6400 representative pixels; (**b**) Contours showing the data intensity of (**a**). The Figures are plotted applying WxTernary [42].

The garnet in this samples has probably grown from a quartz consuming reaction, possibly chlorite + quartz + muscovite = garnet + biotite + H_2O, which can explain for the biotite and quartz inclusions in the core of the garnet (Figures 6 and 7). From the element concentration plots (Figure 7c,e) can be seen that the composition of the biotite in the inclusions is different from the biotite in the foliation: the included biotite is Mg-rich, while the biotite in the foliation is Fe-rich. The inner rim of the garnet showed a continuation of the same reaction, probably at a slightly higher temperature, thus slightly changing the Fe–Mg concentration in biotite and garnet [43] (Figure 7). Ti is largely incompatible in aluminous garnet and remains as ilmenite and rutile inclusions. The foliation of the rock bends around the previously formed garnet and some biotite is formed in the pressure shadows demonstrating a weak dextral sense of shear (Figures 6 and 7). The fractured and healed quartz and anorthite visible in the CL image show that deformation might have been fast and intense, as quartz and anorthite typically are deforming plastically at garnet-forming temperatures [44,45]. Large rims around anorthite show that anorthite precipitated under a differential stress from fluids during metamorphism (darker rims of anorthite visible in the CL image (Figure 7b) are thicker along the horizontal grain axis than the vertical). Ca in garnet may also originate from these fluids.

3.3. Reaction Rims Around Rubies

The ruby-bearing sample 521111 from the Fiskenæsset complex contains the reaction products of the interaction of a tonalitic sheet intruding into an ultramafic rock in contact with anorthosite. As not all of the ruby in the thin section is of gem-quality, we refer to the ruby as corundum. The reaction occurred at amphibolite-facies metamorphic conditions, i.e., post peak-metamorphism (which was at granulite facies conditions) [37,41]. It has been discussed previously whether the alumina of corundum and other aluminous minerals (spinel, kornerupine, cordierite, and sapphirine) are primary or secondary minerals [37,38,41,46]. Reaction temperatures are estimated to be ca. 600 °C, for the southern part of the complex [37,41], but may have been higher nearer to the village of Fiskenæsset where the current sample was collected. with a gradient from lower granulite facies conditions in the western part of the Fiskenæsset complex. The corundum-forming reaction occurred during the main folding phase in the region [37].

Like for the garnet example above, the sample was mapped for its mineralogy, and afterwards recalculated with new mineral lists to create mineral association maps and an element ratio map (Figures 9 and 10). The mineral association maps can be made in the Mineral recipe tab by redefining the colours of the minerals, highlighting two or three phases, while all other mineral phases are displayed in white. As for the element concentration maps, the mineral list can also be redefined to show steps in element ratios in order to display e.g., changes in Si/Al or Fe/Mg ratios in the sample.

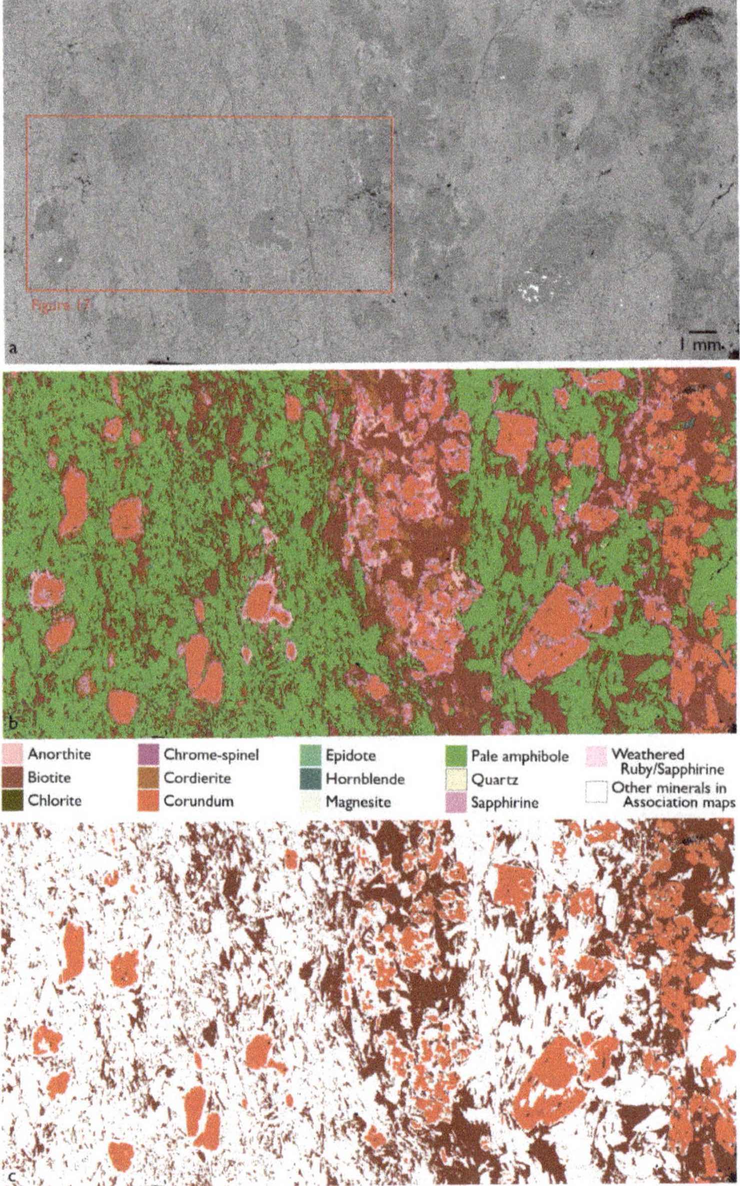

Figure 9. Corundum (Ruby-)bearing rock showing the mineral assembly after the formation of corundum, sapphirine, cordierite, anorthite, pale amphibole and biotite (peak metamorphic reaction products) and the retrograde reaction products magnesite, chlorite, and quartz. (**a**) BSE image of a selected part of the sample. (**b**) Mineral map of nearly the entire thin section of the sample (1.63 million pixels). (**c**) Mineral association map of biotite and ruby. All other minerals are displayed in white.

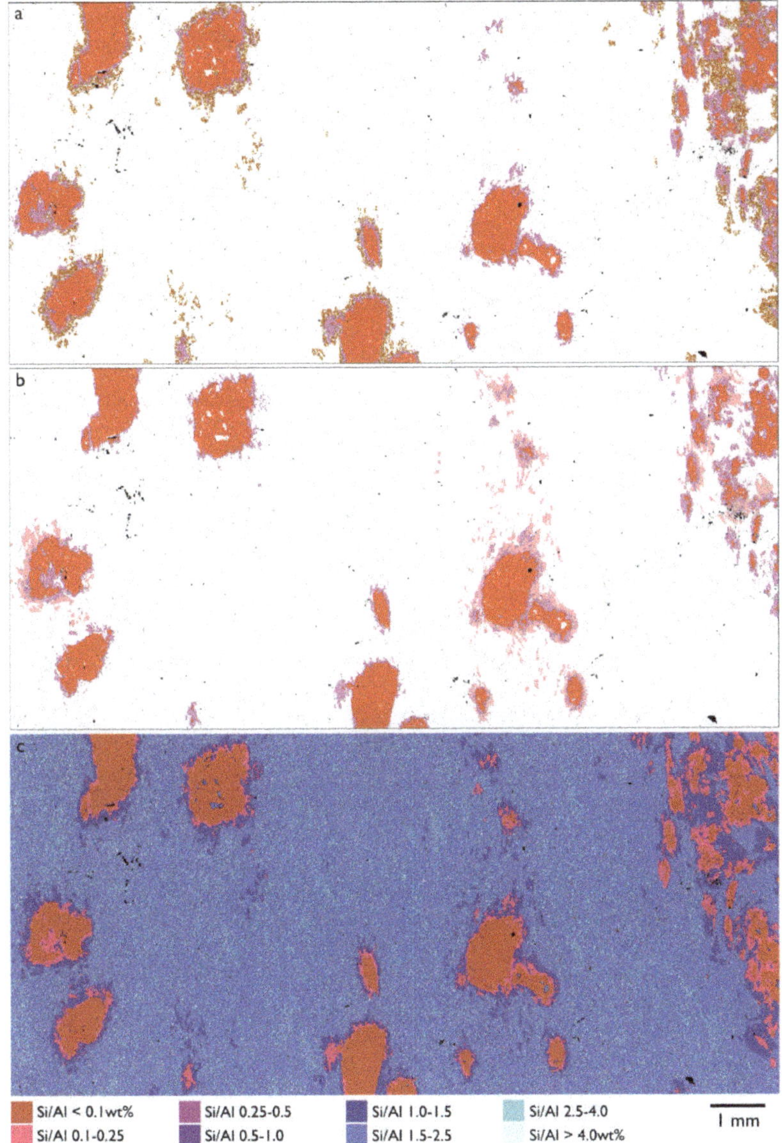

Figure 10. Detail of the sample displayed in Figure 9. (**a**) Corundum-sapphire-cordierite association map. (**b**) Corundum-sapphirine-anorthite association map. All other minerals are displayed in white. (**c**) Element ratio map showing the Si/Al ratios in the sample. wt% indicates that wt% data were used to calculate the ratios. Legend for (**a**,**b**) in Figure 9.

Desilification of the ultramafic rock results in the formation of pale amphibole (mainly gedrite), biotite, and one or more aluminous phases (cordierite, sapphirine and corundum), e.g., by the reactions: olivine + K-feldspar (tonalite) + H_2O = biotite + anthophyllite or olivine + SiO_2(aqueous) (tonalite) = anthophyllite [41] (Figure 9). The intruding tonalite is not peraluminous and can therefore not be the source of aluminium, however anorthite is able to react in a balanced reaction: olivine + anorthite + H_2O (tonalite) = Ca-bearing amphibole + corundum [41]. In the case of sample 521111, where reaction temperatures may have been slightly higher, the corundum mineral is surrounded by two rims of reaction products (Figures 9b and 10a,b). Corundum in the centre, sapphirine around the corundum, and cordierite or anorthite around the sapphirine. The reaction rims are very well developed with the same width all around the corundum minerals, no symplectites or other fine-grained minerals are observed in the reaction rims, which are therefore interpreted to have developed near-simultaneously with the corundum-forming reaction at peak or near-peak metamorphic conditions. Figure 10c indicates that the minerals formed in the reaction rims are increasingly rich in Si compared to Al, ranging from no Si in ruby to more Si than Al in cordierite. This shows that the degree of silica-desaturation of the rock may have fluctuated during the ruby-forming reaction, or that the Al from the original anorthite was consumed towards the end of the reaction series. Cordierite or anorthite as the outer reaction rim, occur systematic (Figures 9b and 10a,b) and may reflect the original hornblenditic and anorthitic layers, respectively, in the anorthosite before intrusion of the tonalite. Biotite is not associated with corundum (Figure 9c), but cordierite seems more strongly associated with biotite, than anorthite (Figure 9b). Here, sapphirine and cordierite are peak metamorphic minerals, ruby started growing just before peak metamorphic conditions, sapphirine and ruby are part of the peak metamorphic assemblage, and cordierite and anorthite grow immediately after initial decompression [42], as was also observed in other parts of the Fiskenæsset complex [37,41]. However, a decrease in temperature could additionally create retrograde sapphirine or plagioclase, as also was described previously for other parts of the Fiskenæsset complex [46].

3.4. Grain Size Distribution of Chromite in Leucogabro

Within the Fiskenæsset gabbro and leucogabbro thin layers of chromitite have been observed, which previously have been investigated for their platinum group elements content [34,47]. Within the Fiskenæsset complex both primary and secondary chromite occur [48]. The chromite in the layer is homogeneous in composition but shows a wide variation in grain size. The chromite grains are situated in the Mg-rich hornblende layers of the metamorphosed gabbro (Figure 11a), while only a few grains are associated with anorthite. The chromitite is highly dominated by chromite (Figure 11b), thus individual chromite grains are touching each other, which makes automated grain size analysis more complex.

In order to measure the grain sizes a modified version of the watershed method, which the image processing tap of the software provides, was applied. After thresholding to select the chromite grains from the rock, the image was eroded in three iterations, followed by dilation, as a modification of the opening function that is routinely applied—this gives better separation of individual grains. The watershedding was set to 40 units difference in order to separate touching grains. After image processing the chromite was selected by mineral classification (minerals tab) and characterized by morphology and chemistry in the same tab.

The association data for sample show that 54% of the chromite is in contact association with hornblende and only 17.1% with anorthite, while 23% is associated with the background (mainly holes and cracks between chromite grains) and the remaining fraction of the association is made up by minor phases including rutile and chrome-spinel. The grain size distribution of the grains is visualised in Figure 11. It shows that individual chromite grains in this layer range in size (Feret mean diameter) from 100 to 1000 micrometer. Part of these grains are clusters of several grains, despite the watershed procedure during the image processing. Chromite settled together with hornblende during the cumulation of the igneous rock. Individual chromite grains were already inter-grown during their crystallization and their grain size and texture have not been affected by metamorphism. The chromite grains in this sample are primary minerals. An analysis of the roundness and orientation of the individual grains (Feret angle) shows that the chromite grains are not systematically flattened during the three tectonometamorphic events that affected the area, showing that the chromite grains are very rigid under those tectonometamorphic conditions (ca. 600 °C, 5 kbar [37,41]).

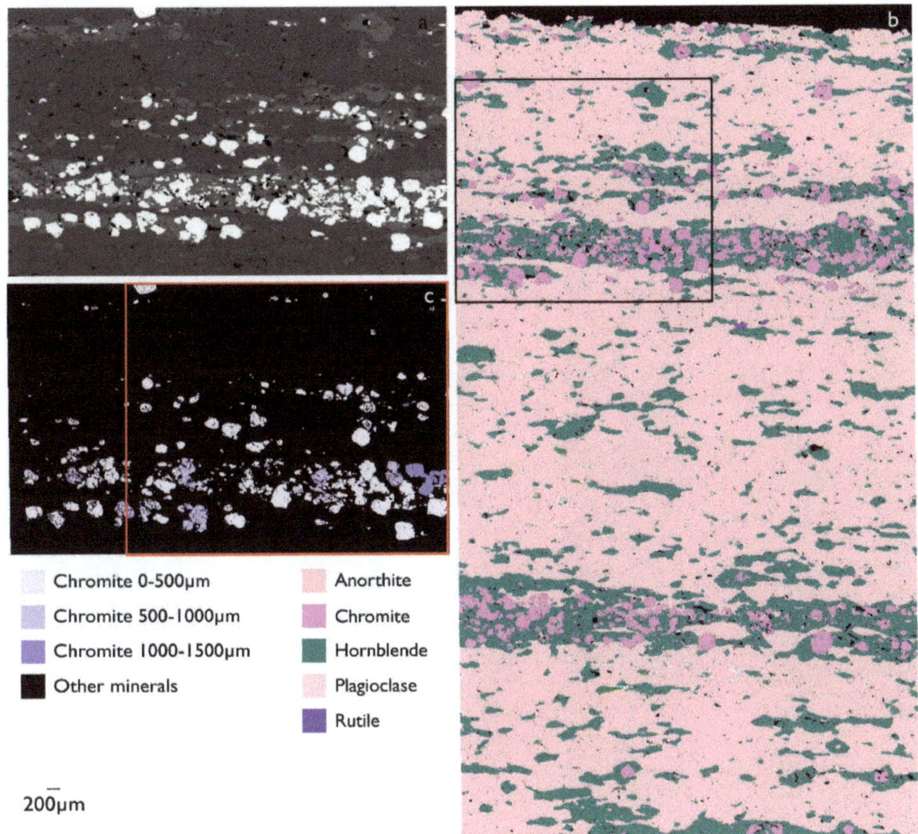

Figure 11. Leucogabbro with bands of chromite, sample 510152. (**a**) BSE micrograph of part of sample, showing chromite grains in white. (**b**) Mineral map of a larger part of the same sample. Chromite is associated with hornblende. 1.63 million pixels. (**c**) Grain size map showing the same part of sample as in (**a**), with colour-coding of the chromite grain sizes. (**a**,**c**) partially overlap with the upper left part of (**b**), indicated with squares in (**b**,**c**).

3.5. Maximum Feret Angle Determination

Sample 508599 was investigated for the size and orientation of the biotite minerals in the thin section, which define the foliation and the isoclinal fold in the sample. The orientation of the biotite minerals can be described with the maximum Feret angle. The maximum Feret angle is the angle between the maximum Feret diameter of a particle and the horizontal axis. The maximum Feret diameter is the longest diameter of irregular shaped particles.

In order to measure these morphological features for biotite in the volcanoclastic sediment, in the image processing recipe the BSE image of the sample (Figure 12A) was thresholded by grey scale value to only investigate the minerals with a bright BSE contrast; these include mainly biotite and a few accessory phases. The biotite grains are now isolated features of single grains or clusters of grains that lie in a darker matrix. The biotite grains can thus be investigated with a spot centroid or feature scan analysis. The mineral list was adapted to show all none-biotite grains in black, while a morphochemical criterion was added to the biotite classification in order to classify the biotite grains in colours according to their Feret angle (Figure 12B). The results are shown in Figure 12, together with the BSE image and the mineral map of the entire thin section.

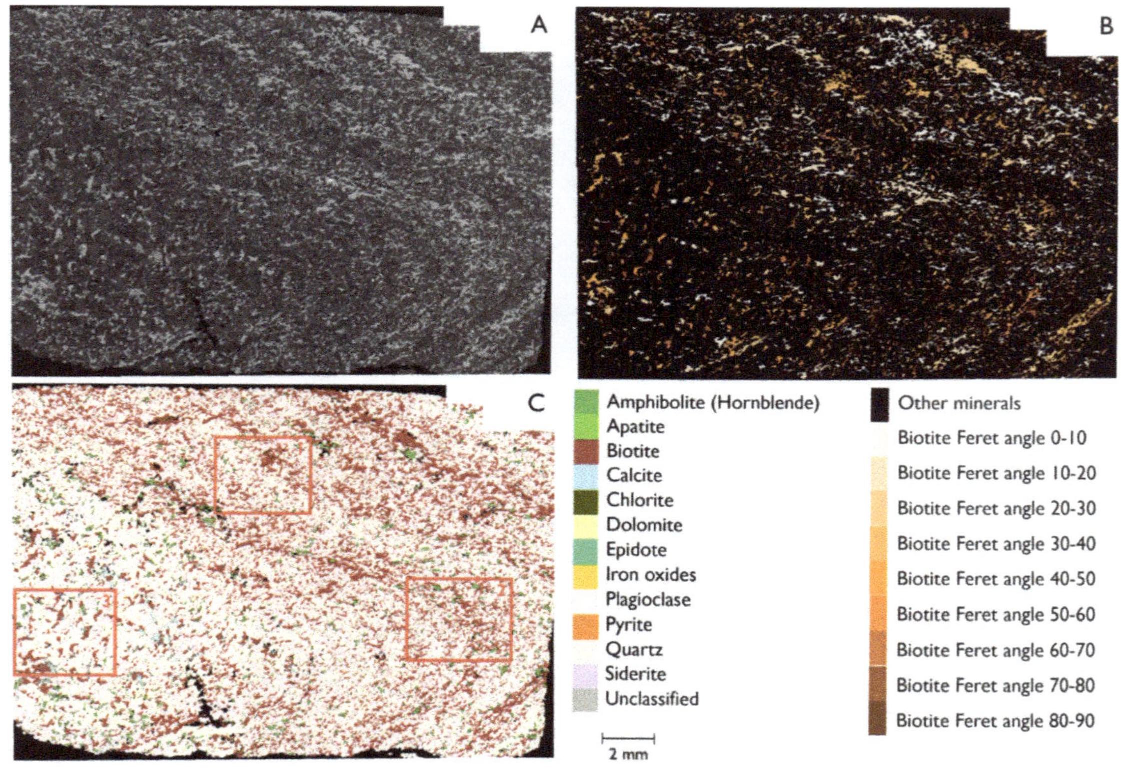

Figure 12. Isoclinal fold in a volcanoclastic sediment containing biotite, amphibole, anorthite and quartz. (**A**) BSE micrograph of the sample. The three main phases (biotite, anorthite and quartz) are light, intermediate and dark grey respectively. (**B**) Maximum Feret angle map over the thin section. All none-biotite minerals are black, while biotite is coloured by its maximum Feret angle. Positive and negative angles are indicated in the same colours. (**C**) Mineral map showing the distribution of the mineral phases in the sample. This mineral map contains 2.04 million pixels. Squares define three smaller areas within the sample, with their own data for surface area and association data.

The BSE image and the Mineral map show a different behavior for the individual minerals in the sample across the isoclinal fold. In the flanks of the fold, biotite occurs in clusters of grains. Individual grains are occasionally oriented randomly, but the larger clusters of grains are oriented into elongated clusters with low Feret angles (preferentially 20–40°; see Figure 13), while most individual biotite grains and smaller clusters have lower Feret angles (mainly 0–20°) and are stacked stair-case-wise to follow the local foliation. Thus, the individual grains follow the main foliation in the area (oriented ca. 0–10° with respect to the horizontal axis), while they form stacked arrays that follow the local foliation (20–40°) in the individual fold flank. In the core of the fold, biotite occurs more scattered, and all Feret angle orientations with an large elongation are observed between −30 and 40° (Figures 12 and 13), while the grains with the largest area have orientations around 40 and −10° (Figure 13), fitting with the orientation of the two flanks mapped in Area 2 (Figure 12). In the least deformed part of the fold (Area 3) biotite grains are smaller and show no preferred orientation for the most elongated grains (Figure 13).

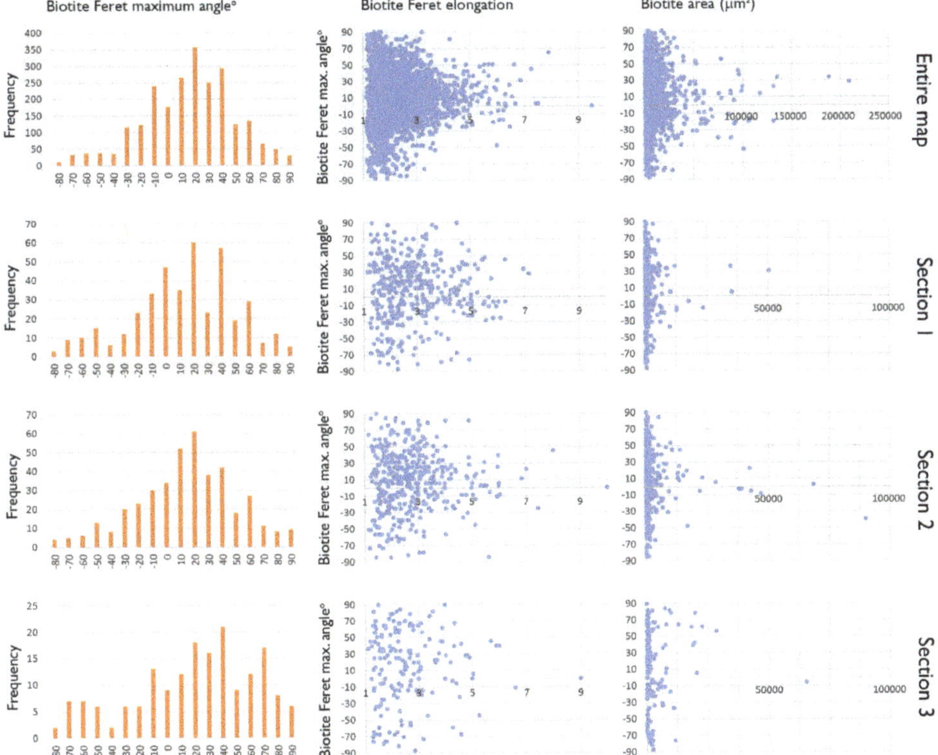

Figure 13. Plots of the maximum Feret angle frequency, and this angle compared to the elongation and area of the biotite grains and clusters of grains. Figures represent the entire map in Figure 12, as well as the three areas outlined in Figure 12C.

In the less extensively deformed parts of the fold, quartz and plagioclase occupy roughly the same percentage of the area (compare least deformed Area 3 to intensively deformed Areas 1 and 2 in Table 4). Both minerals are partially intermixed but prefer to cluster with their own phase. In the flanks of the fold, a more well-defined layering of quartz, biotite and anorthite occurs, where both quartz and biotite are sandwiched between layers of anorthite. In the core of the fold, hardly any quartz minerals are observed. Biotite is associated with anorthite (Figures 12C and 13). This is illustrated for association data for the three areas in Figure 12C (see Table 4). Area 1, derived from the core of the fault, and Area 2, derived from the flank, show that biotite is strongly associated with plagioclase (45.3 area% and 48.7 area%, respectively), while in the least deformed Area 3 this association is only 34.0 area%. Plagioclase in the flanks of fold (Area 2) is strongly associated with biotite (45.3 area%), and the association with quartz is the least of the three areas (34.5 area%). Plagioclase in the core of the fold (Area 1) is forming the sandwiching layer between biotite (37.8 area% association) and quartz (40.9 area%). In the least deformed part of the sample, plagioclase is strongly associated with quartz (51.6 area%) and much less with biotite (21.3 area%), which is partially caused by a much lower area% of biotite in this section.

Table 4. Area% and association data for biotite, quartz and plagioclase in the three areas indicated in Figure 12C. TS = thin section.

Mineral	Area in TS (%)	Association Data (%)		
		Plagioclase	Quartz	Biotite
Area 1				
Biotite	19.8	45.3	35.4	-
Plagioclase	40.2	-	40.9	37.8
Quartz	35.6	46.6	-	34.2
Area 2				
Biotite	23.1	48.7	31.7	-
Plagioclase	40.3	-	34.5	45.3
Quartz	31.1	43.5	-	37.7
Area 3				
Biotite	11.3	34.0	40.8	-
Plagioclase	36.1	-	51.6	21.3
Quartz	45.5	47.7	-	23.7

Folding in the area occurred at ca. 600 °C [37,41]. Under these conditions all three minerals (quartz, anorthite and biotite) deform plastically under most strain rates [44,45]. Quartz, which under these conditions is more plastic than plagioclase, moves to the flanks of the fold, while the more rigid plagioclase is pressed towards the cores of the fold (compare Areas 1 and 2 in Figure 12C and Table 4). Biotite rotates by a combination of recrystallization and mechanical fracturing and healing to accommodate to the main stress in the region and can be used to read the record of the main foliation outside the isoclinal fold (Figure 13), i.e., 0°. However, the movement of quartz and plagioclase also force the biotite to follow the local foliation, leading to a stacking of horizontally oriented minerals into a local foliation of ca. 30° to the horizontal fold axial plane (compare to Figure 13). Continued stress on the rock will remove the evidence for a fold core from the record of the rock, leaving the sample layered by mineral phase and with clusters of flat-lying biotite. We are thus able to study the generation of a schistose layering from an initially little deformed rock in a single sample by studying individual areas of the fold core.

4. Discussion

The five examples of application above show a small set of the vast range of possibilities for modern AQM systems, like Mineralogic, where visualization of the data and the generation of chemical, mineralogical and morphological data are combined. Traditionally, mineral compositions in metamorphic rocks are quantified with optical microscopy or microprobe analyses, while metamorphic textures are investigated with optical microscopy and often unquantified or, mainly in case of deformed samples, quantified by electron back-scatted diffraction. By applying AQM, visualization and quantification can be combined in one tool and executed on the same minerals. The AQM software is creating a false-coloured mineral map, while simultaneously measuring grain morphology and chemistry. The false-coloured map resembles the optical microscope mineral display of metamorphic textures, but also forms the basis to gain quantitative data for the sample. Compared to the microprobe, AQM offers mineral maps and not only element maps. This is combined with the higher analytical speed for EDX analyses compared to the wavelength dispersive spectrometry analyses on the microprobe. The diversity and flexibility in the chemical, textural and mineralogic visualization is demonstrated in this paper as a means of contextualization of what can initially be complex data sets.

AQM is often regarded as a slow analytical tool. However, depending on the information required, analytical speed can be adapted to the optimal combination of mineralochemical precision, map-quality and time. An area of 1.5 × 2 cm can be analysed in 83 min at a 40 μm step size, and provide almost identical mineral concentrations, when compared to runs with smaller step sizes and longer analysis times (Table 3; Figures 3 and 5). A time-saving approach could be to first run a large (here 20 μm) step size line-scan or Mineral map (here 40 μm step size) on the selected sample, then build a solid mineral list and run selected area(s) with a smaller step size to obtain detailed information on specific features e.g., reaction rims. Further, these investigations indirectly showed that the general reproducibility of the data, even while generated at different step-sizes (pixel sizes), is good.

Within the Mineralogic software platform, it is also possible to analyze samples with different acceleration voltages, without changing the mineral list. Samples 521106 and 521111 originate from the same area, were analyzed at 20 and 15 kV respectively, and were recalculated with the same mineral list without any adaptions (Table 1).

The garnet and ruby examples outline unique capabilities the Mineralogic software is able to access when compared to other AQM platforms. Because Mineralogic's EDX software is matrix-correcting and quantifying every spectrum for every pixel in the mineral maps and for each spot or feature scan for every epoxy-embedded particle, the exact chemistry is known with the same precision as for EDX point analyses. Elements can be easily detected with concentrations down to ca. 0.5 wt%, and major and minor elements can be analysed with an accuracy of 1–2 wt%. This is a large improvement compared to spectrum matching AQM platforms, which are not able to detect elements in concentrations of less than 5% [22]. Also, by obtaining the chemistry for every pixel makes the addition of new minerals to the mineral library more certain, and easily adaptable to changes in composition due to slightly different metamorphic conditions (e.g., Mg/Fe ratio dependence on pressure and especially temperature during metamorphism). The ability to measure detailed chemistry of the minerals for every pixel of the mineral maps has been applied to make element concentration maps (Figure 7c–e) and element ratio maps (Figure 10c). But the chemical data can also be exported after analysis, e.g., to obtain a bulk geochemical assay or to obtain raw data for a garnet XMg–XFeMn–XCa ternary diagram (Figure 8). A comparison with whole rock geochemical data and EDS generated data with Mineralogic on a homogeneous sample (Table 2) shows a good agreement in the results between the two methods. The tested sample had slightly higher Ca and lower Si in the analysed thin section than the bulk rock sample and comparable results for the other elements., The discrepancy in Ca and Si probably occurred due to the presence of a calcite vein in the thin section.

The garnet example shows how the zonation into core-inner rim-outer rim could be observed by the garnet elements Fe, Ca, and Mg. These three zones in garnet were each associated with their own set of inclusion minerals, visualized in the mineral map of the sample (Figures 6 and 7a), revealing three stages of the metamorphic history of the garnet. Similarly, the Si/Al element ratio map show how silica-concentration is increasing in the reaction rims around corundum in the sample, causing the formation of sapphirine, cordierite and anorthite, each with a higher Si/Al ratio. Sapphirine, cordierite and anorthite are clearly associated with the ruby, as is visible in the mineral association map for these three minerals (Figures 9 and 10). Mineral mapping of the reaction rim showed that these are a prograde/peak metamorphic feature and not part of the retrograde reaction path [41].

The chromite-bearing gabbro example and the biotite example show the power of morphochemical classifications of the minerals (Figure 11). The ability to precisely select certain minerals and to cluster and colour-code them by morphological features is a very powerful feature in AQM software products. For the chromite-bearing leucogabbro sample it was demonstrated that the Mineralogic software can evaluate and colour-code the grain size distribution of grains in a rock. Image processing tools help to automatically separate near-touching grains (though not the entirely clustered grains). The strong preference of chromite for the hornblende layers is quantified by the mineral association data that AQM software is able to produce.

The biotite sample's Feret analysis is an example of a monomineralic display used to illustrate a morphological criterion. Here, the most important mineral is mapped out only, and colour-coded by its mineral orientation (Figure 12). However, the results are not only displayed, but the data for the analysed particles from the mineral map formed the basis for association data, Feret angle orientation and Feret elongation, grain size (area) as well as modal mineralogy (Figure 13 and Table 4), showing how intensive isoclinal foliation causes mineral layering in the formation of a schistosity.

Even though most scientific investigations cannot exist without the availability of numerical data and the quantification of processes. However, the visualization of results and qualitative observations still remain very important in order to fully understand the data, and often this visualization is the key to gaining new insights through conceptualization of the data.

Here, we show that AQM software in general and especially Mineralogic software is able to generate many different ways of displaying the same sample and thereby highlighting several aspects of same sample. In the examples above we apply BSE and CL images, mineral maps, element concentration maps, element ratio maps, single mineral maps, mineral association maps on paired mineral groups, and morphology maps highlighting grain shape and orientation. Still these range of maps are only a small selection of the range ways the same sample can be displayed. As most humans are visually oriented, this displaying of results makes it easier for us to gain new insights. Visualization of the data adds context and meaning to the data tables. For example: the zonation of the garnets would not be visible without the Fe, Ca, and Mg concentration maps (Figure 7) and the orientation of biotite with respect to the local and main orientation of the foliation was more easily visible with colour coded oriented minerals (Figure 12).

The flexibility of having platform independent, acceleration voltage independent mineral classification, precise chemical measurements allowing for element concentrations and element ratio maps supplement and support the quantified data analysis, and interpretations, and on these topics ZEISS Mineralogic has proven its worth for metamorphic texture investigations, as well as a large number of other areas of investigation.

Supplementary Materials: The following are available online at http://www.mdpi.com/2075-163X/10/1/47/s1, Supplementary 1: Outcrop Images and Mineral Composition; Supplementary 2: Mineralogical Method (Figure S1: Overview of the SEM recipe (left) and the image navigation tool (right); Figure S2: Holder recipe; Figure S3: Calibration recipe; Figure S4: Image processing recipe; Figure S5: Morphology recipe; Figure S6: Mineral recipe set-up; Figure S7: Mining recipe); Table S1: EDX point analyses on garnet standard.

Author Contributions: Conceptualization, N.K., S.G. and S.N.M.; methodology, N.K., S.G. and S.N.M.; validation, N.K., and S.N.M.; formal analysis, N.K., S.N.M.; investigation, N.K., S.N.M.; resources, N.K.; writing—original draft preparation, N.K. and S.N.M.; writing—review and editing, N.K., S.G. and S.N.M.; visualization, N.K. and S.N.M.; supervision, N.K.; project administration, N.K. All authors have read and agreed to the published version of the manuscript.

Funding: This research received no external funding.

Acknowledgments: The authors wish to thank R. Kahn and S. Lode for discussion and suggestions on the use of the Mineralogic software. J.C. Schumacher is thanked for his work on the metamorphic petrology in the Fiskenæsset area. Two anonymous reviewers and the editor provided valuable comments to an earlier version of this publication.

Conflicts of Interest: The authors declare no conflict of interest.

References

1.　Huggins, F.E.; Kosmack, D.A.; Huffman, G.P.; Lee, R.J. Coal mineralogy by SEM analysis. *Scanning Electron Microsc.* **1980**, *1*, 531–540.

2.   Friedrichs, K.H. Electron microscopic analyses of dust from the lungs and the lymph nodes of talc-mine employees. *Am. Ind. Hyg. Assoc. J.* **1987**, *48*, 626–633. [CrossRef] [PubMed]

3.   Heasman, I.; Watt, J. Particulate pollution case studies which illustrate uses of individual particle analysis by scanning electron microscopy. *Environ. Geochem. Health* **1989**, *11*, 157–162. [CrossRef] [PubMed]

4.   Steffen, S.; Otto, M.; Niewoehner, L.; Barth, M.; Brozek-Mucha, Z.; Biegstraaten, J.; Horvath, R. Chemometric classification of gunshot residues based on energy dispersive X-ray microanalysis and inductively coupled plasma analysis with mass-spectrometric detection. *Spectrochim. Acta* **2007**, *62*, 1028–1036. [CrossRef]

5.   Hrtska, T.; Gottlieb, P.; Skála, R.; Breiter, K.; Motl, D. Automated mineralogy and petrology—Applications of Tescan Integrated Mineral Analyzer (TIMA). *J. Geosci.* **2018**, *63*, 47–63.

6.   Pirrie, D.; Rollinson, G.K. Unlocking the applications of automated mineral analysis. *Geol. Today* **2011**, *27*, 226–235. [CrossRef]

7.   Gäbler, H.-E.; Melcher, F.; Graupner, T.; Bahr, A.; Sitnikova, M.A.; Henjes-Kunst, F.; Brätz, H.; Gerdes, A. Speeding up the analytical workflow for Coltan Fingerprinting by an Integrated Mineral Liberation Analysis/LA-ICP-MS approach. *Geostand. Geoanal. Res.* **2011**, *35*, 431–448. [CrossRef]

8.   Graham, S.D.; Brough, C.; Cropp, A. An Introduction to ZEISS Mineralogic Mining and the correlation of light microscopy with automated mineralogy: A case study using BMS and PGM analysis of samples from a PGE-bearing chromitite prospect. *Precious Met.* **2015**, 1–2. Available online: https://www.researchgate.net/profile/Christopher_Brough2/publication/277669986_An_Introduction_to_ ZEISS_Mineralogic_Mining_and_the_correlation_of_light_microscopy_with_automated_mineralogy_a_ case_study_using_BMS_and_PGM_analysis_of_samples_from_a_PGE-bearing_chromitite_prospect/links/ 5570388208aeccd77741818c.pdf (accessed on 15 June 2019).

9.   Keulen, N.; Frei, D.; Riisager, P.; Knudsen, C. Analysis of heavy minerals in sediments by Computer-Controlled Scanning Electron Microscopy (CCSEM): Principles and applications. In *Quantitative Mineralogy and Microanalysis of Sediments and Sedimentary Rocks*; Sylvester, P., Ed.; Mineralogical Association of Canada Short Course: Toronto, ON, Canada, 2012; Volume 42, pp. 167–184.

10.   Elghali, A.; Benzaazoua, M.; Bouzahzah, H.; Bussiere, B.; Villarraga-Gomez, H. Determination of the available acid-generating potential of waste rock, part 1: Mineralogical approach. *Appl. Geochem.* **2018**, *99*, 31–41. [CrossRef]

11.   Sandmann, D. Method Development in Automated Mineralogy. Ph.D. Thesis, Technischen Universität Bergakademie Freiberg, Freiberg, Germany, 2015. Available online: https://pdfs.semanticscholar.org/6c7d/ fc4f887906a3c987f47277981fbec453538f.pdf (accessed on 1 June 2019).

12.   Scheller, S.; Tagle, R.; Gloy, G.; Barraza, M.; Menzies, A. Advancements in Minerals Identification and Characterization in Geo-Metallurgy: Comparing E-Beam and Micro-X-ray-Fluorescence Technologies. *Microsc. Microanal.* **2017**, *23*, 2168–2169. [CrossRef]

13.   Holwell, D.A.; Adeyemi, Z.; Ward, L.A.; Graham, S.D.; Smith, D.J.; McDonald, I.; Smith, J.W. Low temperature alteration and upgrading of magmatic Ni-Cu-PGE sulfides as a source for hydrothermal Ni and PGE ores: A quantitative approach using automated mineralogy. *Ore Geol. Rev.* **2017**, *91*, 718–740. [CrossRef]

14.   Bernstein, S.; Frei, D.; McLimans, R.K.; Knudsen, C.; Vasudev, V.N. Application of CCSEM to heavy mineral deposits: Source of high-Ti ilmenite sand deposits of South Kerala beaches, SW India. *J. Geochem. Explor.* **2008**, *96*, 25–42. [CrossRef]

15.   Gu, Y. Automated Scanning Electron Microscope Based Mineral Liberation Analysis—An introduction to JKMRC/FEI Mineral Liberation Analyser. *J. Miner. Mater. Charact. Eng.* **2003**, *2*, 33–41. [CrossRef]

16.   Olivarius, M.; Rasmussen, E.S.; Siersma, V.; Knudsen, C.; Kokfelt, T.F.; Keulen, N. Provenance signal variations caused by facies and tectonics: Zircon age and heavy mineral evidence from Miocene sand in the north-eastern North Sea Basin. *Mar. Pet. Geol.* **2014**, *49*, 1–14. [CrossRef]

17.   Sylvester, P. Use of the Mineral Liberation Analyzer (MLA) for mineralogical Studies of sediments and sedimentary rocks. In *Quantitative Mineralogy and Microanalysis of Sediments and Sedimentary Rocks*; Sylvester, P., Ed.; Mineralogical Association of Canada Short Course: Toronto, ON, Canada, 2012; Volume 42, pp. 1–16.

18.   Ma, K.; Jiang, H.; Li, J.; Zhao, L. Experimental study on the micro alkali sensitivity damage mechanism in low-permeability reservoirs using QEMSCAN. *J. Nat. Gas Sci. Eng.* **2016**, *36*, 1004–1017. [CrossRef]

19. Ayling, B.; Rose, P.; Petty, S.; Zemach, E.; Drakos, P. Qemscan° (Quantitative evaluation of minerals by scanning electron microscopy): Capability and application to fracture characterization in Geothermal systems. In Proceedings of the 37th Workshop Geothermal Reservoir Engineering 2012, Stanford, CA, USA, 30 January–1 February 2012. SGP-TR-194.

20. Mackay, D.A.R.; Simandl, G.J.; Ma, W.; Redfearn, M.; Gravel, J. Indicator mineral-based exploration for carbonatites and related specialty metal deposits—A Qemscan orientation survey, British Columbia, Canada. *J. Geochem. Explor.* **2016**, *165*, 159–173. [CrossRef]

21. Graham, S.; Keulen, N. Nano-scale automated quantitative mineralogy: A 200 nm quantitative mineralogy assessment of fault gouge using Mineralogic. *Minerals* **2019**, *9*, 665. [CrossRef]

22. Nikonow, W.; Rammlmair, D. Automated mineralogy based on micro-energy dispersive X-ray fluorescence microscopy (micro-EDXRF) applied to plutonic rock thin sections in comparison to a mineral liberation analyzer. *Geosci. Instrum. Methods Data Syst.* **2017**, *6*, 429–437. [CrossRef]

23. Nikonow, W.; Rammlmaier, D.; Meima, J.A.; Schodlok, M.C. Advanced mineral characterization and petrographic analysis by μ-EDXRF, LIBS, HIS and hyperspectral data merging. *Miner. Petrol.* **2019**, *113*, 417–431. [CrossRef]

24. Senesi, G.S. Laser-Induced Breakdown Spectroscopy (LIBS) applied to terrestrial and extraterrestrial analogue geomaterials with emphasis to minerals and rocks. *Earth Sci. Rev.* **2014**, *139*, 231–267. [CrossRef]

25. Gottlieb, P.; Wilkie, G.; Sutherland, D.; Ho-Tun, E.; Suthers, S.; Perera, K.; Jenkins, B.; Spencer, S.; Butcher, A.; Rayner, J. Using quantitative electron microscopy for process mineralogy applications. *JOM* **2000**, *52*, 24–25. [CrossRef]

26. Andersen, J.C.Ø.; Rollinson, G.K.; Snook, B.; Herrington, R.; Fairhurst, R.J. Use of QEMSCAN® for the characterization of Ni-rich and Ni-poor goethite in laterite ores. *Miner. Eng.* **2009**, *22*, 1119–1129. [CrossRef]

27. Schulz, B. Polymetamorphism in garnet micaschists of the Saualpe Eclogite Unit (Eastern Alps, Austria), resolved by automated SEM methods and EMP–Th–U–Pb monazite dating. *J. Metamorph. Geol.* **2016**. [CrossRef]

28. Kern, M.; Möckel, R.; Krause, J.; Teichmann, J.; Gutzmer, J. Calculating the deportment of a fine-grained and compositionally complex Sn skarn with a modified approach for automated mineralogy. *Miner. Eng.* **2018**, *116*, 213–225. [CrossRef]

29. Williams, M.L.; Jercinovic, M.J. Tectonic interpretation of metamorphic tectonites: Integrating compositional mapping, microstructural analysis and in situ monazite dating. *J. Metamorph. Geol.* **2012**, *30*, 739–752. [CrossRef]

30. Windley, B.F.; Garde, A.A. Arc-generated blocks with crustal sections in the North Atlantic craton of West Greenland: New mechanisms of crustal growth in the Archaean with modern analogues. *Earth Sci. Rev.* **2009**, *93*, 1–30. [CrossRef]

31. Friend, C.R.L.; Nutman, A.P. New pieces to the Archaean terrane jigsaw puzzle in the Nuuk region, southern West Greenland: Steps in transforming a simple insight into a complex regional tectonothermal model. *J. Geol. Soc. Lond.* **2005**, *162*, 147–162. [CrossRef]

32. Kalsbeek, F. Metamorphism of Archaean rocks of West Greenland. In *The Early History of the Earth*; Windley, B.F., Ed.; Wiley: London, UK, 1976; pp. 225–235.

33. Henriksen, N.; Higgins, A.K.; Kalsbeek, F.; Pulvertaft, T.C.R. *Greenland from Archaean to Quaternary Descriptive Text to the 1995 Geological Map of Greenland, 1:2,500,000*, 2nd ed.; Geological Survey of Denmark and Greenland Bulletin: Copenhagen, Denmark, 2009; Volume 18, pp. 1–93.

34. Myers, J.S. Stratigraphy and structure of the Fiskenæsset Complex, southern West Greenland. *Grønlands Geol. Undersøgelse* **1985**, *150*, 1–72.

35. Keulen, N.; Næraa, T.; Kokfelt, T.F.; Schumacher, J.C.; Schersten, A. Zircon record of the igneous and metamorphic history of the Fiskenæsset anorthosite complex in southern West Greenland. *Geol. Surv. Den. Greenl. Bull.* **2010**, *20*, 67–70.

36. Polat, A.; Frei, R.; Schersten, A.; Appel, P.W.U. New age (ca. 2970 Ma), mantle source composition and geodynamic constraints on the Archean Fiskenæsset anorthosite complex, SW Greenland. *Chem. Geol.* **2010**, *277*, 1–20. [CrossRef]

37. Keulen, N.; Schumacher, J.C.; Næraa, T.; Kokfelt, T.F.; Schersten, A.; Szilas, K.; van Hinsberg, V.J.; Schlatter, D.M.; Windley, B.F. Meso- and Neoarchaean geological history of the Bjørnesund and Ravns Storø Supracrustal Belts, southern West Greenland: Settings for gold enrichment and corundum formation. *Precambrian Res.* **2014**, *254*, 36–58. [CrossRef]

38. Keulen, N.; Thomsen, T.B.; Schumacher, J.C.; Poulsen, M.D.; Kalvig, P.; Vennemann, T.; Salimi, R. Formation, origin and geographic typing of corundum (ruby and pink sapphire) from the Fiskenæsset complex, Greenland. *Lithos* **2019**. submitted.

39. Szilas, K.; Hoffmann, J.E.; Scherstén, A.; Rosing, M.T.; Windley, B.F.; Kokfelt, T.F.; Keulen, N.; van Hinsberg, V.J.; Næraa, T.; Frei, R.; et al. Complexcalc-alkaline volcanism recorded in Mesoarchaean supracrustal belts north of Frederikshåb Isblink, southern West Greenland: Implications for subductionzone processes in the early Earth. *Precambrian Res.* **2012**, *208*, 90–123. [CrossRef]

40. Keulen, N.; Kokfelt, T.F.; The Homogenisation Team. *A Seamless, Digital, Internet-Based Geological Map of South-West and Southern West Greenland, 1:100,000, 61°30′–64°*; Geological Survey of Denmark and Greenland: Copenhagen, Denmark, 2011.

41. Schumacher, J.C.; van Hinsberg, V.J.; Keulen, N. *Metamorphism in Supracrustaland Ultramafic Rocks in Southern*; Geological Survey of Denmark and Greenland: Copenhagen, Denmark, 2011; Volume 6, pp. 1–29.

42. Keulen, N.; Heijboer, T. The provenance of garnet: Semi-automatic plotting and classification of garnet compositions. *Geophys. Res. Abstr.* **2011**, *13*, EGU 2011-4716.

43. Gublin, Y.L. Garnet–Biotite Geothermometer and Estimation of Crystallization Temperature of Zoned Garnets from Metapelites: I. Reconstruction of Thermal History of Porphyroblast Growth. *Geol. Ore Depos.* **2012**, *54*, 602–615. [CrossRef]

44. Stünitz, H.; Fitzgerald, J.D. Deformation of granitoids at low metamorphic grade. II: Granular flow in albite-rich mylonites. *Tectonophysics* **1983**, *221*, 299–324. [CrossRef]

45. Tullis, J.; Yund, R.A. Dynamic recrystallization of feldspar: A mechanism for ductile shear zone formation. *Geology* **1985**, *13*, 238–241. [CrossRef]

46. Herd, R.K.; Windley, B.F.; Ghisler, M. The mode of occurrence and petrogenisesis of the sapphirine-bearing and associated rocks of West Greenland. Rapp. *Grønlands Geol. Undersøgelse* **1969**, *24*, 1–44.

47. Appel, P.W.U.; Dahl, O.; Kalvig, P.; Polat, A. Discovery of new PGE mineralisation in the Precambrian Fiskenaesset anorthosite complex, West Greenland. *Geol. Surv. Den. Greenl. Rep.* **2011**, *3*, 1–48.

48. Ghisler, M. Pre-metamorphic folded chromite deposits of stratiform type in the early Precambrian of West Greenland. *Miner. Depos.* **1970**, *5*, 223–236. [CrossRef]

Article

Nanoscale Automated Quantitative Mineralogy: A 200-nm Quantitative Mineralogy Assessment of Fault Gouge Using Mineralogic

Shaun Graham [1] and Nynke Keulen [2],*

[1]   Carl Zeiss Microscopy GmbH, ZEISS Group, 50 Kaki Bukit Place, Singapore 415926, Singapore; shaun.graham@zeiss.com

[2]   Department of Petrology and Economic Geology, Geological Survey of Denmark and Greenland (GEUS), Øster Voldgade 10, DK-1350 Copenhagen K, Denmark

*   Correspondence: ntk@geus.dk

Received: 14 August 2019; Accepted: 25 October 2019; Published: 29 October 2019

Abstract: Effective energy-dispersive X-ray spectroscopy analysis (EDX) with a scanning electron microscope of fine-grained materials (submicrometer scale) is hampered by the interaction volume of the primary electron beam, whose diameter usually is larger than the size of the grains to be analyzed. Therefore, mixed signals of the chemistry of individual grains are expected, and EDX is commonly not applied to such fine-grained material. However, by applying a low primary beam acceleration voltage, combined with a large aperture, and a dedicated mineral classification in the mineral library employed by the Zeiss Mineralogic software platform, mixed signals could be deconvoluted down to a size of 200 nm. In this way, EDX and automated quantitative mineralogy can be applied to investigations of submicrometer-sized grains. It is shown here that reliable quantitative mineralogy and grain size distribution assessment can be made based on an example of fault gouge with a heterogenous mineralogy collected from Ikkattup nunaa Island, southern West Greenland.

Keywords: scanning electron microscopy (SEM); automated quantitative analysis (AQM); spectrum quantification; signal deconvolution; fault gouge; 200-nm resolution; grain size distribution; Ikkattup nunaa; mineral maps; submicrometer

---

1. Introduction

1.1. Automated Quantitative Mineralogy (AQM)

Little work has been done in the use of scanning electron microscopy (SEM)-based automated quantitative mineralogy (AQM) in the evaluation of mineralogy and microstructures of fault gouges. Physicochemical processes taking place in these rocks are recorded in the mineralogy and the microstructures of the fault gouge itself. However, the subsequent mineralogical assemblages are incredibly fine-grained, and therefore it is challenging for microanalytical applications to provide a high-resolution quantitative mineralogy of a large area in order to understand these physicochemical processes.

"Automated mineralogy", or more accurately described as automated quantitative mineralogy (AQM), has been available since the early 1980s with technologies such as QEMSCAN [1] and the more recent TIMA-X [2]. The key application area of this technique was to provide an automated and routine analysis capability to quantify the mineralogy and textures of ore mineralogy (modal mineralogy, liberation, association, etc.) to aid mineral processing plant development, optimization, and troubleshooting [1,3].

These automated mineralogy techniques control the electron beam to step across the sample surface, at a user-defined spatial resolution (i.e., 5-μm steps) acquiring unquantified energy-dispersive spectroscopy (EDX) spectrum. This unprocessed (no matrix corrections or spectrum quantification) spectrum is matched to a list of known referenced EDX spectra to provide a mineral name [2,4]. Without a direct measured chemical quantification from that analysis point, spectral fingerprinting on validated known mineral standards using external techniques such as electron microprobe is used [2,4,5].

The EDX spectrum, once acquired, is matched to the referenced spectrum from the mineral to classify the individual analysis. Whilst this technique is widely applied within AQM, it only provides a "crude" assessment of the mineralogy where elemental concentrations exceed 10 wt.% [5]. As such, this historic methodology falls short of providing a direct chemical analysis or correct analytical procedure to provide the necessary quality of analysis or standardized results.

Since the inception of AQM technologies, no significant methodological development of this capability occurred with little technological advancement despite the huge strides made in EDX technology and analytical capability. Multiple examples in the literature point toward the quantitative measurement capability of modern day EDX on major elements, which are able to achieve relative errors of less than 2% on a standardized analysis of polished samples [6,7].

However, the development of Mineralogic provides a step change in the analytical capabilities of AQM with the integration of these new EDX capabilities. This AQM software acquires an EDX spectrum, and performs matrix corrections and peak deconvolution for each analysis point. Minerals are subsequently classified based on stoichiometry values [8,9], providing new analytical capabilities for AQM and opening up new application areas. Concentrations of elements that cannot be analyzed correctly (e.g., H), can be assigned. The chemistry in each pixel is saved by the Mineralogic software, and the pixel's mineralogy can therefore be reinterpreted after the analysis. Thus, each EDX analysis provides one pixel on the false-colored mineral map [9].

In dispersive spectrometry, fundamental electron beam interaction with the sample still poses challenges. All electron beam microanalysis users will be aware of the challenges of carrying out an analysis at low spatial resolutions. As a result of the energy of the primary electron beam (acceleration voltage; kV) hitting the sample, X-rays are generated from an interaction volume, which may consequently be measured by the SEM's energy-dispersive X-ray spectrometer (EDX). The size of this interaction volume is a function of the primary energy of the beam and the mean atomic number of the sample, and it is usually presented as the diameter of the beam inside the sample material projected in the X-Z plane, which is also called the diameter of the footprint of the electron beam [10]. The interaction volume is larger than the diameter of the beam on the landing spot. In geological samples, typically containing silicates, interaction volumes can be in a region of 2–5 µm, with a 20-kV primary electron beam. Therefore, providing a chemical analysis and mineral classification based on an EDX spectrum can be challenging for fine-grained material, due to the "mixing" of compositions of a number of grains. For those grains, the interaction volume can result in the generation of X-rays from multiple minerals at a single point, providing complex mixed mineralogical compositions. Whilst these challenges remain, there is little alternative substitute for the quantification of such mineral assemblages, as gray level alone is not enough to separate and quantify the mineralogy and textures.

In this study, we provide a proof of concept in showing the capability of a new generation of SEM-based AQM to map the mineralogy and investigate the grain size of the fault gouge of varying composition at a 200-nm spatial EDX resolution.

1.2. Geological Background

The sample that was used for this study was collected in southern West Greenland from the northern coast of Ikkattup nunaa Island in the Ikkattoq anlanngua fjord (Figure 1). This locality is situated in the southern part of the Ravn Storø Supracrustal belt, which is intruded by a well-known Meso-Archaean Fiskenæsset anorthosite-leucogabbro complex. The fault rock sample represents an amphibolite intercalated with orthogneiss. The amphibolite is part of the ca. 3.0 Ga Ravn Storø Supracrustal complex, whereas the intruding orthogneiss was dated to 2.88 Ga. Later tectonic activity in the area includes two folding and thrusting events during the Neo-Archaean, during which the Kvanefjeld block to the south collided with the Bjørnesund block (which includes the sample locality and the Fiskenæsset complex to the north of the outcrop). The thrust zones were reactivated in an extensional setting during the latest Neoarchaean or earliest Palaeoproterozoic, and the fault gouge studied here most likely was formed during this extensional event ([11]; and references therein).

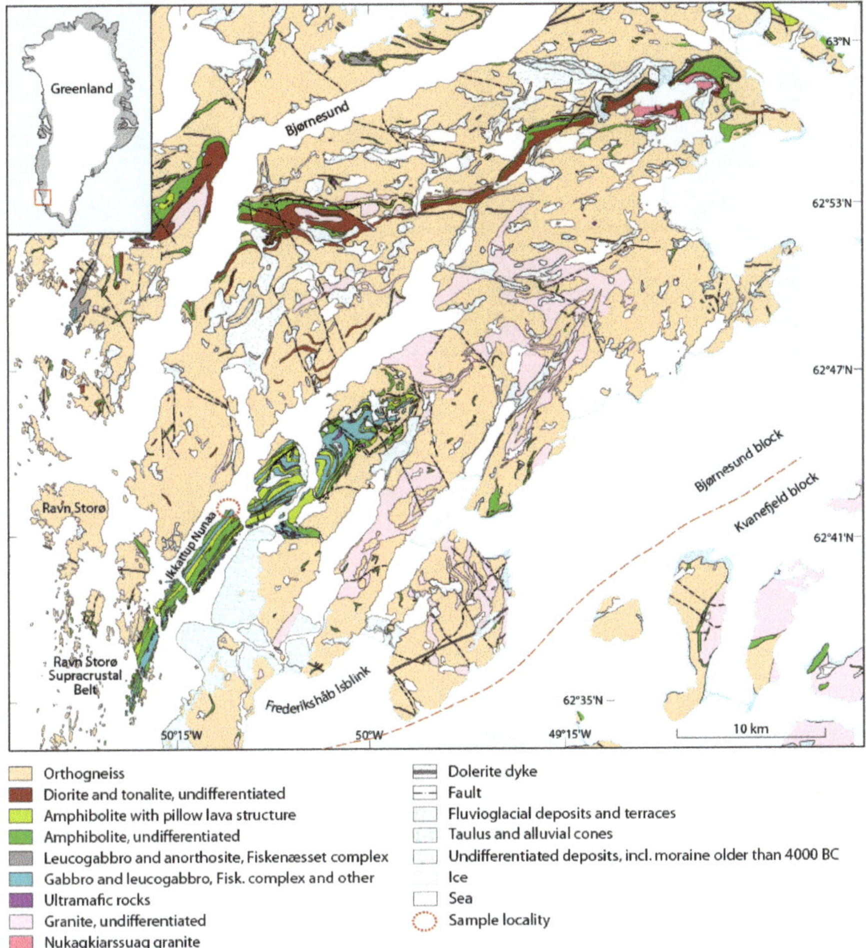

Figure 1. Geological map showing the Bjørnesund region in southern West Greenland. Sample locality, as well as the major tectonic boundary between the Bjørnesund Block and the Kvanefjeld block are indicated [12]. Map adapted from the Geological Survey of Denmark and Greenland (GEUS) [13]; reproduced with permission from GEUS.

2. Materials and Methods

The fault gouge was studied in a polished thin section and shows a mixture of grains that were originally part of the orthogneiss and of the amphibolite. Intensive mixing of the different minerals in the fault gouge prevented healing of the fault gouge to larger grain sizes [14]. Therefore, many comminuted grains of sizes smaller than 200 nm are still present in the sample. However, mineral reactions can be observed: epidote, illite, and chlorite (chamosite) formed to replace the original mineral assembly.

To perform the microanalysis, the thin section across the fault rock was coated with carbon and studied with a Zeiss SIGMA 300VP Scanning Electron Microscope (SEM) equipped two Bruker XFlash 6|30 energy-dispersive spectroscopy detectors (EDX), with 129-eV energy resolution and with the Zeiss Mineralogic automated quantitative mineralogy software platform. The region was selected and imaged to provide a high-resolution back-scattered electrons contrast (BSE) mosaic of the region of interest. In addition, on this region of interest, a 200-nm quantitative mineralogical analysis was carried out using Mineralogic, creating a mineral map (Figure 2).

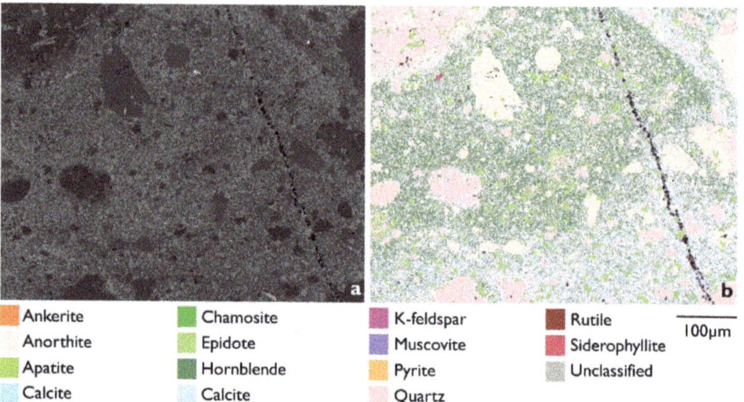

Figure 2. Large area overview image of the scanned sample area. (**a**) Displays the high-resolution BSE image (**b**) Displays the false-colored mineral map and key, displaying the mineralogy across this location.

To analyze the fine-grained fault gouge, we reduced the acceleration voltage of the primary electron beam to 10 kV to reduce the interaction volume [15–17], and therefore to minimize the amount of mixed pixel data, and used the 60-µm aperture providing a 1.8-nA beam current. Despite this reduction of primary electron energy to 10 kV, the interaction volume still has a diameter of up to approximately 1 µm for minerals with the lowest average atomic number (Z) and down to 400 nm for a high Z phase on top of a low Z phase (Figure 3). 200-nm EDX pixel size would greatly provide oversampling and a large volume of mixture pixel signatures.

Despite this, the data below outlines how the analysis and advanced mineral classifications employed in the Mineralogic mineral library were able to deconvolute any mixed signals and provide a reliable quantitative mineralogy and textural assessment. We were able to do this due to the access to fully quantitative EDX classifications for each pixel within the Mineralogic software and the ability to quantify the weight percent contribution of the elements present on a per-pixel basis (see also [9]). This enabled the creation of highly discriminatory mineral classifications whereby "contamination" of the chemical analysis provided by the above described mixed signals were able to be identified and factored into the mineral classifications of the mineral library to enable the correct classification (Figure 4). This capability and deconvolution procedure allowed the creation of true mineral classifications without the need for creating false mixed signal classifications. The mineral classifications could be used on particles down to 200 nm, even though the interaction volume of the beam was larger than 200 nm. Therefore, the obtained resolution of the analysis is 200 nm, which is the applied pixel size for the mineral, despite the larger size of the interaction volume.

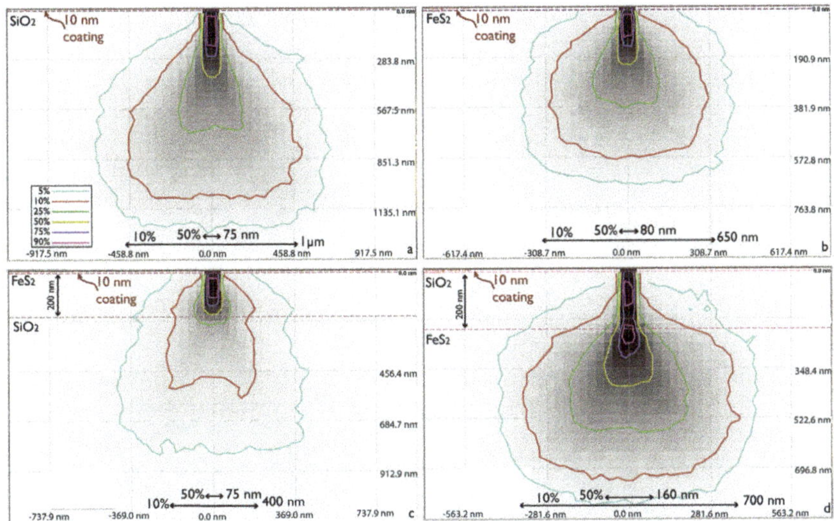

Figure 3. Monte Carlo simulations for the trajectories of 20,000 electrons in quartz (SiO$_2$) and pyrite (FeS$_2$) in carbon-coated samples. Arrows show the width of the interaction volume at 10 kV for 50% and 90% of the electrons. (**a**) Quartz, (**b**) pyrite, (**c**) 200-nm pyrite on top of quartz, (**d**) 200-nm quartz on top of pyrite. Simulations were made with Casino [18,19].

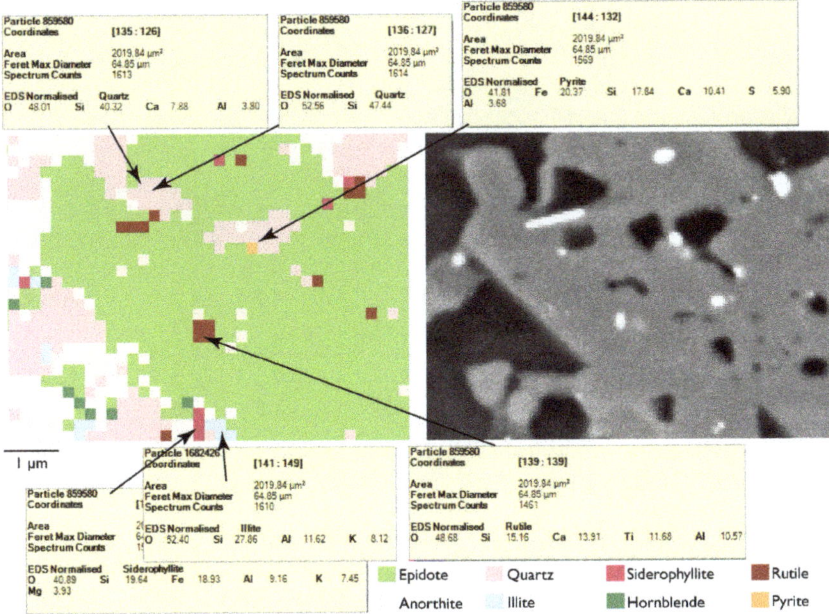

Figure 4. Detailed view of a false-colored mineral map and the associated BSE image. Each pixel represents 200 × 200 nm. Grains of ca. 200 nm can be recognized and classified from the energy-dispersive spectroscopy (EDX) data for that pixel. Yellow boxes display example EDX normalized weight percentages and the interpretation of the operator for the mineralogy.

3. Results

3.1. Mineralogy

The results presented here are for a particular area of interest that is presented and displayed in Figure 2a (backscattered electron micrograph) and Figure 2b with the false-colored mineral map from Mineralogic. The analysis was carried out of a 360,000-μm^2 region where a quantified EDX analysis was taken at every 200 nm, and therefore results in approximately 1.8 million quantified analyses.

Figure 2a displays the BSE data acquired from the region of interest and provides some initially information on the mineralogy based on the average atomic number of the mineral. However, the BSE gray-level intensity is not suitable to quantify mineral abundance and texture in these samples. As Figure 2a,b outlines, minerals such as illite and epidote appear to have almost identical gray-level signatures, as do hornblende, quartz, and anorthite.

However, Figure 2b displays the overlaid false-colored mineral data that was acquired at a 200-nm spatial resolution. The false-color minerals clearly display the mineralogy and the textural variation across this sample.

Combined with the modal mineralogy data (Table 1), these results clearly show how anorthite (c. 39%), illite (c. 20%), hornblende (c. 20%), and quartz (c. 13%) are the dominant mineral in this sample. It is also clear from Figure 2b and Table 1 that all the minerals show predominantly fine-grained sizes (200 nm); however, rare large clasts of quartz and anorthite are observed while hornblende, illite, and epidote mainly consist of a fine-grained matrix.

The 200-nm resolution of the scan was able to identify two distinct mineralogical and textural groups in this region (Figure 5). Textural group 1 is formed by the former amphibolite minerals and their reaction products. It can be found in the center of Figure 2, and is dominated by an epidote- and hornblende-dominated matrix that is populated with large (up to 80 microns in length) remnants of the orthogneiss, mainly consisting of single grains of quartz and anorthite. A second textural group can be found that encapsulates textural group 1 and is dominated by an illite and fine-grained anorthite matrix, which represents the remnants of the orthogneiss. In this textural group, hornblende is a minor phase in the matrix. Figure 5c,d shows that this mineralogical and textural group is dominated by anorthite, illite, and quartz with a small amount of epidote intergrown within this matrix.

3.2. Grain Size Distributions

Whilst the hornblende is relatively granular within this matrix, the epidote is often found as relatively coarse grains up to 20 microns in length (Figure 6). The anorthite grains display bimodal distribution as they are found in the matrix as grains of 1–5 microns in size, and also as much coarser remnant clasts that can be 50 microns in length or more.

The average grain size in textural group 2 is clearly much larger than that in textural group 1, and the matrix is dominated by illite and anorthite with only minor amounts of epidote and hornblende. Quartz grains are more common and of a larger grain size than those within textural group 1 (Table 2).

Table 1. Modal mineralogy data from the selected region of interest displayed in the graphical abstract. This data displays the mineral, its area % occurrence in the sample, the average grain size of that mineral with its standard deviation, and its average composition in wt.%.

Mineral	Area%	Avg Grain Size (µm)	Grain Size Std Dev (µm)	Average Composition
Anorthite	38.78	1.57	3.73	O 40.8; Si 35.62; Al 15.26; Ca 7.29; Fe 0.94; Mg 0.08; K 0.02
Illite	19.91	0.95	1.72	O 37.19; Si 31.84; Al 11.91; Fe 8.64; Ca 4.56; K 3.48; Mg 2.38
Hornblende	19.86	0.85	1.84	O 37.4; Si 31.01; Al 12.71; Fe 9.41; Ca 7.81; Mg 1.46; Ti 0.18; K 0.02
Quartz	13.27	1.11	2.62	Si 54.77; O 43.28; Al 1.95
Epidote	6.12	0.76	1.89	O 36.56; Si 25.6; Ca 16.18; Al 14.49; Fe 7.01; Mg 0.09; Ti 0.07
Siderophyllite	0.25	0.44	0.46	O 34.83; Si 22.99; Fe 19.45; Al 12.26; K 6.23; Mg 4.24
Muscovite	0.09	0.31	0.25	O 50.65; Si 30.4; Al 10.65; K 6.19; Mg 1.52; Na 0.25; Fe 0.18; Ca 0.16
Pyrite	0.08	0.44	0.32	Fe 65.19; S 34.8; Zn 0.01
Apatite	0.07	1.91	3.80	Ca 42.22; O 40.11; P 17.6; Cl 0.07; F 0
Ankerite	0.06	0.32	0.27	O 57.36; Ca 27.14; Fe 12.99; Mg 2.52
Rutile	0.05	0.52	0.34	O 65.7; Ti 34.3
Chamosite	0.02	0.61	1.02	O 33.59; Fe 24.66; Si 22.26; Al 14.33; Mg 5.03; Ca 0.13; K 0.01; Mn 0
K Feldspar	0.02	0.46	0.55	O 37.48; Si 34.62; Al 13.03; K 12.34; Ca 1.74; Fe 0.59; Mg 0.21
Calcite	0.01	0.85	0.86	O 52.63; Ca 44.23; Al 2.42; Fe 0.49; K 0.1; Mg 0.08; S 0.02; Mn 0.02

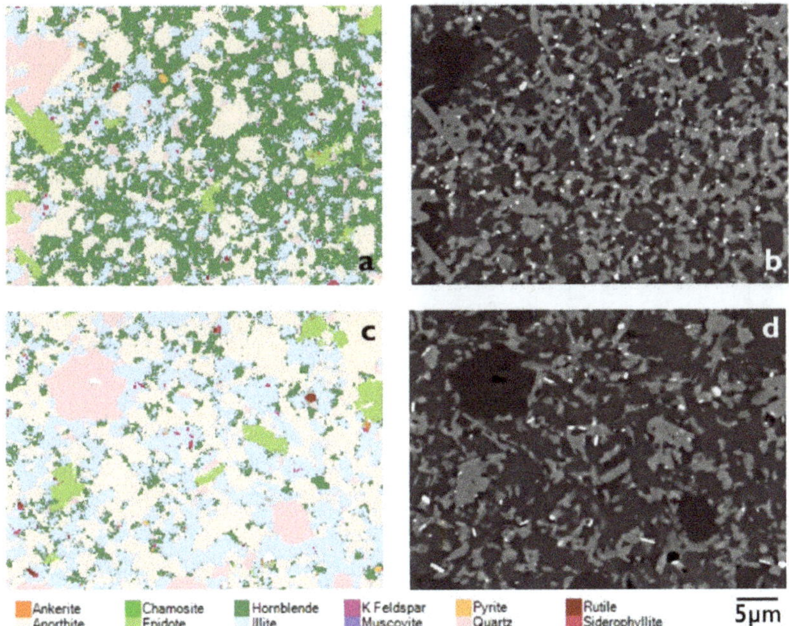

Figure 5. Two details of Figure 2: Combined false-colored mineral map (**a,c**) and BSE images (**b,d**) for textural group 1 (**a,b**) and textural group 2 (**c,d**). The width of view for each field is 46 microns.

Table 2. Modal mineralogy and grain size distribution data for textural groups 1 (TG1) and 2 (TG2) and comparative modal mineralogy and grain size distribution data for TG1 and TG2.

	Area				Average Grain Size				Grain Size Standard Distribution			
Mineral	TG1	TG2	Absolute Difference	Relative Difference (%)	TG1	TG2	Absolute Difference	Relative Difference (%)	TG1	TG2	Absolute Difference	Relative Difference (%)
Anorthite	40.68	39.92	0.76	1.89	1.55	1.62	−0.07	−4.50	3.59	4.21	−0.62	−15.94
Illite	15.14	23.46	−8.32	−43.09	0.87	1.08	−0.22	−22.26	1.26	2.23	−0.98	−55.94
Quartz	9.59	18.86	−9.27	−65.17	1.07	1.21	−0.15	−12.73	2.26	3.27	−1.01	−36.69
Hornblende	26.36	10.31	16.05	87.54	1.02	0.71	0.31	36.15	2.36	1.09	1.28	74.06
Epidote	6.23	5.49	0.74	12.62	0.61	1.40	−0.79	−78.44	1.59	2.71	−1.12	−52.18
Siderophyllite	0.20	0.29	−0.08	−34.21	0.41	0.47	−0.06	−14.16	0.46	0.50	−0.04	−7.59
Muscovite	0.07	0.12	−0.05	−54.68	0.32	0.31	0.02	4.92	0.27	0.25	0.02	7.87
Apatite	0.04	0.09	−0.06	−88.82	1.99	1.83	0.16	8.47	2.81	4.16	−1.35	−38.59
Ankerite	0.06	0.06	0.00	−0.89	0.32	0.33	0.00	−1.10	0.27	0.28	−0.01	−4.21
Pyrite	0.09	0.06	0.03	41.86	0.41	0.48	−0.08	−16.87	0.34	0.26	0.08	25.95
Rutile	0.05	0.05	0.00	−9.21	0.50	0.55	−0.05	−9.64	0.34	0.34	0.00	−0.12
Chamosite	0.01	0.04	−0.03	−127.75	0.41	0.76	−0.35	−60.31	0.68	1.17	−0.49	−52.38
K Feldspar	0.02	0.01	0.00	0.51	0.54	0.40	0.14	29.69	0.77	0.37	0.41	71.40
Calcite	0.01	0.01	0.00	−20.93	0.75	0.81	−0.07	−8.41	0.71	0.78	−0.07	−9.25

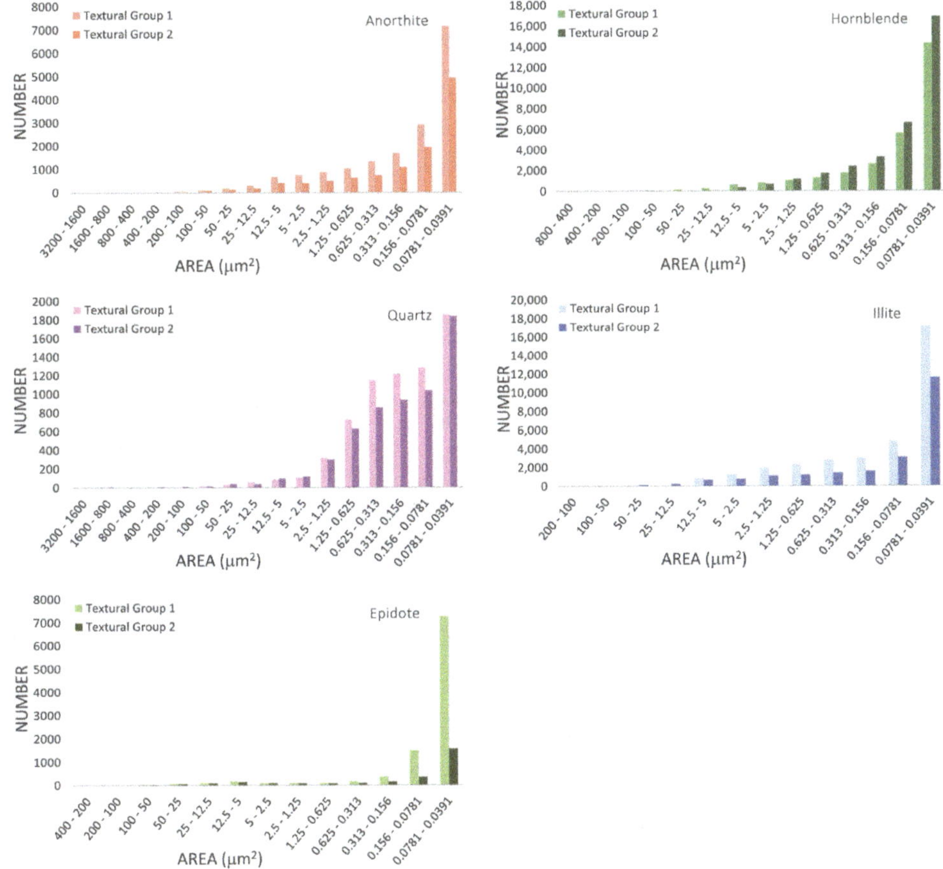

Figure 6. Grain size distribution for anorthite, hornblende, quartz, illite, and epidote in the analyzed fault gouge for textural groups 1 (TG1) and 2 (TG2). The distribution is shown as grain area in μm² and was measured as a 2D grain size distribution. Most grains are in the smallest bin that corresponds to grain diameters down to 200 nm.

The grain size distribution per mineral can be extracted from the Mineralogic output. Figure 6 presents the grain size distribution for one of the originally present grains, quartz, and for one of the reaction products, illite, which formed after the breakdown of feldspar minerals. Both minerals show a considerable decrease in grain size, with most grains in the bin down to 200 nm.

4. Implications

Fine-grained samples can be mapped for their mineralogy with EDX on an SEM with a step size that is considerably smaller than the diameter of the interaction volume of the primary electron beam. Whilst the BSE photomicrograph provides a higher resolution display of the region of interest, it is clearly visible that there is not enough average atomic number contrast available between the mineral phases present to clearly differentiate and quantify the mineralogy only based on the BSE image. Therefore, mineral mapping at a very high resolution is necessary to obtain most of the information on the composition of the fault gouge sample. To be able to do this, a basic understanding of the mineralogy needs to be gathered from the sample (e.g., rock type, metamorphic facies, major minerals). Building a mineral library based on the measured compositions is the initial starting point to base classifications on for a finer pixel spacing (i.e., 200 nm) and the quantification at the nanoscale mineralogy. Whilst analyses at this resolution provided mixed pixels/signals, the ability to quantify the wt.% contribution of the elements allows for the identification and deconvolution of the mixtures and the successful classification of the mineralogy at this scale. Even for minerals with a low Z and thus a high interaction volume, which is several times larger than the size of the smallest particles and larger than the step size, a large part of the chemical information comes from the area where the beam hits the sample. Figure 3 shows that for all the investigated cases, 50% of the electron trajectories remain within a 75–160-nm wide interaction volume area and reach depths of maximum ca. 200–300 nm. Therefore, the mixed signals could successfully be identified and translated into classifications in the mineral library that enable the correct classification of those particles (Figure 4). The latter figure shows that even for 200-nm pixels, and most of the pixels at the grain boundary between two phases, the correct mineral could be determined from the mixed signal. Only one pixel away from the grain boundaries, the signal for each mineral is very near its expected composition. Compare e.g., the two yellow boxes for quartz in Figure 4.

The small step size applied for this analysis allows for an investigation of the fault gouge mineralogy at the nanometer scale with a quantification of the mineralogy to complement the BSE imaging. As EDX analyses are based on chemistry only, and the identification of the smallest grains is dependent on a beam interaction volume of a size larger than the smallest grains in the sample, X-ray diffraction or electron backscatter diffraction analyses might need to be performed in order to establish the crystallography of the minerals with an independent characterization method. This is especially important for clay minerals, which may show interstratifications at a nanometer scale, which might remain unnoticed with EDX analyses only.

The minerals in the amphibolite (anorthite, hornblende) were more intensively comminuted than the minerals in the orthogneiss (quartz, anorthite, and feldspar). Large grains of anorthite and especially quartz are still present in the sample, even though the majority of the grains were reduced to very small grain sizes. Illite is more fine-grained than quartz, which suggests that feldspar was more intensively comminuted than quartz, and that the illite grains did not grow after their formation. This brittle behavior of quartz and feldspar suggests greenschist facies temperature conditions during comminution [20,21]. Therefore, the fault gouge formed after the uplift of the area to upper-crustal levels (greenschist facies conditions) during a late stage of the geological history of the area. This is also confirmed by the observed mineral reactions of hornblende and anorthite to epidote, biotite to chamosite, and feldspar to illite, which all are typically occurring at lower greenschist facies conditions [22–24]. No large veins and no stacks of mica grains (chamosite, muscovite) or illite were observed, suggesting that these minerals grew after deformation, as a result of diagenetic processes (compare to [24]).

5. Conclusions

This study is believed to be the first recorded example of a successful automated quantitative mineralogy analysis of fault gouge to be accurately performed at such a fine pixel detail (200 nm). The ability to quantify the EDS spectrum at each analyzed point into its primary chemical components, and the use of in-built matrix corrections and peak deconvolutions, provide a robust analytical framework to provide high-quality quantitative chemical data that have never before been possible in AQM. Then, this can be used to define mineral species and use the quantified chemical capability to identify and deconvolute the mixed signals at such fine scales, despite the mixed signatures provided from the primary electron beam interaction. Therefore, the ability to map and successfully classify at such high resolutions opens up a wide range of new applications in automated quantitative mineralogy that have never before been possible.

Author Contributions: Conceptualization, S.G.; methodology, S.G.; validation, S.G. and N.K.; formal analysis, S.G.; investigation, S.G. and N.K.; resources, N.K.; writing—original draft preparation, S.G. and N.K.; writing—review and editing, S.G. and N.K.; visualization, S.G. and N.K.; project administration, S.G. and N.K.

Funding: This research received no external funding.

Acknowledgments: The authors would like to thank the three anonymous reviewers for their comments to an earlier version of this paper. The editors are thanked for their careful handling of the manuscript.

Conflicts of Interest: The authors declare no conflict of interest.

References

1. Sutherland, D.N.; Gottlieb, P. Application of automated quantitative mineralogy in mineral processing. *Min. Eng.* **1991**, *4*, 753–762. [CrossRef]
2. Hrtska, T.; Gottlieb, P.; Skála, R.; Breiter, K.; Motl, D. Automated mineralogy and petrology—Applications of Tescan Integrated Mineral Analyzer (TIMA). *J. Geosci.* **2018**, *63*, 47–63.
3. Lotter, N.-O.; Kowal, D.L.; Tuzun, M.A.; Whittaker, P.J.; Kormos, L. Sampling and flotation testing of Sudbury Basin drill core for process mineralogy modelling. *Miner. Eng.* **2003**, *16*, 857–864. [CrossRef]
4. Gottlieb, P.; Wilkie, G.; Sutherland, D.; Ho-Tun, E.; Suthers, S.; Perera, K.; Jenkins, B.; Spencer, S.; Butcher, A.; Rayner, J. Using quantitative electron microscopy for process mineralogy applications. *JOM* **2000**, *52*, 24–25. [CrossRef]
5. Andersen, J.C.Ø.; Rollinson, G.K.; Snook, B.; Herrington, R.; Fairhurst, R.J. Use of QEMSCAN® for the characterization of Ni-rich and Ni-poor goethite in laterite ores. *Min. Eng.* **2009**, *22*, 1119–1129. [CrossRef]
6. Newbury, D.E. Energy-dispersive spectrometry. In *Methods in Materials Research*; Kaufmann, E.N., Abbaschian, R., Barnes, P.A., Bocarsly, A.B., Chien, C.-L., Dollimore, D., Doyle, B.L., Fultz, B., Goldman, A.I., Gronsky, R., et al., Eds.; John Wiley & Sons: New York, NY, USA, 2000; pp. 11d.1.1–11d.1.25.
7. Newbury, D.E.; Ritchie, N.W.M. Is Scanning Electron Microscopy/Energy Dispersive X-ray Spectrometry (SEM/EDS) Quantitative? *Scanning* **2013**, *35*, 141–168. [CrossRef] [PubMed]
8. Holwell, D.A.; Adeyemi, Z.; Ward, L.A.; Graham, S.D.; Smith, D.J.; McDonald, I.; Smith, J.W. Low temperature alteration and upgrading of magmatic Ni-Cu-PGE sulfides as a source for hydrothermal Ni and PGE ores: A quantitative approach using automated mineralogy. *Ore Geol. Rev.* **2017**, *91*, 718–740. [CrossRef]
9. Keulen, N.; Malkki, S.N.; Graham, S. Automated quantitative mineralogy applied to metamorphic rocks. *Minerals* **2019**. Submitted this volume.
10. Goldstein, J.I.; Newbury, D.E.; Michael, J.R.; Ritchie, N.W.M.; Scott, J.-H.J.; Joy, C.C. *Scanning Electron Microscopy and X-Ray Microanalysis*; Springer: New York, NY, USA, 2017; p. 550.
11. Keulen, N.; Schumacher, J.C.; Næraa, T.; Kokfelt, T.F.; Schersten, A.; Szilas, K.; van Hinsberg, V.J.; Schlatter, D.M.; Windley, B.F. Meso- and Neoarchaean geological history of the Bjørnesund and Ravns Storø Supracrustal Belts, southern West Greenland: Settings for gold enrichment and corundum formation. *Precambrian Res.* **2014**, *254*, 36–58. [CrossRef]
12. Windley, B.F.; Garde, A.A. Arc-generated blocks with crustal sections in the North Atlantic craton of West Greenland: New mechanisms of crustal growth in the Archaean with modern analogues. *Earth Sci. Rev.* **2009**, *93*, 1–30. [CrossRef]
13. Keulen, N.; Kokfelt, T.F. The Homogenisation Team, 2011. A Seamless, Digital, Internet-Based Geological Map of South-West and Southern West Greenland, 1:100 000, 61°30′–64°. Copenhagen: Geological Survey of Denmark and Greenland. Available online: http://geuskort.geus.dk/gisfarm/svgrl.jsp (accessed on 9 May 2018).
14. Keulen, N.; Stünitz, H.; Heilbronner, R. Healing microstructures of experimental and natural fault gouge. *J. Geophys. Res.* **2008**, *113*, B06205. [CrossRef]
15. Hombourger, C.; Outrequin, M. Quantitative Analysis and High-Resolution X-ray Mapping with a Field Emission Electron Microprobe. *Microsc. Today* **2013**, *21*, 10–15. [CrossRef]
16. McSwiggen, P. Characterisation of sub-micrometre features with the FE-EPMA. *IOP Conf. Ser. Mater. Sci. Eng.* **2014**, *55*, 012009. [CrossRef]
17. Fournelle, J.; Cathey, H.; Pinard, P.T.; Richter, S. Low voltage EPMA: Experiments on a new frontier in microanalysis—Analytical lateral resolution. *IOP Conf. Ser. Mater. Sci. Eng.* **2016**, *109*, 012003. [CrossRef]
18. Hovington, P.; Drouin, D.; Gauvin, R. CASINO: A new Monte Carlo Code in C Language for Electron Beam Interaction—Part 1: Description of the program. *Scanning* **1997**, *19*, 1–14. [CrossRef]

19. Drouin, D.; Couture, A.R.; Joly, D.; Tastet, X.; Aimez, V.; Gauvin, R. CASINO V2.42—A fast and Easy-to-use modeling tool for scanning electron microscopy and microanalysis users. *Scanning* **2007**, *29*, 92–101. [CrossRef]

20. Stünitz, H.; Fitzgerald, J.D. Deformation of granitoids at low metamorphic grade. II: Granular flow in albite-rich mylonites. *Tectonophysics* **1983**, *221*, 299–324.

21. Tullis, J.; Yund, R.A. Dynamic recrystallization of feldspar: A mechanism for ductile shear zone formation. *Geology* **1985**, *13*, 238–241. [CrossRef]

22. Velde, B.; Suzuku, T.; Nicot, E. Pressure-temperature of illite-smectite mixed-layer minerals: Niger delta mudstones and other examples. *Clays Clay Miner.* **1986**, *34*, 435–441. [CrossRef]

23. Kristmannsdottir, H. Alteration of Basaltic Rocks by Hydrothermal-Activity at 100–300 °C. *Dev. Sedim.* **1979**, *27*, 359–367.

24. Abd Emola, A.; Charpentier, D.; Buatier, M.; Lanari, P.; Monié, P. Textural-chemical changes and deformation conditions registered by phyllosilictes in a fault zone (Pic de Port Vieux thrust, Pyrenees). *Appl. Clay Sci.* **2017**, *144*, 88–103. [CrossRef]

Article

Evaluation of Recyclability of a WEEE Slag by Means of Integrative X-ray Computer Tomography and SEM-Based Image Analysis

Markus Buchmann [1,2,*], **Nikolaus Borowski** [3,4], **Thomas Leißner** [1], **Thomas Heinig** [2], **Markus A. Reuter** [2], **Bernd Friedrich** [3] **and Urs A. Peuker** [1]

[1]   Institute of Mechanical Process Engineering and Mineral Processing, TU Bergakademie Freiberg, Agricolastraße 1, 09599 Freiberg, Germany; thomas.leissner@mvtat.tu-freiberg.de (T.L.); urs.peuker@mvtat.tu-freiberg.de (U.A.P.)

[2]   Helmholtz-Zentrum Dresden-Rossendorf, Helmholtz Institute Freiberg for Resource Technology, Chemnitzer Straße 40, 09599 Freiberg, Germany; t.heinig@hzdr.de (T.H.); m.reuter@hzdr.de (M.A.R.)

[3]   Institute of Process Metallurgy and Metal Recycling, RWTH Aachen University, Intzestraße 3, 52056 Aachen, Germany; nikolaus.borowski@sms-group.com (N.B.); bfriedrich@ime-aachen.de (B.F.)

[4]   SMS group GmbH, Ivo-Beucker-Straße 43, 40237 Düsseldorf, Germany

*   Correspondence: markus.buchmann@mvtat.tu-freiberg.de or m.buchmann@hzdr.de

Received: 10 February 2020; Accepted: 26 March 2020; Published: 30 March 2020

Abstract: Waste of electrical and electronic equipment (WEEE) is one of the fastest growing waste streams globally. Therefore, recycling of the valuable metals of this stream plays a vital role in establishing a circular economy. The smelting process of WEEE leads to significant amounts of valuable metals and rare earth elements (REEs) trapped in the slag phase. The effective manipulation of this phase transfer process necessitates detailed understanding and effective treatment to minimize these contents. Furthermore, an adequate process control to bring these metal contents into structures that make recycling economically applicable is required. Within the present study, a typical slag from a WEEE melting process is analyzed in detail. Therefore, the material is investigated with the help of X-ray computed tomography (XCT) and scanning electron microscopy (SEM)-based mineralogical analysis (MLA) to understand the typical structures and its implications for recycling. The influencing factors are discussed, and further processing opportunities are illustrated.

Keywords: waste of electrical and electronic equipment; X-ray computed tomography; mineral liberation analysis

1. Introduction

Over the course of the last few years, the pyro-metallurgical recycling process of electronic scraps, so-called WEEE (waste of electrical and electronic equipment), developed significantly not only because of governmental restrictions and statutory recycling rates, but also due to valuable and economically interesting amounts of metal contents (e.g., Cu, Au, Ag) [1]. Nevertheless, the industrial processes are tailored to recover only valuable metals from e-waste. In general, when talking about e-waste, valuable metals are concentrated mainly on PCBs (printed circuit boards). The metals found in and on PCBs can be classified into five main groups, namely, base metals (BMs), precious metals (PMs), platinum group metals (PGMs), metals of concern (MCs, hazardous), and rare earth elements (REEs) [2,3].

- Base metals (BMs): Cu, Al, Ni, Sn, Zn, and Fe;
- Precious metals (PMs): Au, Ag, Pd, Pt, Rh, Ir, and Ru;
- Metals of concern (MCs, hazardous): Hg, Be, In, Pb, Cd, As, and Sb;
- Scarce elements (REEs): Te, Ga, Se, Ta, and Ge.

The general pyro-metallurgical recycling process of PCBs (see Figure 1) is based on the combustion of the organic constituents of the material mix fed into the furnace and using the energy of this exothermic reaction to melt the other constituents. Therefore, depending on the desired process and the PCB content in the feed mix, the organic content is used either to substitute fossil energy (organic content ≤20 wt. %. of the feed mix) or to run the entire process autothermal (organic content >20 wt. %. of the feed mix). Thus, an off gas containing combusted organics, flame retardants (bromide and chloride compounds), and parts of the zinc as oxide is produced. Currently, processes are able to recover most of the base metals (Cu, Ni, Sn), as well as the precious metals (Au, Ag, Pd, Pt), contained within the input material in a metal phase at the bottom of the furnace. Above this metal phase, a slag phase will form, containing the ceramic compounds of the feed mix, such as aluminum oxide, silicon dioxide, and magnesium oxide, as well as oxidized metals such as iron and aluminum. Unfortunately, there is always some of the base metal alloy, which is entrapped as droplets inside the slag (~1–2 wt. % of the slag). Within these entrapped droplets, some of the precious metals, as well as parts of lead and tin in their metallic state, are lost to the slag. Depending on the process, atmosphere, and slag chemistry, oxidic forms of lead and tin may also occur. The last group of elements, i.e., the scarce elements, are mainly bound in the slag. However, depending on the chemical reactions inside the furnace, halides may also be formed and lost to the off gas [4]. Although the recycling process is quite well developed and plenty of research was done in that field, the distribution of scarce elements is still not fully understood, and an efficient recovery method is yet to be developed.

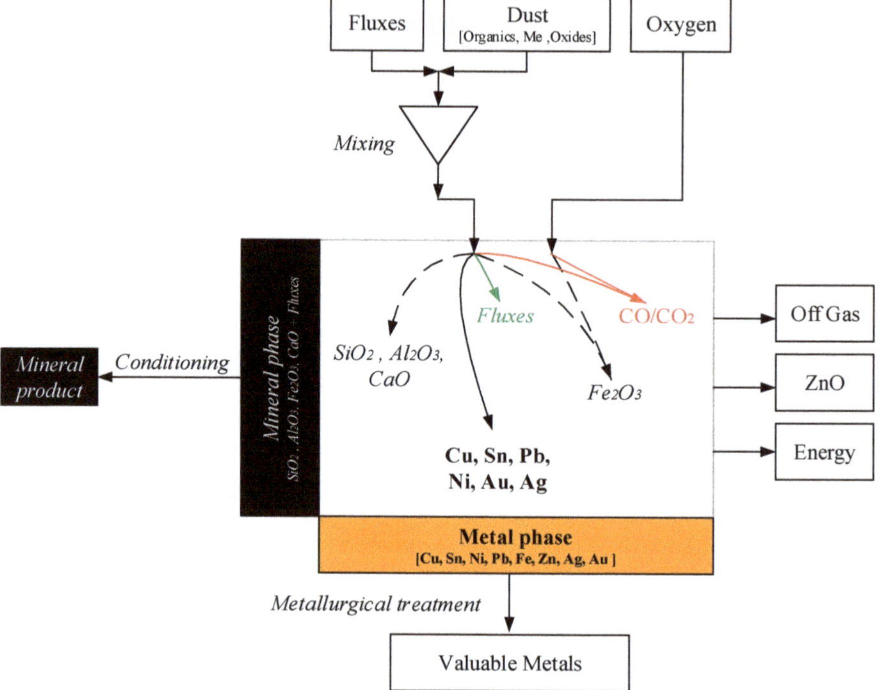

Figure 1. General schematic process of metal recycling from printed circuit board (PCB) scraps [5].

In order to reduce metal losses to the slag, facilitate their recovery, and possibly recover REEs, further investigation must be conducted. In this context, FACTSage™ (Thermafact/CRCT, Montreal, Canada and GTT-Technologies, Aachen, Germany) [6] offers one suitable opportunity to help understand the influence of slag chemistry on the viscosity of the slag, consequently enabling a detailed understanding and manipulation of the settling behavior to reduce these losses. A possible solution, which is investigated within this study, is the targeted concentration of metal parts, as well as scarce elements, in certain phases inside the slag, achieved through controlled cooling rates and slag chemistry. A positive concentration would facilitate further separation, collection, and subsequent recycling of these elements, enabling the recycling of scarce elements to become technically and economically feasible [7,8].

With respect to a sustainable recycling strategy, a detailed understanding of the elemental distribution within the process, the reduction of metal losses to the slag, and the material properties of the product is essential. Therefore, the analysis of the present slag material in terms of formed structures depending on the cooling rate and the enrichment of metals in these structures is of great interest. Two main analysis methods were used in the present study, namely, SEM-based mineralogical analysis (MLA) and X-ray computed tomography (XCT).

SEM-based mineralogy (MLA) is a two-dimensional image analysis method based on scanning electron microscopy coupled with energy-dispersive X-ray spectroscopy [9,10]. It is proven to be a practical method for multidimensional characterization of various phases [11–13]. Using MLA allows for an integrated analysis of shape, size, and distribution of various phases within a specimen which can be a sample block, thin section, or a grain mount. By studying processes such as comminution or physical separation, process parameters can be directly related to product characteristics with respect to particle properties [14–17]. For example, the liberation of valuable phases with respect to grinding fineness, as well as the dominating fracture mechanisms, can be studied [18–20].

XCT enables a non-destructive three-dimensional characterization of specimens. As it is based on the X-ray attenuation of the volume elements (voxel), which is a function of the materials' average atomic number, wavelength, material density, and path length, voxels can have different chemical composition but similar attenuation coefficients [12]. Studying multiphase materials such as ores and slags can, thus, be challenging. Different approaches exist to face this problem. These are direct three-dimensional analysis of the chemical composition using X-ray fluorescence tomography [21,22] or the combination of two-dimensional analysis of the chemical composition with XCT [19,23–25].

The aim of the investigation is to identify if the slag could possibly represent an artificial resource or collecting agent for various valuable metals and rare earth elements. Based on the findings of this study, it is planned to develop an effective and efficient method for the downstream recycling process of the addressed elements. If it is possible to do so, a huge pyro-metallurgical process optimization potential is expected, as this resource can be influenced in various ways and no detailed knowledge is generated so far.

2. Materials and Methods

The slag, which was further investigated in this study, resulted from a previously conducted PCB melting experiment in a TBRC (top-blown rotary converter) of industrial scale (0.5 m^3 melt volume) [5]. Within the experiment, 560 kg of finely ground PCBs were injected into a synthetic slag phase at 1300 °C starting temperature. The main elemental components of this injected material were Cu, Pb, Zn, Al, Ni, and Sn. For more details on the injected material, the reader is referred to Borowski et al. (2018) [5]. The general schematic experimental set-up is shown in Figure 2.

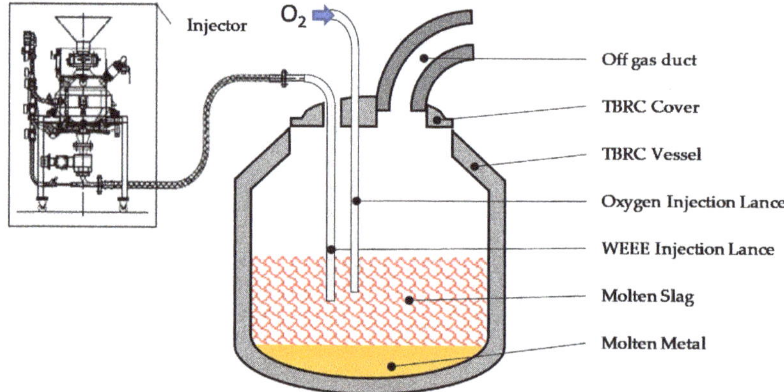

Figure 2. Schematic process of waste of electrical and electronic equipment (WEEE) powder injection into the slag phase.

The experiment was conducted with an initial molten slag bath formed by a composition of SiO_2, Al_2O_3, and CaO, chosen so that the lowest possible liquid temperature was achieved. The powder and fines were injected into the slag phase near the metal slag boundary through the injector (see Figure 3).

Figure 3. Top-blown rotary converter (TBRC) with lances for injecting WEEE powders and oxygen [5].

The mass flow of the injected WEEE powder was adjusted according to the calculated energy requirement for an autothermic melting process between 1350 °C and 1450 °C and was based on the calorific value of the organic content of approximately 30 wt. % in the feed material. After 560 kg of material was continuously injected into the slag phase and 30 min of holding time elapsed, both slag and metal phases were tapped at 1450 °C into a cold (20 °C) cast-iron ladle that was coated with a thin graphite layer (for more detailed information on the smelting process from which the investigated slag originates, see Borowski et al. (2018) [5]). In the ladle (see schematic in Figure 4), the metal was intended to settle at the bottom and to cool relatively quickly while the slag cooled down with different cooling rates from the outside to the center. At the ladle's cold side walls, in a very small zone 1, very high cooling rates of approximately >100 K/s are realized. Here, the slag is intended to form an amorphous, glass-like structure, whereas, in the center of the ladle, in zone 3, a mineral phase with crystalline structures is expected. Here, due to the solidified surrounding slag, the inner core is literally thermal insulated, and it features a very low cooling rate of approximately <1 K/s. In between these two zones, the biggest zone 2, with an "intermediate" cooling rate of approximately 5 K/s (mean cooling rate of the entire zone), is formed.

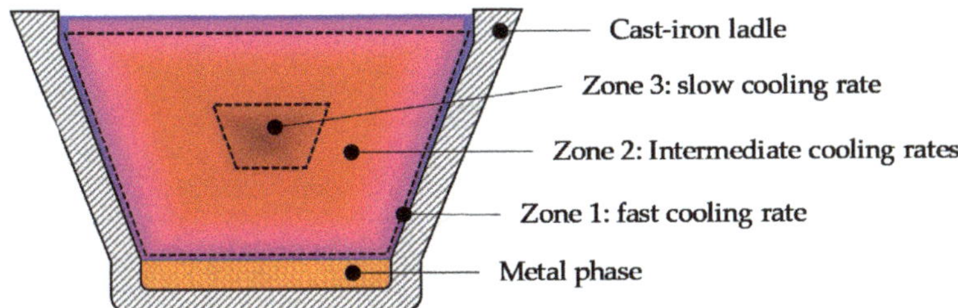

Figure 4. Schematic illustration of the slag ladle with expected cooling zones and cooling rates.

The main difference in metal concentration is expected between the inner (zone 3) and outer phase (zone 1) of maximal and minimal cooling rates. The slags of intermediate cooling rates (zone 2) will probably not differ much in terms of results from the inner phase (zone 3), as the cooling rates are too close to each other.

2.1. X-Ray Computed Tomography

In order to analyze the three-dimensional (3D) structure of the slag, the fragments of samples from the slag ladle (see Figure 4) of zone 1 (H), zone 2 (M), and zone 3 (S) were subsampled by drilling small cores (4 mm in diameter) from them. These cores were then glued on sample holders and scanned using X-ray computed tomography (XCT). The used XCT apparatus was a Zeiss Xradia 510 Versa X-ray microscope (Carl Zeiss AG, Oberkochen, Germany), which combines geometric and optical magnification for increased spatial resolution (Figure 5). The total magnification results from the product of geometric and optical magnification. Detailed information on XCT can be found, e.g., in a review article by Hanna and Ketcham (2017) [26].

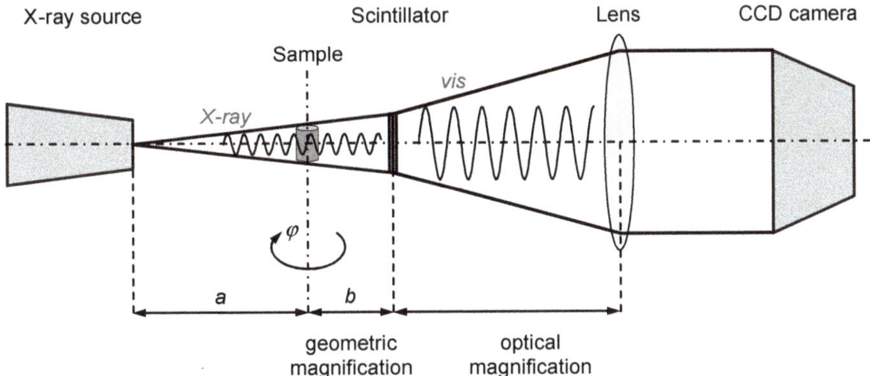

Figure 5. Working principle of an X-ray microscope combining geometric and optical magnification for increased spatial resolution.

XCT scans were done with two sets of parameters to analyze the whole core in a large field of view (FOV, 7.8 μm voxel size) and to additionally get a high-resolution virtual subsample via an interior tomography using a small FOV (2.77 μm voxel size). The latter enabled a detailed view of small structures formed by the complex intergrowth of the different slag phases. The scan parameters are listed in Table 1.

Table 1. Scan parameters from X-ray computed tomography (XCT) analysis.

Scan	Total	Detail
Source position (mm)	−20	−14
Detector position (mm)	160	20
Lens	0.4×	4×
Acceleration voltage (keV)	110	110
Electrical power (W)	10	10
Filter (Zeiss Standard)	HE1	HE1
Camera binning	2	2
Pixel size (µm)	7.8	2.77
Number of projections	801	2001
Scan angle (°)	360	360
Scan time (hh:mm)	1:18	2:23

The projection images were reconstructed using the scanner's proprietary software Scout&Scan Reconstructor (Version 11.1.8043, Carl Zeiss AG, Oberkochen, Germany). The analysis of the final image stack was done with the software VGStudio MAX 3.3 (Volume Graphics GmbH, Heidelberg, Germany).

2.2. Scanning Electron Microscopy-Based Mineralogical Analysis

In preparation for the investigation via MLA, the samples were crushed and ground down to a particle size of <500 µm. Subsequently, the material was screened into four different size fractions. Samples of these size fractions were analyzed separately by MLA. The preparation of the samples is shown in Figure 6.

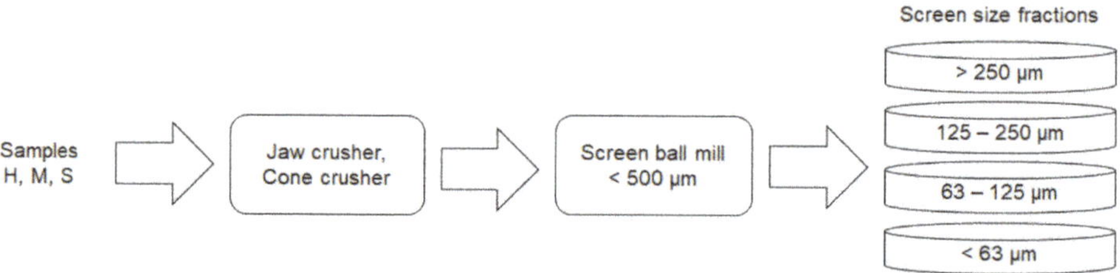

Figure 6. Preparation of samples H, M, and S for analyses via SEM-based image analyses; after a sequence of jaw and cone crushing, the material is milled down to an upper particle size of 500 µm. The material is then screened into four size fractions.

The samples were analyzed by means of a mineral liberation analyzer (MLA). Grain mounts of the particle samples were prepared by mixing 3 g of material with the same volume of graphite and epoxy resin. The grain mounts were polished and subsequently carbon-coated with a Leica (Baltec) MED 020 vacuum evaporator (Leica Microsystems, Wetzlar, Germany). The MLA consisted of an FEI Quanta 650F (Thermo Fisher Scientific, Waltham, MA, USA) field-emission SEM (FE-SEM) equipped with two Bruker Quantax X-Flash 5030 (Bruker, Billerica, MA, USA) energy-dispersive X-ray detectors (EDX). Identification of mineral grains by MLA involves backscattered electron (BSE) image segmentation and collection of EDX spectra of the particles and grains distinguished in GXMAP (grain-based X-ray mapping) mode. In this mode, EDX spectra of each particle are collected in a dense grid and further classified, using a list of mineral spectra collected by the user. The GXMAP measurement mode was applied to all samples [10]. Data processing and evaluation was done with the software package MLA Suite 3.1.4.686. The scan parameters for measurement are listed in Table 2.

Table 2. Scan parameters from mineral liberation analyzer (MLA) analysis. BSE—backscattered electron.

GXMAP Parameters	Total
Voltage (kV)	25
Probe current (nA)	10
Horizontal field width (µm)	1000
BSE calibration	253
Resolution (px)	500×500
Pixel size (µm/px)	2
GXMAP trigger	25
Step size (px)	6
Minimum grain size (px)	4

3. Results

After cooling the tapped slag, the discarded slag showed that the produced metal settled properly (see Figure 7B) at the bottom of the ladle and could, as intended, be easily separated from the slag. The slag, as shown in Figure 7, also showed the expected structures. The outermost slag shell was amorphous and glass-like and looked like a black–green basalt glass. The slag of zone 2 showed no amorphous phases. Here, small metallic inclusions were very easily visible. In the innermost zone 3, after 24 h of cooling (approximately 80 °C), the slag mostly showed the same structure to that in the intermediate zone, but with a more pronounced crystallinity.

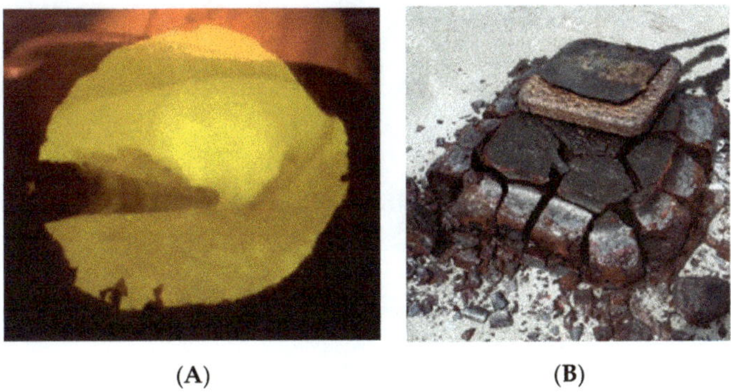

(**A**)   (**B**)

Figure 7. (**A**) Molten slag phase inside the TBRC during the melting process; (**B**) cooled slag and metal block [5].

After cooling and crushing, the slag was subsequently analyzed. Therefore, several samples from different positions inside the ladle were taken, and the mean composition of the slag was determined. Table 3 shows the values for the main components of the slag and most important components in terms of viscosity and cooling rate.

Table 3. Average composition and liquid temperature of the slag.

Slag	SiO_2	Al_2O_3	CaO	Fe_xO_y	CuO	PbO	T_{liq} (°C)
wt. %	36	31.6	19.2	22.0	5.4	0.5	1350–1400

The slag material used in further investigations was individually sampled from different positions of the three cooling zones. One sample (H) originated from zone 1, the outer region of the ladle with a high cooling rate. The two other samples (M and S) were sampled from more central areas of the slag inside the ladle, whereby sample M was taken from zone 2 and sample S was taken from the most central zone 3. All taken samples from the designated cooling zones were analyzed by X-ray computer tomography (XCT) and SEM-based image analysis (MLA).

3.1. Results of Computer Tomography

Results of XCT analysis are given in Figure 8. In the top row of Figure 8, the projection images of the three different samples H, M, and S are shown to give an overview of the structure within the measured volume. The second row shows the cross-sections through the scanned volumes. The red lines visible in the projection images indicate the position of these cross-sections within the volume. The grayscale represents the X-ray attenuation, which is a function of the materials' average atomic number, the wavelength of the X-ray, the material density, and the material thickness. Increased absorption by a region results in a brighter appearance in the image, with pores appearing as dark regions.

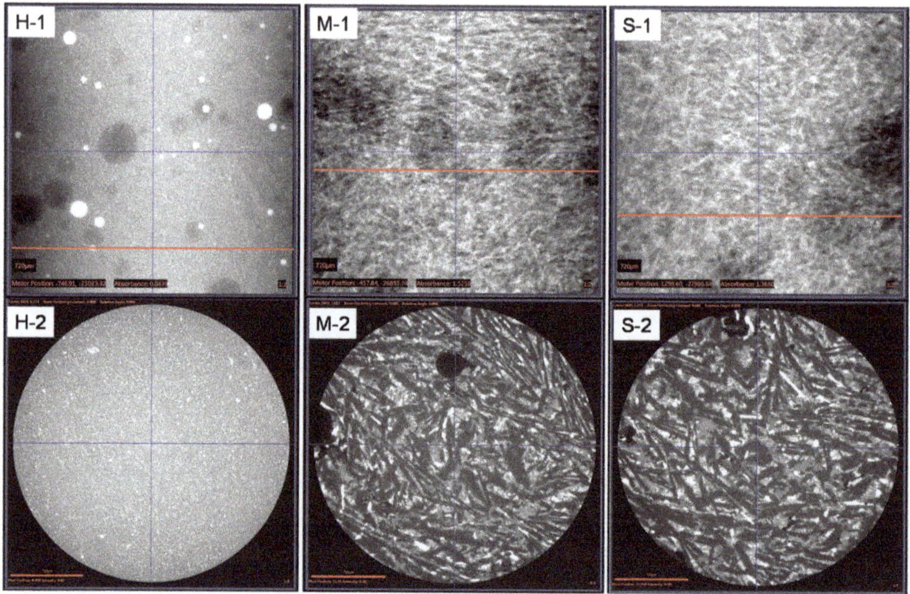

Figure 8. Results from XCT of samples H, M, and S. Top row: projection images of the analyzed volumes giving an overview of the structures and the size scale; bottom row: reconstructed cross-sections from the position of the red line in the projection images.

The morphological analysis of the metal-rich phases (bright regions in the images of Figure 8) showed that they appear as plate-like to needle-like shapes. Their smallest dimension was often below 100 µm. Mechanical liberation by comminution will, therefore, be challenging, and it requires grinding at least down to an upper particle size of roughly the above given number. The visualization of the metal-rich phases shown in Figure 9 gives an idea of their morphology.

Comparing the projection image of sample H and a slice through the reconstructed volume shown in the first row of Figure 8 demonstrates a problem related to two-dimensional analysis. Pores and spherical strong absorbing phases are clearly visible in the projection image but lacking in the chosen section. Therefore, the strong absorbing phases were analyzed in order to measure their size distribution and volume percentage. This helped to estimate whether or not results from subsequent two-dimensional analysis may be misleading. Figure 10 shows a rendering from the inclusion analysis. Spherical inclusions up to approximately 700 µm in diameter were found. These inclusions only accounted for ~0.4% of the sample volume and can, therefore, be neglected with respect to the processing of the slag.

Figure 9. Visualization of the morphology of the metal-rich phases of sample S.

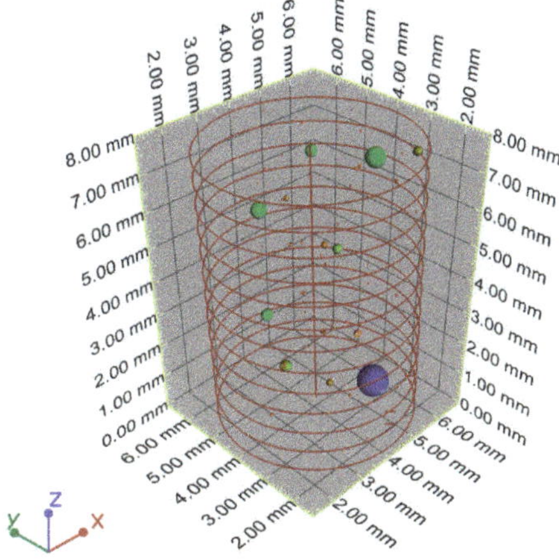

Figure 10. Rendering of spherical metallic inclusions contained in sample H.

3.2. Results of SEM-Based Mineralogical Analysis

In addition to the analyses via XCT, the material was analyzed with MLA according to Section 2.2. Based on these MLA observations, it was shown that the particles in sample H had a different habitus than the particles of the samples M and S, which is in good agreement with the results of the XCT analysis. The Cu-rich phases displayed a variety of shapes, i.e., forming melting droplets and drop aggregates, veins, or dendritic crystals. Those various Cu-containing phases were present in all samples and had to be compared based on concentration, content of Cu, structure, and recoverability by various methods. The main reason for the previous grinding of the sample to <500 μm is explainable by the anisotropic structure of the investigated sample. Another reason is the available size for a typical grain mount, where bigger particles would cause an insufficient representativity and, at the same time, would lead to a disproportional measurement effort. Furthermore, it would help get first insight into the liberation behavior of the valuable material.

An overview of the particles analyzed via MLA of the different samples is shown in Figure 11.

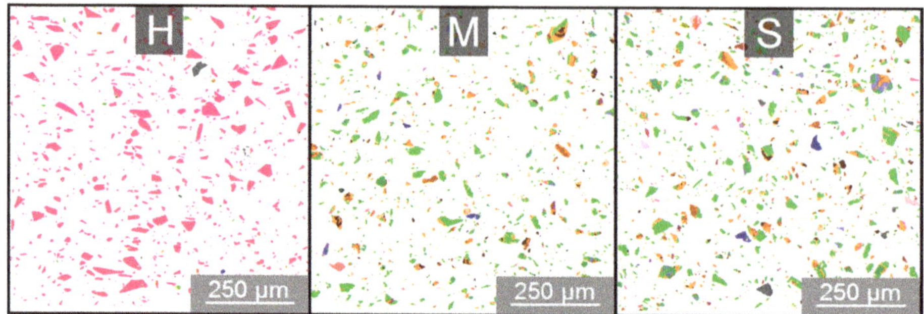

Figure 11. Overview of the three slag samples H, M, and S: MLA false color images display the difference between sample H and samples M and S.

From the false color images in Figure 11, it can be seen that the elemental content and structure of the particles of sample H were significantly different from samples M and S. The samples M and S can be considered as similar regarding particle composition and properties.

The mass recovery of the size fractions gives an overview of the different breakage behavior for the various slag samples. As displayed in Table 4, the mass recoveries for H, M, and S showed significant differences. The fine fraction <63 μm increased from H to M and to S. The coarsest fraction >250 μm showed the inverse trend.

Table 4. Mass recovery of the different size fractions of the samples H, M and S.

Size Fraction (μm)	H (%)	M (%)	S (%)
>250	31.9	30.0	15.2
125–250	30.3	22.5	23.2
63–125	18.1	18.7	22.8
<63	19.8	28.7	38.8

The breakage characteristics of the slag seemed to change between the outer and the inner region. More and more fines appeared below 63 μm. One explanation could be the finer intergrown structure of the samples from the inner regions, compared to the mainly homogeneous structure of sample H. Furthermore, the decreasing brittle character of the inner sample led to more abrasion. Nevertheless, the results showed the effect on the breakage characteristics of the cooling rate, which could, therefore, be adjusted to a certain desired behavior.

As Cu was the main valuable metal with significant content within the investigated slag samples, various aspects regarding Cu are discussed in the following sections. It has to be stated that Cu was present in two different forms in the slag material: Cu-bearing slag phase (Cu-rich) and pure metal (Cu-metal). This statement is evaluated in more detail in the following section with various representations focusing on this aspect. The total content of Cu in the three different samples is shown in Table 5. The Cu content is shown for the different size fractions, and an average value is indicated.

Table 5 illustrates that M and S showed a much higher Cu content than H. The Cu content of H was in the range of 3 wt. %, whereas M and S showed approximately 18 wt. % Cu. Furthermore, there was no enrichment of Cu in certain size fractions, resulting in a particle size-independent content. Additionally, the Cu present as pure metal droplets is shown in Table 6. The portion of Cu present as metal represents the amount of Cu that is recyclable in a certain way. The rest of Cu that is present as a slag phase is very likely lost or difficult to gain for a further recycling process. Fortunately, the content of Cu-metal increased from H over M to S. Sample S showed the highest content of Cu-metal with again no enrichment in a specific size class being visible according to Table 6.

Table 5. Total Cu content of the different size fractions and average value of samples H, M, and S.

Size Fraction (µm)	H (%)	M (%)	S (%)
>250	2.9	17.9	18.1
125–250	2.9	18.7	18.3
63–125	3.1	18.7	18.4
<63	3.5	18.4	17.7
Average	3.1	18.4	18.1

Table 6. Cu-metal content of the different size fractions and average value of the samples H, M, and S.

Size Fraction (µm)	H (%)	M (%)	S (%)
>250	0.4	7.2	8.7
125–250	0.4	7.8	8.7
63–125	0.5	8.0	8.6
<63	0.6	7.7	8.4
Average	0.4	7.7	8.5

Another aspect of interest concerning a potential recycling process for Cu is the size of these metal droplets within the sample. Therefore, their size distribution is shown in Figure 12. The size distributions of M and S were in the same range, while H showed a much finer size distribution. Nevertheless, all three size distributions were in a rather fine range for a potential mechanical separation process. Furthermore, it needs to be stressed that the required liberation size is around 1/10 of the grain size assuming random fracture [27]. Slags are known to be close to a pure random fracture due to their brittle behavior caused by amorphous phases and rough interfaces [18]. Upgrading Cu-rich phases by mechanical separation processes will, thus, be challenging with respect to the required liberation size.

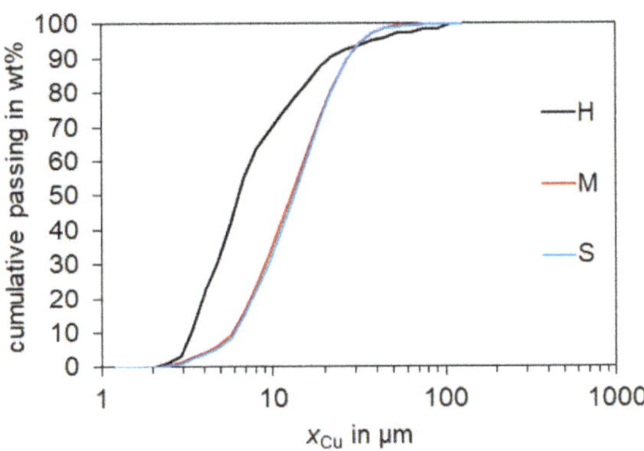

Figure 12. Cu-metal grain size distribution of the three different sampling areas.

A further important characteristic is the shape of the Cu-metal droplets within the different slag structures. From the results of XCT analyses of Section 3.1, it was already concluded that the structure of sample H was different from that of samples M and S. For further investigations, representative examples of the particles analyzed by MLA for sample H (see Figure 13) and for samples M and S (Figure 14) are analyzed in more detail. The shape of the Cu grains in sample H showed a circular shape. In connection with 3D analysis via XCT, it can be concluded that these Cu-metals in sample H were spheres formed by droplets. These spheres (brown color in Figure 13) showed no connection to the surrounding amorphous phase. The amorphous phase in sample H shown in Figure 13 is represented by the pink color, and it consisted mainly of the elements Ca (~12%), Al (~15%), Fe (~2%), Cu (~2%), Si (~27%), and O (~40%).

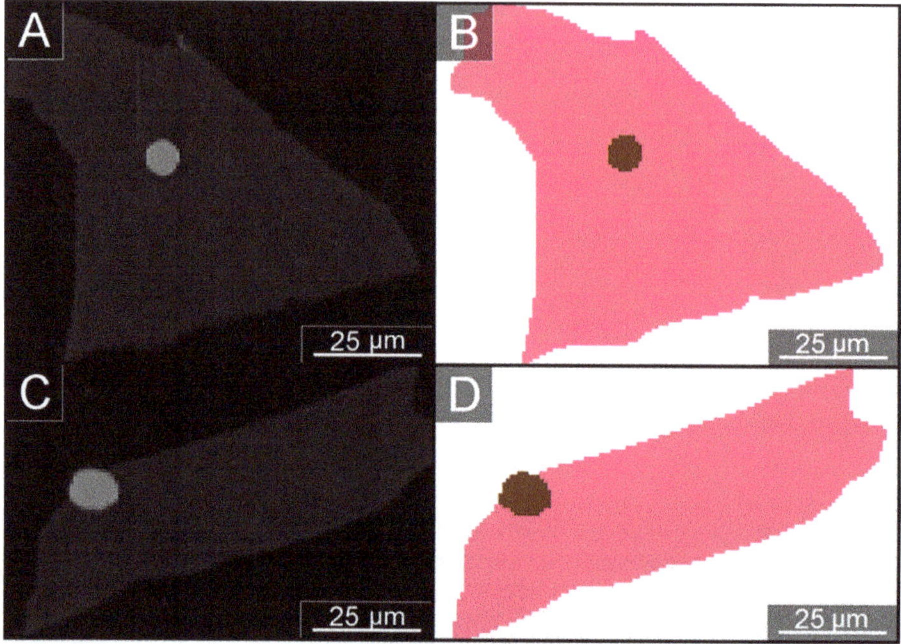

Figure 13. Typical particles with copper inclusion of sample H: (left, **A** and **C**) grayscales from BSE images; (right, **B** and **D**) associated false color images indicating different phases: pink (amorphous Al slag phase), brown (Cu-metal).

These metal droplets in sample H originated from the turbulent mixing during the smelting process. The metal droplets were transported to the slag phase. The cooling rate in this range was high enough to inhibit settling of the Cu droplets back to the metal phase at the bottom of the furnace. This theory is further confirmed as the Cu droplets showed a sharp boundary to the surrounding slag phases of sample H. A very different observation can be made in samples M and S displayed in Figure 14. In these samples, particles containing Cu comprised Cu-metal droplet aggregates in association with Cu-bearing slag phases.

There are two main areas visible in the false color image of Figure 14. Firstly, there is a more or less homogeneous, needle-like green region (the dark phase in Figure 8 and on the left side of Figure 14 A,C). The second main area is more heterogeneous, marked by brownish colors showing a variety of different phases. The Cu-rich phase consisted of a matrix phase consisting of phases b and c, a dendritic structure (phase d), and the Cu-metal visible as round droplets (see phases in Table 7). The main elemental contents of these dominant phases are shown Table 7. The Cu-metal phase (phase e in Table 7) always showed an agglomeration of spherical droplets within the Cu-rich matrix phases (b and c). The phase d indicated a dendritic structure with a quite high Cu content of approximately 49%. In the whole sample (M and S), the Al-rich (a in Table 7) phase summed up to approximately 50%, the Cu-rich phase summed up to approximately 35% (sum of phases b–d), and the Cu-metal represented 9%.

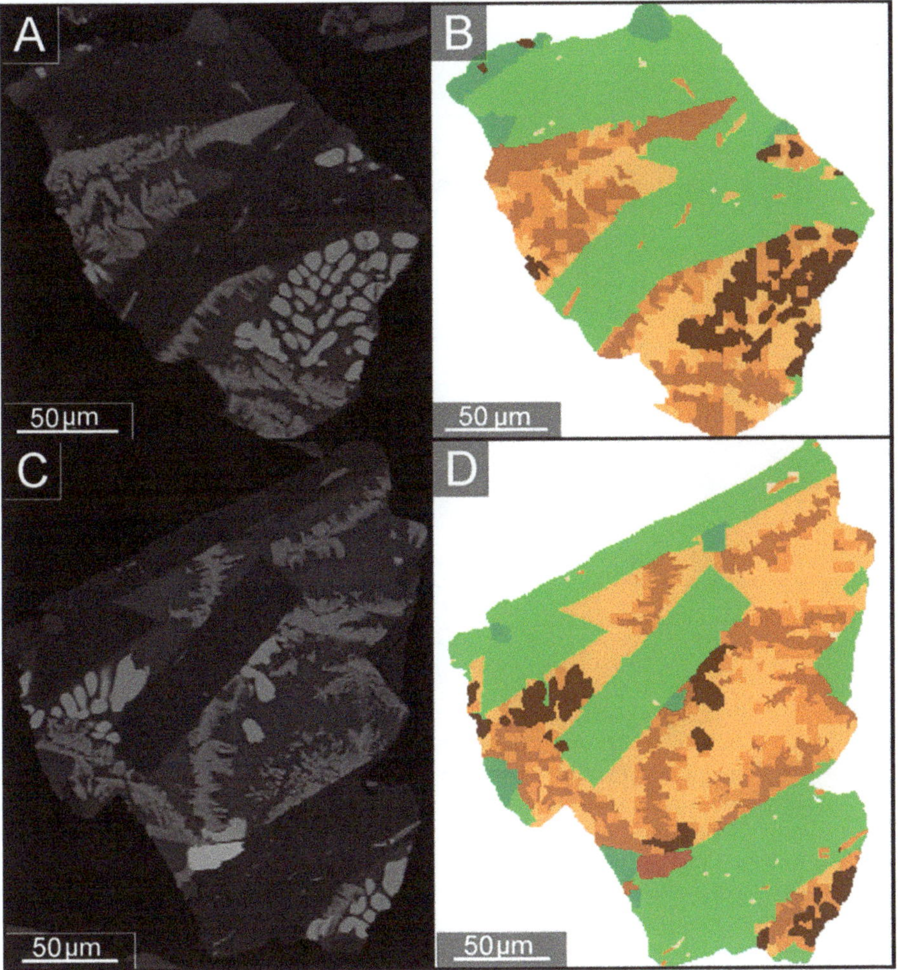

Figure 14. Typical particles with different phases of the samples M and S: (left, **A** and **C**) grayscales from BSE images; (right, **B** and **D**) associated false color images indicating different phases: green (Al-rich slag phase), brownish colors (Cu-rich phases) (see Table 7).

Table 7. Main elemental content of the dominant phases of samples M and S.

Phase (see Figure 14)		Cu (%)	Fe (%)	Al (%)	Ca (%)	Pb (%)	Si (%)	O (%)
a		<1	<1	24	15	0	20	39
b		15	<1	11	4	3	23	37
c		17	4	12	3	3	23	34
d		49	19	12	0	0	0	16
e		100	-	-	-	-	-	-

The Cu-metal droplets always appeared within the Cu-rich matrix (b and c). This fact supports the hypothesis of the occurrence of a transportation process in connection with a formation of Cu-rich dendrites (phase d) and an agglomeration of Cu droplets (phase e). These processes are in fact a function of the cooling rate. In conjunction with the increasing proportion of Cu-metal droplets with more distance to the outer region of the slag (see Table 6), the cooling rate seemed to influence the appearance of the Cu-metal droplets and the characteristics of the surrounding phases.

The deportment of the element Cu, which is the major valuable element in the present material, into the different phases of the investigated slag structures is an important factor for the assessment of its recyclability. Figure 15 shows the deportment of Cu into the different main phases of the three samples H, M, and S.

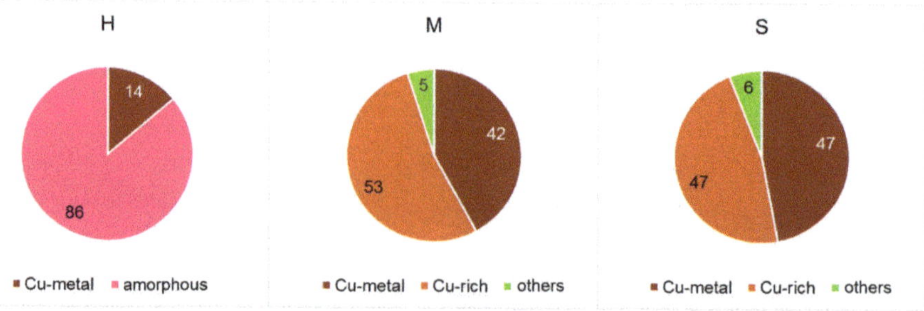

Figure 15. Deportment of Cu as a percentage in the three different samples (H, M, and S) between the different main phases: Cu-metal, Cu-rich, and amorphous.

For sample H, the majority of Cu was found in the amorphous phase. Only 14% of the total Cu content within this sample appeared as the metallic phase. This ratio underlines the low potential for the recycling of the material of this outer zone (H). In contrast to that, the ratios for sample M and S were much more promising with regard to potential recycling, e.g., via mechanical separation processes. For these samples, the amount of Cu present as the metallic phase was much higher (42% and 47%, respectively). The Cu-rich phase in Figure 15 summarizes the phases b, c, and d from Table 7.

For a potential mechanical processing of the material to enrich the Cu-metal phase via separation processes, the liberation of this phase is of major interest. Therefore, in Figure 16, the particle-based information from MLA is binned into property classes of particle size (x_P in µm) and liberation of the Cu-metal phase (L in vol. %, indicating the proportion of Cu-metal in a particle). In the individual bins of x_P and L, the theoretical recovery of Cu-metal phase ($r_{Cu\text{-metal}}$) is indicated by number and the color scale. The focus of these considerations is only on samples M and S, due to the low potential of sample H for further recycling steps.

The recovery values ($r_{Cu\text{-metal}}$) of the Cu-metal phase in Figure 16 summed up to 100% for each sample. According to Figure 16, the distribution in the different x_P and L classes of the Cu-metal phase was more or less the same for samples M and S with only minor differences. The majority of the Cu-metal was in the range of 40 µm to 320 µm, showing a rather low liberation ($\leq$50%). This material needs to be further liberated by additional comminution steps. These comminution steps have to be controlled (e.g., via combination with a suitable classification process) to avoid an overgrinding of the Cu-metal phase, not to increase the amount in the lower particle size range. Already more than 20% of the Cu-metal is in a range below 40 µm. Despite some fractions showing a satisfying liberation ($L > 70\%$), this relatively low size range is already challenging for conventional mechanical separation processes (e.g., density separation via shaking tables).

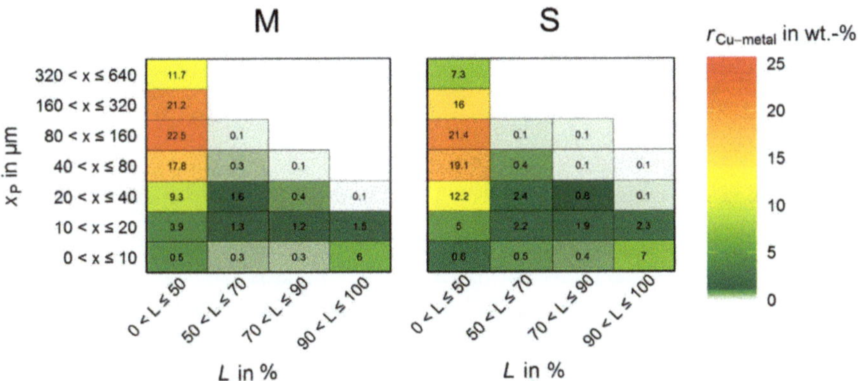

Figure 16. Theoretical recovery ($r_{\text{Cu-metal}}$) of the Cu-metal phase into size (x_P) and liberation (L) classes for the two main products M and S.

3.3. Other Elements

The above evaluations focused on the element Cu, due to its relatively high content. There are other interesting elements present in the investigated material. Table 8 shows the averaged content of some other interesting elements in samples M and S.

Table 8. Averaged content of various elements in samples M and S measured by MLA.

Content	Al	Fe	Cr	Sn	Pb	Zn	Ni
%	17	5	2	2	1	1	0.5

The appearance and deportment of the elements of Table 8 are decisive for the potential recycling of these materials. For samples M and S, the majority of these elements were distributed between the Cu-rich phases (mainly phases b, c, and d of Table 7) and the Al-rich phases (mainly phase a of Table 7). Some elements were present in other phases with significant amounts. Table 9 gives an overview of the deportment of these elements within samples M and S.

Table 9. Deportment of various elements between the Cu-rich, Al-rich, and other phases in samples M and S.

Element	Cu-Rich (%)	Al-Rich (%)	Other Phases (%)
Al	29	70	1 (present as Al metal phase)
Fe	59	21	20 (present as iron oxide)
Cr	43	50	7 (distributed between various minor phases)
Sn	30	-	70 (present as SnO_2)
Pb	99	-	1 (distributed between various minor phases)
Zn	21	62	17 (present as ZnO)
Ni	20	78	2 (in minor phases)

Obviously, the majority of Al was found in the Al-rich phases; nevertheless, 29% was found in the Cu-rich phases, and 1% was present as Al metal inclusions. For Fe, the main amount was found in the Cu-rich phases, whereas only 21% was in the Al-rich phases, and 20% was present as a more or less pure oxide phase. The situation for Sn was totally different, as no Sn was found in the Al-rich phase, while 30% was present in the Cu-rich phases, and the majority appeared as isolated SnO_2 in the sample. Nearly all Pb was in the Cu-rich phases. For Zn and Ni, the majority was found in the Al-rich phases. In contrast to Cu, the other elements had no significant appearance in a pure metallic form, while some minor metallic occurrences were found only for Al (1%). This analysis of the deportment of various elements gives a first overview of the different behavior of these elements. A more detailed evaluation of these elements and other trace elements requires the application of additional analysis methods [28].

4. Discussion

The slag structure is an important factor for the potential recycling of various valuable metal contents. By adjustment of the process parameters, the structure can be tailor-made for optimizing the recovery of different valuable materials [29].

Firstly, it has to be stated that the results from XCT and MLA are in good agreement. In both cases, clear differences between sample H on the one hand and sample M and S on the other hand were visible. There were two main occurrences of Cu-metal within the slag structure. The first represents the presence of single Cu droplets within the amorphous matrix in the outer region of the TBRC. These droplets appeared detached from the surrounding amorphous slag structure (see Figure 13). In this region, the cooling rate was rather fast, whereby the droplets could not sink back into the metal phase and were entrapped within the glass-like slag structure. This area is of less importance, as this zone 1 (see Figure 4) was determined to sum up to <1 wt. % of the total mass in the process. Nevertheless, if possible, a further reduction of this zone could be beneficial.

The main slag material showed a structure as identified in samples M and S. Although these two samples originated from different zones of the furnace (zones 2 and 3, see Figure 4), no significant differences between their phase structures were identified. There were two main patterns present in these samples: an Al-rich slag phase and a Cu-rich phase. These main areas showed significant differences regarding their structure and elemental contents. With respect to the solidification order of these main phases, the following assumptions can be made: the Al-rich phase (a) represents the material that solidifies first, followed by the Cu-metal phase. The metal droplets then accumulate as loos agglomerates within the Cu-rich matrix (b and c). Meanwhile the Cu-rich dendritic structures are formed, and phase b and c finally solidify.

The distribution of various valuable metals (e.g., Au, Ag, Ge, In) between the different phases of the slag, their deportment, and in which phases they accumulate have to be investigated further [30–32]. Special focus should also be on the effect of the applied slag system [33] and the adjustment of the components to improve the breakage properties with respect to a downstream comminution process [18,34].

By adjusting the cooling rate, the occurrence of the described Cu droplets has to be minimized. The slag should be kept in a certain temperature window to allow these droplets to settle back onto the metal face at the bottom of the furnace. As these isolated droplets predominantly appear in the outer regions of the furnace and the Cu content is already lower compared to other regions, the handling and adjustment of this structure is of less importance. By lowering the cooling rate at the same time, the Cu structure in the inner regions of the slag can also be altered. The target is a better recyclability of the Cu contents of these materials. A detailed understanding of the occurrence of the main phases and their formation process is necessary for a better understanding of potential parameters to optimize the processability. The structures have to be adjusted to the requirements of the following processes. A potential downstream grinding and separation cascade possibly implies increasing the size of the Cu-metal agglomerates.

In the end, there are two main aspects to improve the recyclability of potential valuable metals within the slag structure. The first point, which is only indirectly linked with the improvement of recyclability, is to reduce these metal contents in the slag by allowing the metal droplets to sink back into the metal phase on the bottom of the furnace. The second point is the adjustment of the slag structure to improve the performance of potential downstream processes [35]. Both require a detailed understanding of the process–parameter–slag–structure relationship and the proposed investigation procedure is able to reveal and quantify the critical effects. Additionally, other observation techniques should complement the presented approach to get the full picture [36]. Finally, it has to be stated that the found structures within the slag analyzed via XCT and MLA strongly depend on the applied reactor technology and the resulting flow patterns. These dependencies have to be investigated further.

5. Conclusions

The applied analysis methods enabled the characterization of the present material structures of the investigated slag samples. Through the synergetic effect of the two imaging techniques (SEM-based image analysis and X-ray CT), it was possible to identify potential valuable phases and assess the recyclability of these contents. The application of these imaging techniques improves the understanding of the size, shape, and association of the present phases. The Cu-metal contents within the slag samples represent the main valuable component. For a sustainable recycling of these contents, the slag structure has to be optimized and adjusted to the requirements of following processes with the help of various process parameters (e.g., cooling rate). The detailed characterization of the slag structure by imaging processes in combination with fluid flow modeling approaches via computational fluid dynamics represents the basis for an adjusted process optimization and further improvement of recyclability. The main conclusions from the present study in short are as follows:

- The outer amorphous layer of the slag is not suitable for recycling of the metal contents and needs to be minimized.
- The phase structures (e.g., size and shape) showed no differences for sample M (zone 2) and S (zone 3).
- The actual size range of the Cu metal inclusions is too small for mechanical separation processes. The grain size needs to be increased to improve the recyclability; otherwise, these contents need to be lowered in the slag.
- Other elements, e.g., Fe, Sn, Ni, Zn, Pb, and Cr, were not found to be present as pure metals in the slag; rather, they were in other oxide or silicate phases.
- An interdisciplinary approach of different fields (i.e., metallurgy, mineral processing, and analytics) is necessary to improve the recyclability of such complex slag materials.

Author Contributions: M.B. conceptualized the paper. N.B. was responsible for the metallurgical tests. T.H. was responsible for sample analysis with a mineral liberation analyzer (MLA) including quality control, data processing, and data evaluation. T.L. was responsible for sample analysis with X-ray computed tomography (XCT) including quality control, data processing, and data evaluation. M.B. and T.L. were responsible for the grinding tests of the material. M.B., N.B., T.L., and T.H. analyzed, interpreted, and collated all results. M.A.R., B.F., and U.A.P. interpreted the results, supervised the work, and supported the preparation of the manuscript. All authors have read and agreed to the published version of the manuscript.

Funding: The authors would like to thank the German Research Foundation (DFG) for funding of the X-ray microscope (Zeiss Xradia 510 Versa) INST 267/129-1.

References

1. Worrel, E.; Reuter, M.A. *Handbook of Recycling*; Elsevier: Waltham, MA, USA, 2014; p. 595.
2. Hageluken, C. Improving metal returns and eco-efficiency in electronics recycling—A holistic approach for interface optimisation between pre-processing and integrated metals smelting and refining. In Proceedings of the 2006 IEEE International Symposium on Electronics and the Environment, Scottsdale, AZ, USA, 8–11 May 2006. [CrossRef]
3. Van Schaik, A.; Reuter, M.A. Chapter 22: Material-Centric (Aluminium and Copper) and Product-Centric (Cars, WEEE, TV, Lamps, Batteries, Catalysts) Recycling and DfR Rules. In *Handbook of Recycling*; Worrel, E., Reuter, M.A., Eds.; Elsevier: Waltham, MA, USA, 2014; pp. 307–378.
4. Diaz, F.; Flerus, B.; Nagraj, S.; Bokelmann, K.; Stauber, R.; Friedrich, B. Comparative analysis about degradation mechanisms of printed circuit boards (PCBs) in slow and fast pyrolysis: The influence of heating speed. *J. Sustain. Metall.* **2018**, *4*, 205–221. [CrossRef]

5. Borowski, N.; Trentmann, A.; Brinkmann, F.; Stürtz, M.; Friedrich, B. Metallurgical effects of introducing powdered WEEE to a molten slag bath. *J. Sustain. Metall.* **2018**, *4*, 233–250. [CrossRef]

6. Bale, C.W.; Bélisle, E.; Chartrand, P.; Decterov, S.A.; Eriksson, G.; Gheribi, A.E.; Hack, K.; Jung, I.H.; Kang, Y.B.; Melançon, J.; et al. FactSage Thermochemical software and databases, 2010–2016. *Calphad* **2002**, *26*, 189–228. [CrossRef]

7. Obiso, D.; Kriebitzsch, S.; Reuter, M.; Meyer, B. The importance of viscous and interfacial forces in the hydrodynamics of the top-submerged-lance furnace. *Metall. Mater. Trans. B* **2019**, *50*, 2403–2420. [CrossRef]

8. Akashi, M.; Keplinger, O.; Shevchenko, N.; Anders, S.; Reuter, M.A.; Eckert, S. X-ray radioscopic visualization of bubbly flows injected through a top submerged lance into a liquid metal. *Metall. Mater. Trans. B* **2019**, *51*, 124–139. [CrossRef]

9. Sutherland, D.N.; Gottlieb, P. Application of automated quantitative mineralogy in mineral processing. *Miner. Eng.* **1991**, *4*, 753–762. [CrossRef]

10. Fandrich, R.; Gu, Y.; Burrows, D.; Moeller, K. Modern SEM-based mineral liberation analysis. *Int. J. Miner. Process.* **2007**, *84*, 310–320. [CrossRef]

11. Schach, E.; Buchmann, M.; Tolosana-Delgado, R.; Leißner, T.; Kern, M.; van den Boogaart, G.; Rudolph, M.; Peuker, U. Multidimensional characterization of separation processes—part 1: Introducing kernel methods and entropy in the context of mineral processing using SEM-based image analysis. *Miner. Eng.* **2019**, *137*, 78–86. [CrossRef]

12. Buchmann, M.; Schach, E.; Tolosana-Delgado, R.; Leißner, T.; Astoveza, J.; Kern, M.; Möckel, R.; Ebert, D.; Rudolph, M.; van den Boogaart, G.; et al. Evaluation of magnetic separation efficiency on a cassiterite-bearing skarn ore by means of integrative SEM-based image and XRF–XRD data analysis. *Minerals* **2018**, *8*, 390. [CrossRef]

13. Buchmann, M.; Schach, E.; Leißner, T.; Kern, M.; Mütze, T.; Rudolph, M.; Peuker, U.A.; Tolosana-Delgado, R. Multidimensional characterization of separation processes—part 2: Comparability of separation efficiency. *Miner. Eng.* **2020**, *150*, 106284. [CrossRef]

14. Sousa, R.; Simons, B.; Bru, K.; de Sousa, A.B.; Rollinson, G.; Andersen, J.; Martin, M.; Machado Leite, M. Use of mineral liberation quantitative data to assess separation efficiency in mineral processing—Some case studies. *Miner. Eng.* **2018**, *127*, 134–142. [CrossRef]

15. Charikinya, E.; Robertson, J.; Platts, A.; Becker, M.; Lamberg, P.; Bradshaw, D. Integration of mineralogical attributes in evaluating sustainability indicators of a magnetic separator. *Miner. Eng.* **2017**, *107*, 53–62. [CrossRef]

16. Evans, C.L.; Wightman, E.M.; Manlapig, E.V.; Coulter, B.L. Application of process mineralogy as a tool in sustainable processing. *Miner. Eng.* **2011**, *24*, 1242–1248. [CrossRef]

17. Lastra, R. Seven practical application cases of liberation analysis. *Int. J. Miner. Process.* **2007**, *84*, 337–347. [CrossRef]

18. Leißner, T.; Hoang, D.H.; Rudolph, M.; Heinig, T.; Bachmann, K.; Gutzmer, J.; Schubert, H.; Peuker, U.A. A mineral liberation study of grain boundary fracture based on measurements of the surface exposure after milling. *Int. J. Miner. Process.* **2016**, *156*, 3–13. [CrossRef]

19. Reyes, F.; Lin, Q.; Cilliers, J.J.; Neethling, S.J. Quantifying mineral liberation by particle grade and surface exposure using X-ray microCT. *Miner. Eng.* **2018**, *125*, 75–82. [CrossRef]

20. Little, L.; Mainza, A.N.; Becker, M.; Wiese, J.G. Using mineralogical and particle shape analysis to investigate enhanced mineral liberation through phase boundary fracture. *Powder Technol.* **2016**, *301*, 794–804. [CrossRef]

21. Egan, C.K.; Jacques, S.D.; Wilson, M.D.; Veale, M.C.; Seller, P.; Beale, A.M.; Pattrick, R.A.; Withers, P.J.; Cernik, R.J. 3D chemical imaging in the laboratory by hyperspectral X-ray computed tomography. *Sci. Rep.* **2015**, *5*, 15979. [CrossRef]

22. Jacques, S.D.; Egan, C.K.; Wilson, M.D.; Veale, M.C.; Seller, P.; Cernik, R.J. A laboratory system for element specific hyperspectral X-ray imaging. *Analyst* **2013**, *138*, 755–759. [CrossRef]

23. Furat, O.; Leissner, T.; Ditscherlein, R.; Sedivy, O.; Weber, M.; Bachmann, K.; Gutzmer, J.; Peuker, U.; Schmidt, V. Description of ore particles from X-ray microtomography (XMT) images, supported by scanning electron microscope (SEM)-based image analysis. *Microsc. Microanal.* **2018**, *24*, 461–470. [CrossRef]

24. Furat, O.; Leissner, T.; Bachmann, K.; Gutzmer, J.; Peuker, U.; Schmidt, V. Stochastic modeling of multidimensional particle properties using parametric copulas. *Microsc. Microanal.* **2019**, *25*, 720–734. [CrossRef]

25. Reyes, F.; Lin, Q.; Udoudo, O.; Dodds, C.; Lee, P.D.; Neethling, S.J. Calibrated X-ray micro-tomography for mineral ore quantification. *Miner. Eng.* **2017**, *110*, 122–130. [CrossRef]

26. Hanna, R.D.; Ketcham, R.A. X-ray computed tomography of planetary materials: A primer and review of recent studies. *Geochemistry* **2017**, *77*, 547–572. [CrossRef]

27. Ferrara, G.; Preti, U.; Meloy, T.P. Inclusion shape, mineral texture and liberation. *Int. J. Miner. Process.* **1989**, *27*, 295–308. [CrossRef]

28. Cook, N.; Ciobanu, C.; Ehrig, K.; Slattery, A.; Verdugo-Ihl, M.; Courtney-Davies, L.; Gao, W. Advances and opportunities in ore mineralogy. *Minerals* **2017**, *7*, 233. [CrossRef]

29. Lennartsson, A.; Engström, F.; Samuelsson, C.; Björkman, B.; Pettersson, J. Large-Scale WEEE recycling integrated in an ore-based Cu-extraction system. *J. Sustain. Metall.* **2018**, *4*, 222–232. [CrossRef]

30. Anindya, A.; Swinbourne, D.R.; Reuter, M.A.; Matusewicz, R.W. Distribution of elements between copper and FeOx–CaO–SiO$_2$slags during pyrometallurgical processing of WEEE. *Miner. Process. Extr. Metall.* **2013**, *122*, 165–173. [CrossRef]

31. Anindya, A.; Swinbourne, D.R.; Reuter, M.A.; Matusewicz, R.W. Distribution of elements between copper and FeOx–CaO–SiO$_2$slags during pyroprocessing of WEEE: Part 2—Indium. *Miner. Process. Extr. Metall.* **2013**, *123*, 43–52. [CrossRef]

32. Klemettinen, L.; Avarmaa, K.; O'Brien, H.; Taskinen, P.; Jokilaakso, A. Behavior of tin and antimony in secondary copper smelting process. *Minerals* **2019**, *9*, 39. [CrossRef]

33. Avarmaa, K.; Yliaho, S.; Taskinen, P. Recoveries of rare elements Ga, Ge, in and Sn from waste electric and electronic equipment through secondary copper smelting. *Waste Manag.* **2018**, *71*, 400–410. [CrossRef]

34. Chris Pistorius, P.; Kotzé, H. Role of silicate phases during comminution of titania slag. *Miner. Eng.* **2009**, *22*, 182–189. [CrossRef]

35. Sarrafi, A.; Rahmati, B.; Hassani, H.R.; Shirazi, H.H.A. Recovery of copper from reverberatory furnace slag by flotation. *Miner. Eng.* **2004**, *17*, 457–459. [CrossRef]

36. Thyse, E.L.; Akdogan, G.; Olivier, E.J.; O'Connell, J.H.; Neethling, J.H.; Taskinen, P.; Eksteen, J.J. 3D insights into nickel converter matte phases: Direct observations via TEM and FIB SEM tomography. *Miner. Eng.* **2013**, *52*, 2–7. [CrossRef]

Article

An Improved Evaluation Strategy for Ash Analysis Using Scanning Electron Microscope Automated Mineralogy

Andrea C. Guhl [1], Valentin-G. Greb [1], Bernhard Schulz [2] and Martin Bertau [1,*]

[1] TU Bergakademie Freiberg, Institute of Chemical Technology, Leipziger Straße 29, 09599 Freiberg, Germany; andrea.guhl@chemie.tu-freiberg.de (A.C.G.); valentin.greb@alba.info (V.-G.G.)

[2] TU Bergakademie Freiberg, Institute of Mineralogy, Brennhausgasse 14, 09599 Freiberg, Germany; bernhard.schulz@mineral.tu-freiberg.de

* Correspondence: martin.bertau@chemie.tu-freiberg.de

Received: 31 March 2020; Accepted: 20 May 2020; Published: 25 May 2020

Abstract: Sewage slush ashes are materials composed of polyphase particles. Ashes are fine-grained with many amorphous components, and analytical techniques such as X-ray diffractometry cannot recover all the properties. For sewage sludge ash, treatment often focuses on phosphate recovery. Automated mineralogy techniques were applied in order to study phosphate associations and their behavior towards chemical processes. This work shows the distribution of phosphate content in sewage sludge ash and identifies the main recovered phosphate phases in acid leaching. Data interpretation was focused on the target material, phosphate. The approach documents spectra labelling with respect to one target component, phosphorus. This is a tool for assessing sewage sludge ashes towards their phosphate recovery potential and highlights issues processing needs to address.

Keywords: sewage sludge ashes (SSA); phosphate; recycling; recovery; SEM-automated mineralogy; mineral liberation analysis (MLA)

---

1. Introduction

In 2014, the European Commission published a document on critical raw materials, including phosphate [1]. In conjunction with the circular economy initiative of the European Commission, the waste management sector started working on commodity recovery from materials previously regarded as disposable. Thereby, the term "resources" was widened from ores to include industrial ashes. Several countries are incinerating sewage sludge, as previous disposal choices are no longer an option legally [2]. The resulting sewage sludge ash (SSA) is now the product workers are focusing P-recovery techniques on (e.g., [3–8]). SSA is not only complex in composition, but also diverse (depending on provenance, incineration parameters and other factors). In order to optimize recovery techniques, the P-content (target) needs exploring. Here, strategies from primary resources may help. This paper follows previous work [9–11] and adds insights from scanning electron microscopy-based automated mineralogy (SEM-AM) work, a technique from primary resources, to this waste management issue. A number of systems are available among this group, such as QEMSCAN, Mineralogic, TIMA (TESCAN Integrated Mineral Analyzer) and MLA (Mineral Liberation Analyzer).

Knowing both particle morphology as well as composition is an advantage in devising technologies to extract a targeted component from a diverse material. Addressing these issues, the MLA by FEI, Inc. was chosen to study SSA particles with an SEM-AM system [12]. This offers a complete phase analysis (definitive array of minerals), as well as information about the X-ray amorphous components. Since current approaches to evaluating these datasets aim at identifying minerals among the EDX spectra [13–15], and SSA do not consist of known minerals only, the evaluation of datasets needed some adjustments [16]. Here, we present an approach to complex material investigation using SEM-MLA.

2. Materials and Methods

2.1. Materials

SSA investigated originates from a sewage sludge mono-combustion plant that was kindly provided by "Wirbelschichtfeuerungsanlage Elverlingsen GmbH", 58791 Werdohl-Elverlingsen, Germany. All chemicals were supplied by Merck KGaA, Darmstadt, Germany. Reactants used were HCl, reagent grade 36.5% (used 493.2 g and diluted to 1 L in a volumetric flask with distilled water to prepare 5.0 M) and NH_4Cl. Epo-TEK by Epoxy Technology Inc., Billerica, MA, USA was supplied by Logitech Ltd., Glasgow, UK.

2.2. Methods

The methods used are acid digestion, thermochemical treatment of SSA and MLA investigation. SSA was subjected to digestion. Digestion was carried out by adding inorganic acid (HCl (5 N)) for 5 min at ambient temperature. The amount of solids was 5 g sample and 20 g acid. This was done in a beaker while stirring permanently. Following digestion, cake washing with distilled water was carried out using 2 × 50 mL. The resulting residue was dried at 110 °C for 2 h in a drying cabinet.

Then, acid digestion was undertaken after thermochemical pre-treatment: SSA was mixed with NH_4Cl at a ratio of 1:1 and heated in a kiln. A first temperature rest was kept for 2 h at 250 °C, then another rest was kept for 2 h at 350 °C. Finally, the material was heated to 1000 °C in order to remove the remaining heavy metal volatiles. A continuous inert gas flow of N_2 was applied at 15 L/h. Afterwards, the material was processed using inorganic acid (HCl (5 N)) at a digestion time of 5 min without additional heating and a liquid/solid ratio of 4:1. Both processes were accompanied by sampling for SEM-MLA investigation.

Three powder samples were collected: (i) untreated SSA; (ii) digested, rinsed and dried SSA residue; (iii) thermochemically treated, digested, rinsed and dried SSA residue. These samples were prepared and subjected to the grain mount preparation procedure. Preparation included mixing the dried sample powder 1:1 with graphite (for particle separation) and stirring this mix into epoxy in Teflon containers with 30 mm in diameter. Epoxy blocks were generated, whetted, polished and carbon coated for measurement. The samples were introduced into the vacuum chamber of the SEM and beam acceleration was set at 25 kV with a current of 10 nA; horizontal frame width (HWF) set to 373 pixels. For the acquisition of the grey-scale BSE image, Cu was selected as the BSE grey-scale standard (materials with a Z value of Cu or higher appear as 256 (very bright), all other material turn out darker, respectively). Spectral data were obtained using the GXMAP procedure, that is, particles are automatically detected and covered with a grid of measurement points [12]. Particles smaller than 20 pixels were identified by a single measurement point. Fractions of the sample are given in area-% in this work, since the density of SSA material is difficult to assess. Keeping proportions representative, densities (even of known minerals) were set to one, so "% of the sample" refers to area-% (data were gathered from a polished surface, the actual measurement was that of an area). Proportions of the elemental contents in spectra are weight-%, thus, shares of phosphorus within the material are wt%.

In this study, sewage sludge ashes are analyzed with respect to their P-content and treated SSA checked for residual P-content. For data evaluation, a new approach to sorting and handling EDX spectra is applied. Similarly, workers have previously adapted data evaluation strategies, such as Ayling et al., 2012 [17] devising a new species identification protocol. A similar target-element focused evaluation using QEMSCAN to investigate eudialyte residue has been reported by Ma et al., 2018 [18]. The target component labelling looks at the P-content according to each spectrum and allows for a target material grouping. Conventionally, a mineral grain or particle is classified by comparing the collected EDX spectra to a set of reference spectra labelled by mineral names. EDX spectra of SSA are complex, with a multitude of peaks for Si, Al, Ca, Fe, S and P at various counts (97.6 area-% of the material is comprised of these and oxygen). Other elements found are Ba, Cl, Cr, Cu, K, Mg, Mn, Na and Ti. Due to small grain sizes and transitions between grains of different compositions, there are also some "mixed" spectra which display a combination of peaks from elements in adjacent grains. These conditions, in addition to amorphous, artificially produced material phases, render an interpretation of the spectra as a mineral impossible and do not allow for spectra labelling with mineral names. Using EDX-elemental analysis for quantifying the Si, Al, Ca, Fe, S and P-contents, 53 spectra were collected from the samples, which were then used as reference spectra for classifying the samples. For such reference spectra, generic labelling [16] is a viable approach.

Subsequently, these reference spectra are sorted by their P-content. Spectra groups summarizing the spectra within a range of P-contents (in wt%) are then named descriptively, "<1% P" or "5–10% P", for example (Table 1). This step is called the target component grouping. There is an interesting tendency of elemental complexity (number of elements detected) decreasing as P-content rises. In phosphate recycling, recovering a few other elements in addition to P is desirable as it reduces the need for product purification. The procedure differs from the other ways of grouping by sorts of minerals [16], as it focuses on a single elemental component in amorphous particles. All software operations used MLA datasuite 3.1.4.686.

Table 1. Target component grouping—number of spectra and elemental complexity in each group.

Group	Number of Spectra	Comment
<1% P	13	often mixed spectra
1–5% P	9	On average, composed of 9.66 distinct elements
5–10% P	4	On average, composed of 9.75 distinct elements
10–15% P	4	On average, composed of 7 distinct elements
>15% P	2	On average, composed of 6 distinct elements
3 key	3	completely removed by acid digestion, not present in residues; on average, 7.33 distinct elements

Non-P-containing material components are quartz and other silica-rich species, calcium sulphate spectra, iron oxides and residuals/erroneous identification. The GXMAP measurements were classified against the reference EDX-ray spectra list using the MLA 3.1.4 software version. All spectra were collected with 11,000 counts minimum, with an error level of 0.5% or less. Thereby, this method tries to meet the general accuracy challenges of EDX with statistical significance. Less than 0.6% of the sample area remained classified as unknown, with no X-ray counts, or as unspecified due to a low number of counts. Not all phases have been detected in amounts of relevance. These were grouped as others. The arbitrary cut-off value (below—others, above—relevant compositional phase) is 0.4 area-%. For "BaCaO", the percentage is <0.3 area-% of the sample, represented by <1400 grains. All other spectra are below these values, and thus, only secondary in relevance. As an area was measured, values of percentages refer to area-% unless otherwise noted.

3. Results

While the MLA software package offers several options for exploring data, not all are suited to study SSA. The locking operation offers insights into preferential association of material phases, e.g., how a certain component forms particles with other material phases. This encompasses preferential partner species in multi-material particles as well as the degree of liberated (mono-material) particles. Another function, calculated assay, gives a rough estimate of the elemental composition of the whole sample. This is used as a relative function to see trends, in the case here, P-content of SSA (6%) dropped to 0.5% in acid residue after.

The other operation deemed applicable here is chemical composition of the material as based on EDX spectra. Unlike X-ray diffractometry (XRD) measurements, which are only applicable to the crystalline portion of the sample, this technique yields information about the whole sample (including amorphous material). In addition, XRD measurements of this material suffer from peak shifting due to the considerable portion of amorphous material and hardly any clearly developed peaks. Initial endeavors to fit EDX-derived material phases to XRD-derived mineral phases were only able to determine quartz. The reflexes of known phosphorus minerals (whitlockite, apatite, etc.) are not developed in the XRD spectra. This is interpreted as one of two modes: (1) P-phases in this SSA are present as a yet unknown mineral, or (2) the P-content of SSA is largely present within the amorphous part of the material. As Table 1 lists, phosphate-bearing components found in SSA contain more than just Ca, P and O. For P-recovery, secondary resource users are interested in phases with low complexity and as few elements as possible. During phosphoric acid production, these other elements would necessitate purification, an elaborate and costly process. As P-content rises, the complexity of phases decreases. Thus, it might be interesting to focus on high P portions of SSA in recovery in order to lower purification cost. The general appearance of this SSA prior to chemical treatment is reflected in the BSE images in Figure 1. In the following, particles are discussed. Particles in this material are either inherited—they existed prior to combustion—or a product of incineration processes.

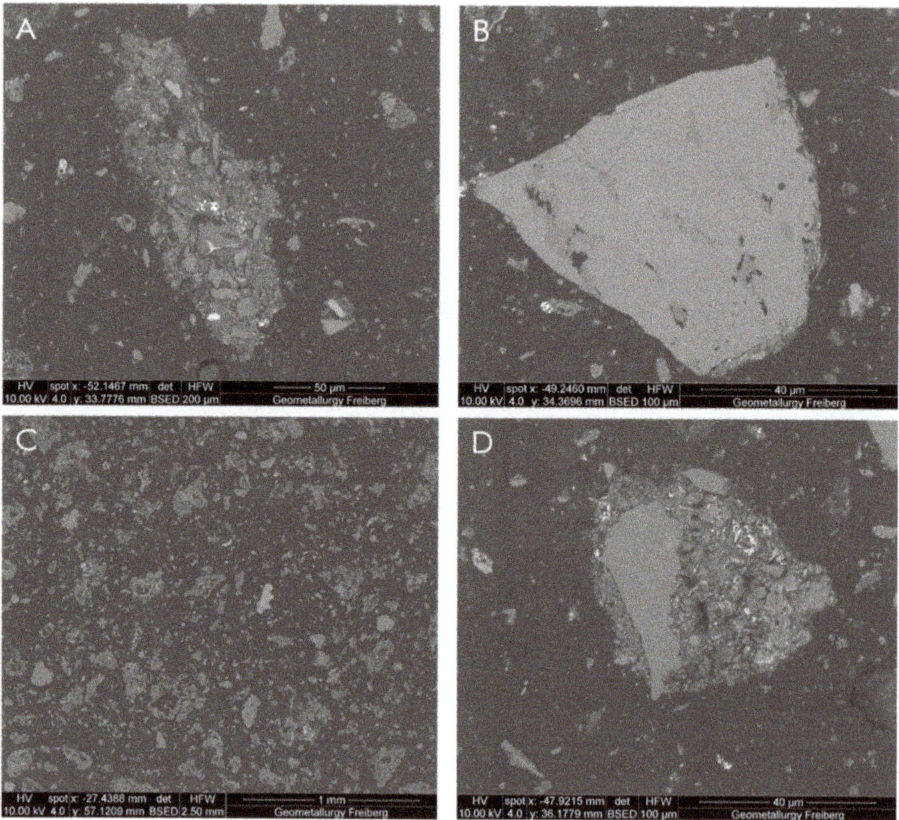

Figure 1. BSE images of sewage sludge ash mixed with graphite for particle separation and stabilized with epoxy for analysis. (**A**) agglomeration of fine flakes; (**B**) inherited mineral grain, presumably from fluidized bed, bottom right of the grain shows P-rich material condensing onto grain; (**C**) general overview of SSA and (**D**) agglomeration around inherited grain.

Using the target component grouping subsequent to a generic labelling approach [16], the material composition of three samples is shown. Figure 2 compares the initial ash, digestion residue after HCl treatment, and digestion residue after thermochemical pre-treatment.

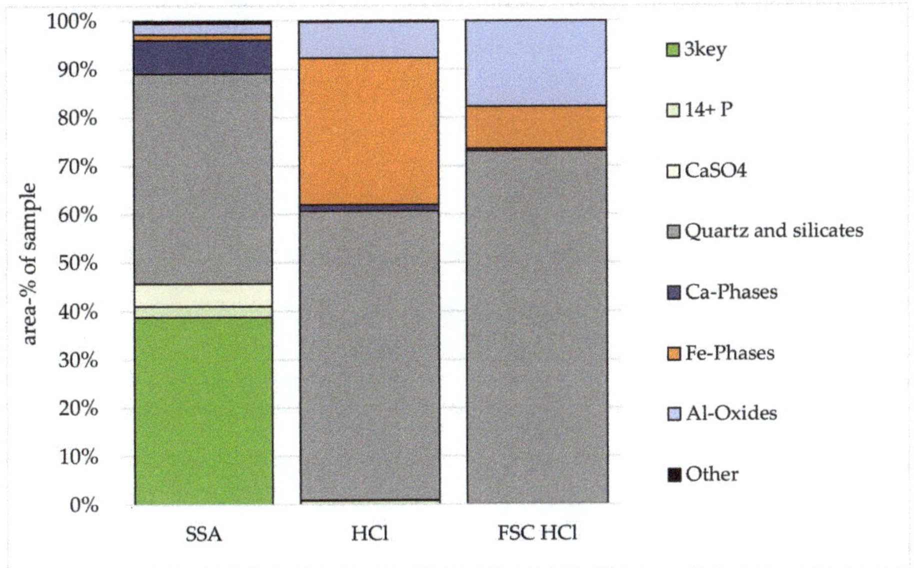

Figure 2. Composition of untreated SSA before and after chemical digestion by different acids (generic labelling); SSA: untreated, original sewage sludge ash, HCl: residue after SSA digestion with HCl; FSC-HCl: residue of HCl digestion of thermochemically treated SSA. Shares of material given in area-%.

When untreated and treated samples are compared, some of the material is missing. This material has been fully digested and moved to the liquid part. Analysis of the raw phosphoric acid showed other elements as well [11]. Wherever material has been removed in significant amounts, the proportion of other components changes. Upon closer investigation, three "key" spectra were identified. They represent the main P-bearing components of this SSA and are fully digested using HCl.

3.1. Insights Gathered by Generic Labelling

Using generic labelling [16], the spectra names are CaFeSiAlPO, CaPAlSiMgFeKO and PCaAlSiFeMgO. They contain 6, 12 and 16 wt% P, respectively, as determined by EDX analysis, and account for about 40 area-% of the material. By generic labelling, the first two of these would have ended up in the Ca-dominant group, and only the third as part of the P-rich material. Thus, when tracking P deportment, it is helpful to label spectra according to their P-content rather than the most abundant element.

3.1.1. CaSO4

SSA contains calcium sulphate. HCl digests this initial SSA component. This material is commonly found around the edges of angular silicate particles, forming narrow rims (max ~30 μm) around grains. In terms of particle formation within the thermal process, this is interpreted as a result of condensation onto already formed particles. According to the mineral grain size distribution function of the MLA dataview software, this material has a typical D_{50} of ~54 μm (grainsize at 50% of the cumulative passing grain size distribution curve).

3.1.2. Quartz and Other Silicates

Quartz is used as a fluidized bed in sewage sludge combustion, and is the only mineral reliably detected by XRD in these SSA. Since the acid used is unable to digest quartz, the initial amount of quartz should be preserved. For HCl digested residue, the compositional variance is within the normal variation of this material—a mode between 26 and 29 area-% (original SSA: mode of quartz 27 area-%). The spectra grouped as "other silicates" all contain Si as the major component, meaning no other measured element is more abundant within this spectrum. Still, within this group, the measured Si content varies between 7 and 32 wt% and averaging 23 wt% of the mode. "Other silicates" reach a slightly lower D_{50} (~98 μm) than quartz (~140 μm). "Other silicates" tend to be porous, rounded, amoeboid particles whereas quartz particles are more angular and lack pores.

3.1.3. Ca-Dominant Material

Material, in which Ca is the main component, is reduced from 7 area-% in SSA to 1 area-% in digestion residues. Spectra in this group contain an average of 27 wt% Ca. This material reaches a D_{50} of about 50 μm. It is found both as part of narrow rims around larger quartz and other silicate particles as well as porous, amoeboid, particles in conjunction with other silicates or phosphates. This material rarely forms particles with Al- or Fe-dominant material, an observation which encourages the interpretation that the thermal reorganization of SSA material is incomplete: phosphates are precipitated using Al- and Fe-salts during the wastewater treatment process. Since Al- and Fe-ions are unwanted in the recovered phosphoric acid, one goal of sewage sludge incineration for P-recovery is reorganizing the P-content into calcium-phosphates. However, the majority of spectra in this group are not only Ca-dominant but also contain Si, Al, and Fe.

3.1.4. High P-Material (>14 wt% P)

The residue still contains some remnants of P-rich material. Nonetheless, the content was greatly decreased in comparison to the original material. Less than 1.4 area-% of these phases remain, mostly the spectrum named as FePMgO. This residual P-content is rendered as acid-accessible by thermochemical pre-treatment, as was concluded by its absence in thermochemically pre-treated residue. Using the initial generic labelling, this material would have ended up in other groups, the aforementioned spectrum in Fe-rich material, for example. At a D_{50} of 90 μm, these particles are among the larger ones within the material, and they do form particles on their own, but often with the fully recoverable 3 key phases, silicates and Ca-dominant material. One of the spectra in here is Fe-P-O, a product of the precipitation of phosphates using Fe-salts during the wastewater treatment process. This material on its own has a D_{50} of 40 μm and usually forms particles with one of the 3 key spectra or other high P-material, a product of the incomplete thermal reorganization of phosphates during incineration.

3.1.5. Fe- and Al-Rich Material

These material components are formed during phosphorus removal in sewage treatment plants. Using iron and/or aluminum salts, phosphorus compounds precipitate and are preserved during incineration. Some of these only slightly soluble compounds even survive acid treatment, which is why these components appear relatively enriched in the residues. The residue shows an enrichment in Fe-phases compared to SSA. Residues will always represent an enrichment in phases which are hard to digest. As already mentioned, this SSA is contaminated with heavy metals and does not comply with the German fertilizer ordinance [19]. To address this issue, thermochemical pre-treatment was carried out. It was found that this treatment also improved P-recovery [9], which is why this process step was studied in the present experiments testing another acid. Less than 0.1 area-% of the three P-rich (14+ P) spectra remain. Fe-rich material was removed during pre-treatment, thus, reduced contents of this group are not surprising (by calculated assay, Fe content was roughly halved, 13.4 to 6.35 wt%). All apparent increases in components in these residues are relative increases as more material has been digested. Particle sizes for these groups vary; at a D_{50} of ~100 μm, the Al-rich material is much coarser than Fe-oxides (D_{50} of ~45 μm). Fe-rich material typically appears as grains within larger, host particles of almost all other groups, the two exceptions being $CaSO_4$ and high P-material. Al-rich material follows the same trend.

3.1.6. HCl Digestion of Thermochemically Pre-Treated SSA

Comparing the pre-treated residue to the untreated residue, the Al-rich material and silicate material proportions rise. Fractions of Fe-rich, Ca-rich and residual high P-material decrease, the latter almost disappearing (0.02 area-%). P-recovery was more successful and pre-treated digestion residue contains less P (by calculated assay, about 0.5% less—1.08 wt% P remain). This highlights the need for an improved data evaluation—if as much P as possible is to be recovered, the current approach does not necessarily help process evaluation.

3.2. Insights Gathered by Target Grouping of EDX-Spectra

When applying the target grouping (P), compositions appear slightly different (Figure 3). All spectra are now sorted according to their P-content. If there was no P detected in the spectrum, it was labelled as before, by dominant (most abundant element within the spectrum). Thus, values for $CaSO_4$ are not affected by the new grouping procedure, since it does not contain P.

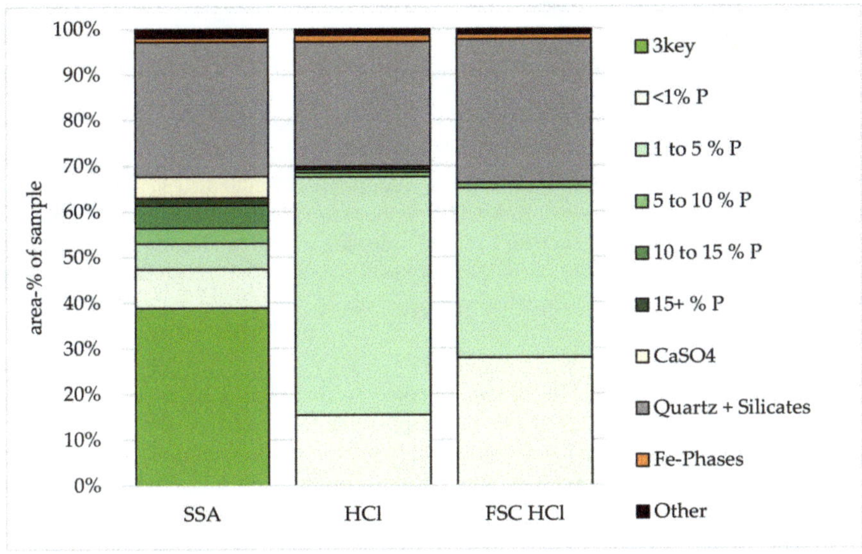

Figure 3. Composition of SSA before and after digestion with different inorganic acids evaluated with the target phase grouping procedure.

3.2.1. Quartz and Silicates

The distribution of quartz and silicates (quartz + silicates) remains in the same range as quartz in generic grouping. Here, generic labelling is expected to yield lower values, as the generic strategy distinguishes between "other silicates" and "quartz". The target grouping combines the two, wherever they are P-free. Hence, whenever values for target quartz + silicates are lower than generic "quartz" and "other silicates" values, the Si-dominant material is P-enriched.

3.2.2. P-Containing Material

Residue content of P-phases digestion with HCl is only significant for the <5 wt% P spectra (groups < 1 and 1–5 wt% P). For spectra with a P-content of more than 15 wt%, the contents are reduced in residues (as compared to original SSA content). Although it looks like the fraction has the same extent as in the original material, taking into account that at least ~40 area-% of SSA were removed, reveals that these values represent a decrease beyond proportion. These material phases must have been particularly acid-accessible. In other words, among the P-containing groups, the original material has the largest share in 5–10 wt% P. Digestion shifts proportions, and in residues, the largest group among the P-containing groups is 1–5 wt% P. While the conventional grouping showed that high P-phases are severely depleted, target phase grouping highlights that the mid-level P-containing groups (quantitative relevance in the material) are also shifted towards lower P-values. Therefore, digestion acts on more P-containing species than just the high P-containing ones.

Phosphate content in SSA was depleted by 86 wt% [11]. This is shown by XRF analyses, since MLA studies cannot supply this information. There is the "Calculated Assay" operation within the MLA software in complex materials such as SSA, however, these values are better understood as a proxy.

The 10 to 15 wt% P-phase group is very accessible by the acid; they have been reduced from ~34 to <1.8 area-%.

3.2.3. Fe-Rich Material

Compared to the original material, Fe-phases have been depleted. Comparing these findings with generic labelling, some of the Fe-phases (generic) must contain phosphorus. Thus, they are not part of Fe-phases group (target phase). Target phase grouping thereby highlights that iron phosphates (generated in phosphate precipitation in wastewater treatment plants) are highly stable and not easily acid-accessible.

3.2.4. Residues of Pre-Treated Digestions (Target Grouping)

Pre-treatment led to an interesting shift in material phases. While the calculated assay values for P increase from SSA to pre-treated SSA (relative increase, as $CaSO_4$ is removed), the share of material components containing more than 10 wt% P decreases. At the same time, the amount of key spectra rises by about 10 area-%. Thus, the improved P-recovery after thermochemical pre-treatment is explained: thermochemical pre-treatment led to the formation of accessible material (see also [11]).

3.2.5. HCl Digestion

Comparing the pre-treated residue to the untreated residue, silicate and <1 wt% P proportions rise. Fractions of all other materials decrease. This implies that HCl digestion performs even better on pre-treated ash. This is also reflected in the images of treatment residues (Figure 3).

3.2.6. Generic Labelling

Prior to digestion, the P-phases were preferentially associated (30% binary) to quartz. In the residues (after digestion of pre-treated SSA), quartz phases are preferentially associated with other silicates, but mostly liberated. As they were mostly liberated in the original material, there is not much change here. Digestion is assumed to have worked as expected and >14 P-material was almost completely leached.

3.3. Target Evaluation

The fraction of liberated grains decreases for quartz + silicates by about 20%. In all P-containing phases, locking increases drastically (up to 98%) in all inorganic acid residues. Contrasting, in the original material, these groups were much more liberated, by up to 55%. The same is true for Fe-phases, although the original degree of locking was higher in this group (75% up to 97%). Comparing locking values for quartz + silicates of pre-treated digestion residues with non-pre-treated digestion residues, these grains reveal reduced degrees of locking in residue (successful digestion of grains previously associated with quartz + silicates) for HCl. Thus, pre-treatment enables access to complex particles where access is desired. Quartz + silicates locking values for their preferred partner (1–5 wt% P-phases) show a decrease in locking values (pre-treated ash yields a residue with less complex quartz + silicates particles).

Locking increases in all phases between original SSA and residue. This lends room to the hypothesis of complex mechanisms leading to ternary$^+$ particles. Keeping in mind that about 40 wt% of the material were already extracted, it is not surprising to find complex particles in the undigested residue.

4. Discussion

Within the literature studying SSA for P-recovery, it has been mentioned that the phosphate content is present as whitlockite [3–7]. The data presented here cannot confirm this. Whitlockite is easily acid-accessible, but none of the recovered material (as represented by the spectra that are present in the original material, but missing from the residue) show good agreement with the material studied here. By generic labelling, whitlockite would appear as PCaMgFeO, and at 20% P-content, fall into the >15 % P group of our target grouping. Neither an appropriate spectrum nor the majority of the P-content of this material in the >15 P group are found here. This is in agreement with other SSA studies [20].

This work suggests grouping EDX-spectra found in SEM-AM investigations of SSA for P-recovery by P-content. As shown, grouping them by their main component is less helpful when assessing the elemental deportment of phosphates and recovery success. Both options, however, show important characteristics for P-recovery. With regards to incineration, the reorganization of P-phases from those generated during P-precipitation in wastewater treatment to those present in ashes does currently not favor a clean recovery of calcium-phosphates. Thus, studies such as these can help assess different combustion regimes and their suitability towards P-recovery. With regards to P-recovery processes, an SEM-AM investigation reveals why recent works of P-recovery always had to deal with Al, Fe, and other macro elements in their process [8,21].

Particle sizes of the groups identified in this work are subject to the error of density settling of material during sample preparation and should be taken as "relative to each other"—this effect should affect all samples alike. Sawing up prepared epoxy blocks of the sample and turning the bands 90°, then analyzing them could show the extent of this error. A comparative study could potentially quantify this effect.

The generic labelling used in [9] allowed for the sorting of spectra by their main component, however, oftentimes material phases found in SSA are very complex and the "main" component is only "major" over other components by a few percent. Thus, the focus on P—as the target constituent which should be extracted—expresses itself in the target grouping. This data evaluation shows the extent to which remnant P-phases are still present in residues and the quality of SSA; a larger proportion of material containing relatively high levels of P is better suited than high levels of overall P-content. This P-recovery quality assessment cannot be done by XRF analyses, which determine the overall P-content of the material or XRD studies, where signals are overlain by a large scatter of random directions due to the large amorphous part of SSA. EDX data as obtained from a polished surface from a multitude of mapped grains, are a suitable method. To accentuate, three spectra were found which are exclusively present in the untreated material. These three spectra, named "key", were completely digested by HCl, regardless of thermochemical pre-treatment of the digestion. In addition, this is the reason why thermochemical treatment resulted in an improved P-recovery: thermochemically treated SSA contained more key material, i.e., was able to generate key material. While other P-containing material phases varied in their response to treatment and thermochemical pre-treatment, these three phases always vanished completely, highlighting the P-recovery success of acid digestion.

It has been shown that generic labelling data evaluation [9] is limited when analyzing the P-contents of residues. A reduction in P-content in the residue is shown—the proportion of spectra containing less P rises, while the fraction of spectra containing more P decreases. Only half of this is illustrated by the conventional grouping: here, high P-phases are seen to vanish, but the rise of reduced P-species is not documented. The shift from 5–10 wt% P to 1–5 wt% P by simple digestion and to <1 wt% P in digestion after pre-treatment is obvious. This may be interpreted as the result of one of two processes: the material is removed, and thus, the relative proportion of remaining components rises—so, there is no actual enrichment; or the extraction mechanism by acid digestion only recovers a part of the P-content and not the full amount. In the latter case, it would be likely for new, mildly P-bearing spectra to appear in the residue but not in SSA. Since the same list of spectra was applied to all samples, this would either show certain spectra only present in residues or a significant increase in "unknown" material in residues. Neither is the case. Thus, there are shifts, triggered by the processes, but no generation of distinct phases. While the benefit of thermochemical pre-treatment is also visible using conventional grouping, it looks only marginal: digestion residues contain <5 area-% of the spectra with high P-content, and pre-treated show none. Using the target grouping, it is obvious that thermochemical pre-treatment affects more parts of the sample than specifically the heavy metal content only. The aforementioned shift in P-phase contents becomes visible. Fe-oxides are extracted by thermochemical pre-treatment.

As has been outlined in mathematical approaches (e.g., [22,23]), it is difficult to give estimates on uncertainty and error based on automated SEM mineral liberation analyses. The primary signal of MLA measurements is a backscattered electron image and energy dispersive spectra composed of data from numerous detector channels. The latter signal is based on ~12,000 counts gained within 10 ms. There is no specific time or channel resolution which could allow uncertainty and error estimates based on peak-to-background ratios and standard deviations for counting rates for distinct peaks. In consequence, for further estimates, the mineral mode appears as the most prominent parameter in comparison to the particle and grain sizes and their shape geometries.

During the preparation of the grain mount block, particle separation may be induced by the stirring of the particles into the liquid epoxy and subsequent gravitational subsidence of high-density particles during epoxy hardening. These effects can lead to heterogeneous particle distributions in the surface of a polished block, an effect that is particularly prominent for sample materials (mineral mixtures) with large differences in particle sizes and/or densities. As all samples thus prepared are affected by this bias towards particle distribution, it is suggested to study more than one sample—SSA and digestion residue, for example, or two types of SSA.

There are also uncertainties related to spectra classification. The EDX spectra classification algorithm for the MLA software packages follows the principle of best match along a scale of reliability between 1×10^{-10} (absolute conformance) to 1×10^{-100} (no conformance), as outlined by [10]. The phosphorus-bearing particles display a comparably complex pattern of X-ray emission lines, with many peaks and sub-peaks that are marked by considerable interference. Due to the complex X-ray spectra characteristics, there is considerable risk that EDX spectra are not at all classified, if classification is carried out at a high reliability value. The classification algorithm allows no alternative assignment to another EDX reference spectrum or to another mineral in the list. Due to this principle, the spectra which cannot be classified by the higher reliability scale value will remain as unknown and increase the mode of unknown grains [11]. For the study presented here, the sample EDX spectra were thus classified by the reliability values of 1×10^{-10} (high degree of conformance) and 1×10^{-25} (fair degree of conformance). The latter reliability value is applied to process samples, in an effort to reduce the amount of unknowns below 0.1 area-% mode. Applying a reliability value of 1×10^{-25} to the samples of the third case study, the modes of unknowns remained low, ranging between 0.53 and 0.16 area-%.

5. Conclusions

This study defined the three P-containing phases in SSA which are completely recovered by HCl acid digestion. This implies that the recoverable P-content of SSA studied (sludge burned at 850 °C) lies not in a single mineral phase, but in multiple element associations. Any recovered phosphorus product from SSA will inevitably require costly purification. At the same time, acid digestion does not recover the full P-content. Legal obligations to recover the full content are likely to result in economic actors not pursuing P-recovery from SSA at all. Thus, several routes are probable: SSA not being used for P-recycling (too costly), legal requirements being adjusted or SSA being optimized—perhaps the P-content can be reorganized in easily recoverable phases by optimizing the combustion regime.

Workers with SSA combusted at higher temperature report less elementary complex phases, and our work shows that higher P-phases are composed of fewer elements. Thus, the impurities in recovered phosphoric acid that need to be stripped off are reduced in complexity. Future work could address SSA from different wastewater treatment plants incinerated at the same furnace to shed light on P-deportment. This would enable the study of provenance effects and should help to differentiate between incineration- and provenance-based effects. Another possible route would be P-recovery prior to incineration, yet this work focused on developing a new spectra naming/grouping strategy for SSA, which is recommended to use during future work for advancing this particular area of waste management.

P-recovery from SSA should, ideally, start with exploring a combustion regime that optimizes P-phase shift towards recoverable P-phases with fewer elements. This study showed that higher P-content in material phases is accompanied by fewer elements. Future work could focus on improving combustion for an efficient P-recovery from SSA.

Author Contributions: Following an idea of M.B., conceptualization by B.S. and M.B. was followed by V.-G.G. carrying out experiments. A.C.G. evaluated data and drew conclusions and proposed spectra labelling. Draft preparation and writing were done by A.C.G., M.B. and B.S. All authors have read and agreed to the published version of the manuscript.

Funding: This research was funded by the German federal ministry for economic affairs and energy, BMWi FKZ 03EFGSN097 and the German federal ministry of education and research, BMBF FKZ 033R099E. These were granted to M.B. and his institute.

Acknowledgments: The authors would like to thank Wirbelschichtfeuerungsanlage Elverlingsen GmbH for providing the samples. Furthermore, we are grateful for Sabine Gilbricht of the Geometallurgy Laboratory at the Department of Economic Geology and Petrology, TU Freiberg, for kindly supporting the MLA analyses. The authors would like to express gratitude towards the three anonymous reviewers whose comments improved this work.

Conflicts of Interest: The authors declare no conflict of interest.

References

1. European Commission. Communication from the Commission to the European Parliament, the Council, the European Economic and Social Committee and the Committee of the Regions on the Review of the List of Critical Raw Materials for the EU and the Implementation of the Raw Materials Initiative. Available online: https://eur-lex.europa.eu/legal-content/EN/TXT/?uri=CELEX%3A52014DC0297 (accessed on 25 May 2020).

2. Werther, J.; Ogada, T. Sewage sludge combustion. *Prog. Energy Combust. Sci.* **1999**, *25*, 55–116. [CrossRef]

3. Fowlie, P.J.; Stepko, W.E. *Sludge Incineration and Precipitant Recovery*; Technical Report No 72-3-4; Pollution Control Branch, Ontario Ministry of the Environment: Toronto, ON, Canada, 1978.

4. Liu, Y.; Kumar, S.; Ra, C.; Kwag, J.-H. Magnesium ammonium phosphate formation, recovery and its application as valuable resources: A review. *J. Chem. Technol. Biotechnol.* **2012**, *88*, 181–189. [CrossRef]

5. Montag, D. Phosphorrückgewinnung bei der Abwasserreinigung—Entwicklung eines Verfahrens zur Integration in Kommunalen Kläranlagen. Ph.D. Thesis, RWTH Aaachen University, Aachen, Germany, 2008.

6. Oliver, B.; Carey, J. Acid solubilization of sewage sludge and ash constituents for possible recovery. *Water Res.* **1976**, *10*, 1077–1081. [CrossRef]

7. Scheidig, K.; Lehrmann, F.; Mallon, J.; Schaaf, M. Klärschlamm-Monoverbrennung mit integriertem Phosphor-Recycling. In *Energie aus Abfall*, 1st ed.; Thome-Kozmiensky, K.J., Beckmann, M., Eds.; TK Verlag: Neuruppin, Germany, 2013; pp. 1039–1046.

8. Ottosen, L.; Jensen, P.E.; Kirkelund, G.M. Phosphorous recovery from sewage sludge ash suspended in water in a two-compartment electrodialytic cell. *Waste Manag.* **2016**, *51*, 142–148. [CrossRef] [PubMed]

9. Greb, V.G.; Guhl, A.; Weigand, H.; Schulz, B.; Bertau, M. Understanding phosphorus phases in sewage sludge ashes: A wet-process investigation coupled with automated mineralogy analysis. *Miner. Eng.* **2016**, *99*, 30–39. [CrossRef]

10. Greb, V.-G.; Fröhlich, P.; Weigand, H.; Schulz, B.; Bertau, M. Phosphatrecycling aus Klärschlammaschen—Warum Phosphorsäure der Königsweg ist. In *Rohstoffwirtschaft Und Gesellschaftliche*

     Entwicklung—Die Nächsten 50 Jahre, 1st ed.; Kausch, P., Matschullat, J., Bertau, M., Mischo, H., Eds.; Springer: Heidelberg, Germany, 2016; pp. 154–196.

11. Greb, V. Rückgewinnung Von Phosphor Aus Klärschlammaschen Durch Sauren Aufschluss Unter Berücksichtigung Einer Thermischen Vorbehandlung. Ph.D. Thesis, TU Bergakademie Freiberg, Freiberg, Germany, 2018.

12. Fandrich, R.; Gu, Y.; Burrows, D.; Moeller, K. Modern SEM-based mineral liberation analysis. *Int. J. Miner. Process.* **2007**, *84*, 310–320. [CrossRef]

13. Smith, D.G.W.; De St. Jorre, L. The MinIdent Data Base—Examples of applications to the Thor Lake rare metal deposits. In Proceedings of the Abstracts with Program, 14th General Meeting of International Mineralogical Association, Stanford, CA, USA, 13–18 July 1986; p. 234.

14. Gu, Y. Automated scanning electron microscopy based mineral liberation analysis an introduction to JKMRC/FEI mineral liberation analyser. *J. Miner. Mater. Charact. Eng.* **2003**, *2*, 33–41.

15. Lastra, R. Seven practical application cases of liberation analysis. *Int. J. Miner. Process.* **2007**, *84*, 337–347. [CrossRef]

16. Schulz, B.; Merker, G.; Gutzmer, J. Automated SEM Mineral Liberation Analysis (MLA) with Generically Labelled EDX Spectra in the Mineral Processing of Rare Earth Element Ores. *Minerals* **2019**, *9*, 527. [CrossRef]

17. Ayling, B.; Rose, P.; Zemach, E.; Drakos, P.; Petty, S. QEMSCAN (Quantitative Evaluation of Minerals by Scanning Electron Microscopy): Capability and application to fracture characterization in geothermal systems. *AGU Fall Meet. Abstr.* **2012**, 1158. [CrossRef]

18. Ma, Y.; Stopic, S.; Gronen, L.; Friedrich, B. Recovery of Zr, Hf, Nb from eudialyte residue by sulfuric acid dry digestion and water leaching with H_2O_2 as a promoter. *Hydrometallurgy* **2018**, *181*, 206–214. [CrossRef]

19. DüMV, Verordnung über das Inverkehrbringen von Düngemitteln, Bodenhilfsstoffen, Kultursubstraten und Pflanzenhilfsmitteln (Düngemittelverordnung–DüMV), 5 December 2012, BGBl. I, 2482 last amendment by article 1 of the ordinance of 27 May 2015, BGBl. I, 886. Available online: https://www.gesetze-im-internet.de/d_mv_2012/D%C3%BCMV.pdf (accessed on 25 May 2020).

20. Pavón, S.; Guhl, A.C.; Pätzold, C.; Bertau, M. Verfahren zur Behandlung von Flugasche und Ascherückstand zur Phosphorsäuregewinnung. In Proceedings of the 1st Annual abonocare®Conference, Leipzig, Germany, 18 March 2020.

21. Fang, L.; Li, J.-S.; Donatello, S.; Cheeseman, C.; Wang, Q.; Poon, C.S.; Tsang, D.C. Recovery of phosphorus from incinerated sewage sludge ash by combined two-step extraction and selective precipitation. *Chem. Eng. J.* **2018**, *348*, 74–83. [CrossRef]

22. Leigh, G.; Sutherland, D.; Gottlieb, P. Confidence limits for liberation measurements. *Miner. Eng.* **1993**, *6*, 155–161. [CrossRef]

23. Tolosana-Delgado, R.; Mueller, U.; Boogaart, K.G.V.D. Geostatistics for Compositional Data: An Overview. *Math. Geol.* **2018**, *51*, 485–526. [CrossRef]

Review

SEM-Based Automated Mineralogy and Its Application in Geo- and Material Sciences

Bernhard Schulz [1,*], Dirk Sandmann [2] and Sabine Gilbricht [1]

[1] Institute of Mineralogy, Economic Geology and Petrology, TU Bergakademie Freiberg, Brennhausgasse 14, D-09599 Freiberg, Germany; sabine.gilbricht@mineral.tu-freiberg.de

[2] ERZLABOR Advanced Solutions GmbH, Halsbrücker Str. 34, D-09599 Freiberg, Germany; d.sandmann@erzlabor.com

* Correspondence: bernhard.schulz@mineral.tu-freiberg.de; Tel.: +49-3731-39-2668

Received: 30 September 2020; Accepted: 9 November 2020; Published: 11 November 2020

Abstract: Scanning electron microscopy based automated mineralogy (SEM-AM) is a combined analytical tool initially designed for the characterisation of ores and mineral processing products. Measurements begin with the collection of backscattered electron (BSE) images and their handling with image analysis software routines. Subsequently, energy dispersive X-ray spectra (EDS) are gained at selected points according to the BSE image adjustments. Classification of the sample EDS spectra against a list of approved reference EDS spectra completes the measurement. Different classification algorithms and four principal SEM-AM measurement routines for point counting modal analysis, particle analysis, sparse phase search and EDS spectral mapping are offered by the relevant software providers. Application of SEM-AM requires a high-quality preparation of samples. Suitable non-evaporating and electron-beam stable epoxy resin mixtures and polishing of relief-free surfaces in particles and materials with very different hardness are the main challenges. As demonstrated by case examples in this contribution, the EDS spectral mapping methods appear to have the most promising potential for novel applications in metamorphic, igneous and sedimentary petrology, ore fingerprinting, ash particle analysis, characterisation of slags, forensic sciences, archaeometry and investigations of stoneware and ceramics. SEM-AM allows the quantification of the sizes, geometries and liberation of particles with different chemical compositions within a bulk sample and without previous phase separations. In addition, a virtual filtering of bulk particle samples by application of numerous filter criteria is possible. For a complete mineral phase identification, X-ray diffraction data should accompany the EDS chemical analysis. Many of the materials which potentially could be characterised by SEM-AM consist of amorphous and glassy phases. In such cases, the generic labelling of reference EDS spectra and their subsequent target component grouping allow SEM-AM for interesting and novel studies on many kinds of solid and particulate matter which are not feasible by other analytical methods.

Keywords: scanning electron microscope; mineral liberation analysis; raw materials; resource technology; granular material; petrology

---

1. Introduction

During the last decade, software developments in Scanning Electron Microscopy (SEM) provoked a notable increase of applications to the study of solid matter. The mineral liberation analysis for the optimisation of mineral processing of metal ores is the economic and thus important drive for innovations which led to various SEM application software versions. These combine the assessment of the backscattered electron (BSE) image to the directed steering of the electron beam for energy dispersive X-ray spectroscopy (EDS) to various measurement routines of Automated Mineralogy applications. The term Automated Mineralogy (AM) does not have a proper definition and is used in somewhat varying meanings. However, as artificial materials can be analysed by this technology too, Automated Materials Characterisation may be used in addition. The term Auto-SEM-EDS can delineate the instrumental combination. Despite a wide distribution of SEM instruments in material research, geosciences and industry, the potential of SEM-based Automated Mineralogy (SEM-AM) is still under-utilised. Characterisation of primary ores, and the optimisation of comminution, flotation, mineral concentration and metallurgical processes in the mining industry by generating quantified reliable data is still the major application field of SEM-AM [1–8]. However, there arises interesting economic and scientific potential beyond the classical fields. Geometallurgy, ore fingerprinting and applications in petrology are still closely related topics [9–14]. Slags, pottery, stoneware and artefacts can be studied in an archaeological context for recognition of provenance and trade routes, but also for the better understanding of their production processes [15–18]. Soil and solid particles of all kinds are objects in forensic science [19–21]. SEM-based Automated Mineralogy allows novel insight in the fields of process chemistry and recycling technology [22–24].

Here, we refer to the main and principle hard- and software components and their combinations to the advanced SEM-based Automated Mineralogy, beyond the pioneering compilations [2,14,25–32]. Potentials and limits of the SEM-AM technology are presented in case studies dealing with metamorphic petrology, ore fingerprinting, slags and firing experiments on archaeological pottery.

2. Principles and Limits of SEM-Based Automated Mineralogy

2.1. Basic Measurement Routines

Common to all SEM-AM systems is the combination of a hardware platform and a specific image analysis and processing software. Almost every scanning electron microscope (SEM) can be used as a hardware platform for AM. For the use in AM, the SEM needs additional internal main boards and must have a high vacuum operation mode. The required pressure range is on the order of 10^{-5} to 10^{-7} Pa. Tungsten cathodes and field emission guns are offered as the electron sources. As long-term stability of the electron beam is required for automated measurements, the field emission guns can be recommended, but tungsten cathodes are economically favourable. SEMs for automated mineralogy are often equipped with two or more EDS spectrometers to increase the count rate of X-rays and subsequently the speed of analysis. A large sample chamber is advantageous to allow a larger number of samples to be analysed in one measurement session. The SEM should have a very accurate stage movement which allows a precise positioning in small intervals. An excellent BSE detector is crucial for good analysis results. The BSE image quality and especially BSE image stability is a critical factor for SEM-AM analysis, as the image is used, in combination with the EDS spectrum, for mineral or phase discrimination. To allow constant BSE image grey levels, fixed working distances must be set prior to measurement. This ensures that a specific mineral or phase always has the same BSE image grey level, as long the image calibration is constant. Usually, the BSE image grey level can be calibrated with reference materials with different BSE grey levels, as gold (very bright), copper (intermediate) and quartz (dark grey). When copper or quartz are used for the BSE image grey level calibration, then particulate materials with dark or intermediate BSE grey levels as many industrial ashes or slags (reported below) are better resolved in the images.

The BSE image analysis and processing software controls the automated to semi-automated measurements and allows thorough data processing and comprehensive results extraction. After the collection of a BSE image for a given frame or area, several steps of image processing are performed. As an example, the FEI-MLA software (version 3.1.4, FEI Company, Hillsboro, OR, USA) versions involve:

(1)   Background removal
(2)   De-agglomeration
(3)   Clean-up of undersized particles and/or of particles touching the frame boundary
(4)   Segmentation of internal particle structures

For the background removal, different values of BSE grey level can be defined. The de-agglomeration step uses specific particle shape factors like circle ratio, rectangular ratio and a combined circle and rectangular ratio to determine if particles are agglomerated. Subsequent routines are performed for their separation. The clean-up function allows the deletion of undersized particles prior to the EDS measurement. Removal of particles touching the frame boundary will avoid artefacts in grain size distribution measurement. At the final segmentation step, the grain boundaries in a particle are determined based on BSE grey level differences. This step also removes any polishing artefacts from the image such as cracks or holes in the particles. A high-quality polished surface is essential for the segmentation procedure. The image processing routines offer variable software functions for the handling of these image analysis procedures. In addition, these image processing steps can be arranged in variable sequences and different weighting in dependence of the sample properties. The design of the image processing routines mostly targets the fast treatment of granular samples of ore and gangue minerals as encountered in mineral processing studies. In non-granular samples with closed surfaces such as petrographic thin sections or plates, most of the image processing steps cannot be performed or will need a specific and adapted software solution. Four principal measurement routines, starting with the collection of a BSE image with a calibrated grey tone level, can be outlined for SEM-AM technology (Figure 1):

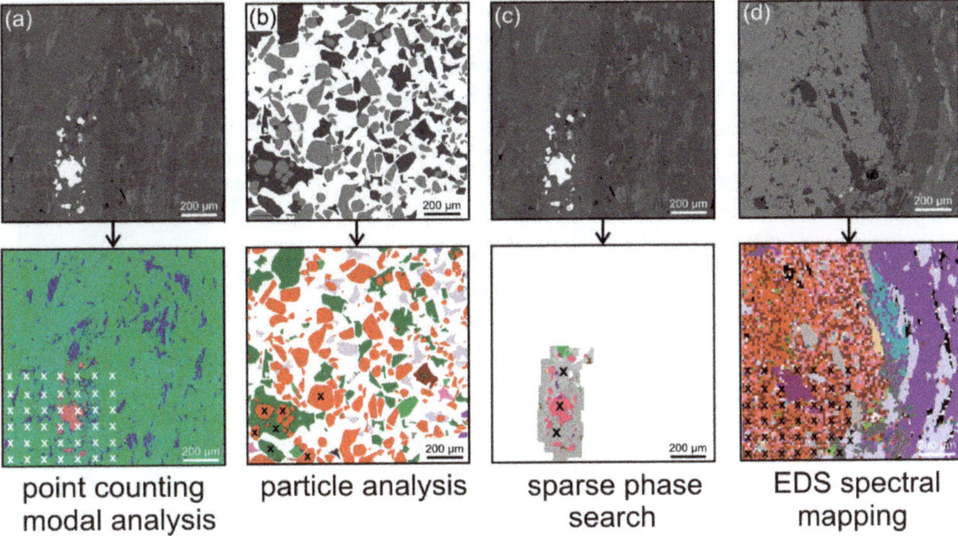

Figure 1. SEM Automated Mineralogy measurement methods in BSE (upper row) and classified EDS images (lower row) of one measurement frame. Points of X-ray analyses are marked by white or black crosses; only some points are shown. (**a**) EDS point counting method for quantification of modal composition. (**b**) EDS particle analysis method for fast measurement of numerous particles with presentation of their geometrical and liberation parameters. (**c**) EDS sparse phase search method for detection and location of rare phases in grain mounts or thin sections. The peripheral mineral associations are analysed within a pre-set µm-scale range of distance. (**d**) EDS spectral mapping combines BSE image analysis with EDS mapping. Each grain with a distinguishable BSE grey level is mapped by numerous single EDS analysis points and visualised as colour-coded pixels, e.g., the red-coloured pixels signal a garnet grain.

(1) The classical point counting method in which mineral or phase identification is determined by one EDS analysis at each counting point (Figure 1a). This measurement uses the BSE imaging merely to discriminate particle matter from background and then collects one EDS spectrum from each grid point across the sample. This method only produces modal mineralogy information, such as the percentages of the mineral or phase components of the sample. The point counting can also be implemented to produce a line scan measurement mode that produces traditional linear intercept data. EDS spectra are taken at a step size of one pixel in the x-direction and a user-defined y-interval determines the line spacing. Typically, $\sim10^4$–10^6 EDS spectra are taken at a fixed step size (for example 10 µm) per sample. The grid step size can be chosen in dependence on the required resolution. This routine does not gather particle sizes and shapes. However, visual information can be taken from the BSE image which is stored during measurement. A summary of the mineral modal compositions in area % and in wt% can be recalculated by introducing an area to each point of the grid and average densities listed in mineral databases for each sort of EDS spectrum, respectively. A calculated chemical assay can be derived from the mineral mode, densities and quantified mineral chemical compositions for each EDS spectrum. The calculated assays represent an approximation of the bulk rock composition in the analysed sample surface area and may differ from the bulk composition gained from a larger volume of material.

(2) Particle analysis by EDS (Figure 1b) has been developed for fast automated characterisation of grain mounts with up to 10^6 particles, as for example encountered in milled products from mining and mineral processing. Some versions of this method use the BSE images of segmented particles to analyse each segment with a single EDS spectrum, as a particle can be composed by several different grains. The particle analysis uses only a comparably small number of EDS analyses to define all particles in a sample. The geometrical parameters of particles and grains as well as their liberation characteristics are extracted from the acquired data. This measurement routine allows the quantification of sizes, geometries, associations and liberation of particles with different chemical compositions within a bulk sample and without previous phase separations. In contrast, this is not possible by conventional sedimentological particle size and size distribution analysis methods.

(3) Sparse phase search (Figure 1c) combines a backscattered electron (BSE) grey tone value trigger and single spot EDS spectral analysis of grains. Grains are selected to be measured by EDS when they match the grey tone level of the defined and pre-set BSE grey level range. These specific grains of interest can then be measured either with a single EDS analysis or the spectral mapping technique (reported below). Not only the sparse phase grain of interest but also its surrounding minerals or the complete particle are analysed within a pre-set µm-scale range of distance. This is also appropriate for massive rock applications, such as in thin sections and drill cores, where mineral liberation data is not relevant. When the BSE grey level trigger is set to high and bright values, this enables the detection of rare phases with such BSE grey levels, such as gold or platinum group minerals. The method also allows grain counts, phase area estimates, and the recording of geometrical grain parameters of sparse phase grains of interest. However, it does not provide bulk mineralogy information, as only selected particles in the sample are analysed. The selectivity of the sparse phase search measurement is designed to efficiently measure liberation of trace minerals in tailings and low-grade feed ores. The method also can be followed by further routines, as a subsequent imaging and spectral mapping at a higher magnification.

(4) Spectral mapping by EDS (Figure 1d) combines BSE image analysis with EDS spectral mapping. It employs a narrow grid (= mapping) of single EDS spectra on both granular and non-granular samples. The mapping can be performed by defined BSE image grey level triggers or by EDS spectrum triggers. Grains which are not selected for EDS mapping will be analysed with a single EDS spectrum. The method is applied where fine details of mineral intergrowths are of interest, especially in particles where grains with similar BSE grey levels and thus similar average atomic numbers (z-numbers) but different chemical compositions are associated and thus cannot be segmented. Such features are not applicable by the particle analysis method. In contrast to the point counting method, the geometrical parameters of particles and grains are fully extracted from the BSE image and the related EDS data.

2.2. Software and Hardware Solutions for Measurement Routines

The image analysis and processing software controls the automated to semi-automated measurements and allows thorough data processing and comprehensive results extraction. Most of the AM providers bundle their image analysis software to their own SEM hardware, such as TESCAN, ZEISS and FEI (now part of Thermo Fisher Scientific). Others allow the usage of their software at any hardware, such as the vendors of EDS spectrometers as Oxford Instruments and Bruker. Summaries of software suites, in common use at present, are described in the following.

An important feature which is poorly explained in the software manuals is the classification of the EDS spectra as a mineral or phase. Such a spectrum is generated by the digital pulse processing of the signals from an X-ray-collecting silicon drift detector and expressed by normalised counts/s per channels along the keV scale (Figure 2). Classification means the comparison of such a single sample EDS spectrum to an extensive list of reference EDS spectra. The FEI-MLA software allows a list of 250 reference spectra; other AM systems can handle more spectra or derivational specific parameters. According to this comparison, the sample EDS spectrum is assigned to the matching reference EDS spectrum. For example, a sample EDS spectrum is assigned to "Albite" when it matches a reference EDS spectrum which has been labelled as "Albite" from the reference EDS spectra list. This list of reference EDS spectra can be compiled by various procedures. Reference EDS spectra can be gained from the sample and/or known reference materials and furnished with corresponding mineral names and/or by generic labels derived from elemental analysis [33,34]. It is also possible to build up a list by constructing synthetic reference EDS spectra by suitable software.

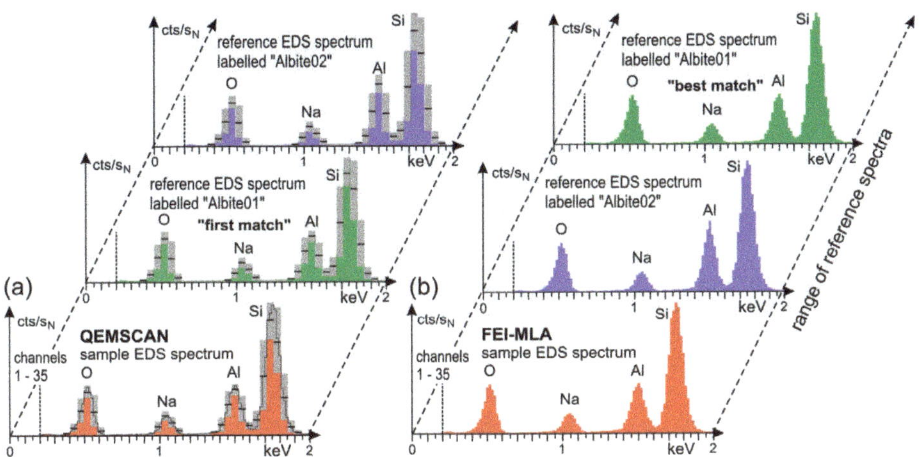

Figure 2. EDS spectra classification modes. Channels 1–35 along the keV scale are not considered and values > 2 of keV scale are omitted for presentation. Scales have been adjusted for visualisation. (**a**) FEI-QEMSCAN software: The sample spectrum is segmented by embracing several channels and compared to a list of reference spectra. Note slightly different counts for Al in the reference spectra. The reference spectra have user-defined minimum and maximal counts, and positions in the keV scale for each segment. The comparison to the range of reference spectra is performed by the "first match principle". (**b**) FEI-MLA software: A Chi-Square test is performed for comparison of the sample EDS spectrum to each of the reference spectra in the list. The comparison to the list of reference spectra is performed by the "best match principle", within user-defined probability limits.

QEMSCAN or Quantitative Evaluation of Minerals by SCANning Electron Microscopy (FEI Company, Hillsboro, OR, USA) is the oldest of all AM solutions, but the sale is terminated now. However, QEMSCAN systems are still in use all over the world. The software suite is bundled to the SEM hardware platforms by the FEI Company. The data acquisition has a strict X-ray centric approach. This means that, from each analysis point, both X-ray spectra and BSE grey value are acquired in a measurements grid (comparable to classical point counting). Thus, the level of detail of the analysis is dependent on the distances in the grid of analysis points. While measured, EDS spectra are acquired and saved for each analysis point. The EDS spectra of the measurement are compared against a Species Identification Protocol (SIP). For the SIP, the count rate information from single channels along the keV scale is condensed (Figure 2a). Two methods of SIP to EDS spectrum comparison can be used. The first (and older one) is the peak intensities method, which compares the exact energy ranges of the elements and the number of X-ray photons in the energy range. The second (and younger one) is the concentrations method which simulates the acquired EDS spectrum by a synthetically computed spectrum. Following this, the elemental composition of the synthetic spectrum is calculated. The concentrations of elements present in the calculation for the spectra are compared against the reference calculations in the SIP. For mineral identification, QEMSCAN uses a first match principle (Figure 2a). This means that, when a matching entry in the SIP list is found, the following entries are not considered anymore [35]. The software for data processing and results extraction is powerful and complex but requires an expert user.

Five types of measurement modes can be applied in QEMSCAN measurement software (version 5.3.2., FEI Company, Hillsboro, OR, USA), BMA (Bulk Mineral Analysis), PMA (Particle Mineral Analysis), SMS (Specific Mineral Search), TMS (Trace Mineral Search) and FieldImage/FieldScan. The BMA mode is a one-dimensional line scan measurement mode which collects EDS data from the x-direction in an x–y coordinates grid only. Before EDS acquisition, the background is distinguished from the particles using a BSE threshold. Only analysis points falling on a particle are measured. As the full sample area is not measured, but lines of intersection on the particles, BMA is a very fast measurement mode. However, only modal mineralogy, calculated assay, estimated particle and grain sizes and mineral association data are obtained. PMA mode is an EDS mapping measurement mode which analyses the full sample area. A custom mapping resolution can be set by defining the interval of analysis points. A false colour mineral map is stored for every single particle, but not a full field map as in FieldImage mode. Before EDS acquisition, the background is distinguished from the particles using a BSE threshold, so that particles are analysed, but not the background. In PMA mode, all particles are analysed, unless a specific filter criterium, for example, the upper particle size was set. From the PMA measurement mode, the full range of results provided by the QEMSCAN software can be extracted. This includes quantitative modal mineralogy, grain sizes, mineral association data, liberation and locking statistics. The SMS and TMS measurement modes are search modes which can be used to find minerals of interest which are present in low amounts or in traces in the sample material. Only particles containing matching mineral grains will be analysed but not the full sample area. Typical examples for areas of application of the SMS mode are the investigation of sulphides in tailings or the search for penalty element-bearing minerals in concentrates. A typical example for an area of application of the TMS mode is the search for precious metal grains in the sample material. The SMS and TMS modes use a custom BSE grey value as a threshold for searching. For these measurement modes, the analytical procedure is similar to PMA. As only a subpopulation of particles in the sample is analysed, SMS and TMS do not provide bulk mineral information. However, the list of results is similar to PMA, with respect to the mineral selective acquisition of data. The FieldImage mode (also known as FieldScan) is an EDS mapping measurement mode which is typically applied to thin sections or coarse-grained particles. It is comparable with the classical point counting procedure at an optical light microscope. FieldImage analyses the full sample area. From each field, a false colour mineral map is stored, so that a composite image of the sample can be produced after measurement. Before EDS acquisition, the background is distinguished from the particles using a BSE threshold, so that only measurement points falling on particles are analysed. The FieldImage mode allows the quantification of the modal mineralogy but also investigating textures and fabrics of larger samples [25,29,36]

By the MLA or Mineral Liberation Analysis software (FEI Company MLA) for each analysis point, a full EDS spectrum is stored during analysis. After completion of the measurement, the spectra are compared (= classified) against a list of reference EDS spectra. The comparison is performed using a Chi-Square difference test. Here, the full spectrum pattern is compared against all the reference spectra pattern. The sample spectrum classification is based on a best match principle. This means that the sample EDS spectrum will be assigned according to the reference EDS spectrum with the highest matching score or probability according to the Chi-Square difference test (Figure 2b). The matching score or probability of match can be selected and adjusted. When a sample EDS spectrum has no matching reference EDS spectrum within the given probability of match, it will be classified as "unknown". As the reference EDS spectra can be directly gained from the sample and receive then their labelling from the operator, the compilation of the reference spectra list is comparably time-saving, simple and user-friendly. This considerably facilitates the characterisation of complex artificial materials as slags or ashes. The composition could be qualitatively estimated by an EDS analysis at the same location where the reference EDS spectrum has been gained. However, as not every single elemental peak is compared one-to-one, complex spectrum pattern, showing many peaks or several neighbouring peaks, can cause misidentification. To overcome this issue, advanced classification rules can be used to allow, force or deny specific elements.

Four types of measurement modes can be applied in the FEI-MLA measurement software (version 3.1.4., FEI Company, Hillsboro, OR, USA), the point counting modal analysis XMOD, a particle analysis routine labelled as XBSE, a sparse phase search routine labelled as SPL, and an EDS spectral mapping mode, named as GXMAP. These principal measurement methods are completed by subroutines, which are designed for the automated collection of reference EDS spectra. For the sparse phase measurement method, subroutines including double zoom or combinations with spectral mapping are also available.

Another AM solution software is TIMA-X (TESCAN Integrated Mineral Analyzer TIMA) (TESCAN ORSAY HOLDING, a.s., Brno, Czech Republic). TIMA-X is based on TESCANs MIRA or VEGA SEM platforms. The software was introduced in 2012. Depending on the analysis mode, the sequence of analytical steps is different. Four acquisition modes are available for TIMA-X. In the 'High Resolution Mapping' mode (resembling EDS spectral mapping), from every measurement point, the BSE signal is taken first. For points above the background BSE threshold, the EDS spectrum is acquired subsequently. BSE and EDS spectra are collected in the same regular grid. Thus, it is very precise but slow and can be used for analysis of samples having complex textures or when a high mapping resolution is required. The 'Dot Mapping' mode starts like 'High Resolution Mapping', by collecting the BSE signal in a high-resolution grid. This is followed by the low-resolution EDS spectrum acquisition in a somewhat wider grid than the BSE grid. Again, the BSE signal is used to exclude areas of background, such as epoxy resin, from analysis. The BSE signal is not only used for background removal, but also for particle segmentation. This mode is faster than the 'High Resolution Mapping' mode, but still enables precise and detailed analysis results. In the 'Point Spectrometry' mode (resembles particle analysis), a high-resolution BSE image grid is collected to detect areas of background and identify areas of constant BSE image grey levels (homogeneous segments) in the particles. Each segment is analysed using a single EDS spectrum acquisition point in the centre of the segment. The whole segment will be assigned to the mineral which corresponds to the composition of the single EDS spectrum point. It should be noted that minerals having similar BSE image grey level, but different chemical composition exist. Thus, the 'Point Spectrometry' mode can lead to wrong results if such minerals occur together in a sample. This mode is very fast, but less precise and should be used only if a good BSE image grey level contrast among the phases is ensured. The 'Line Mapping' mode starts by acquiring the high-resolution BSE image grid for background and particle segment identification. The EDS spectrum points are placed in short regular distances on horizontal lines. The distance between these lines are larger. After EDS spectrum acquisition, the lines are divided into linear sections using BSE and EDS signal information. This mode has a high analysis speed but does not provide the full set of quantitative textural data, due to the one-dimensional type of analysis. TIMA-X can be operated in four measurement analysis types, in particular, 'Modal Analysis', 'Liberation Analysis', 'Bright Phase

Search' and 'Section Analysis'. The 'Modal Analysis' can be used only in 'High Resolution Mapping' acquisition mode and is suitable especially for large area samples. 'Liberation Analysis' can be used in all four acquisition modes mentioned above. This analysis type is most suitable for particulate materials and delivers modal information, process mineralogy-related parameters such as mineral liberation as well as textural information. The 'Bright Phase Search' analysis type can be used to search for rare phases. The BSE signal, chemical composition, or both can be used as a threshold for searching. This analysis type can be used in all acquisition modes, except 'Line mapping'. 'Section Analysis' can be used in all four acquisition modes and is designed for the analysis of thin sections. The classification of the sample EDS spectra is performed already online during the measurement. The mineral or phase identification is performed by comparison of the sample spectra against a library. The library consists of mineral definition rules which can be either determined automatically from standard spectra or are calculated from a theoretical composition or are user-defined. Multiple information can be used here in combination such as BSE image grey levels, EDS spectral counts and/or their ratios. In addition, information from other detectors such as cathodoluminescence (CL) or secondary electrons (SE) can be used for mineral identification. The elemental composition of the minerals can be computed from the EDS spectra directly for library entry, if the theoretical stoichiometric composition does not satisfy. This produces first results already by the time when the analysis is finished. The TIMA-X software has an integrated quantitative EDS analysis which allows the calculation of the elemental composition of minerals directly from TIMA-X measurement data [14,37].

The SEM-AM software (Carl Zeiss Microscopy Ltd., Cambridge, UK) routines labelled as Mineralogic Mining are bundled to the SEM instruments ZEISS SIGMA and EVO. The MinSCAN Ruggedized SEM platform allows the application even in a mine-site environment. Mineralogic Mining features four measurement modes: mapping, spot centroid, feature scan and line scan. A high-resolution BSE image is always taken first. There is a measurement routine which uses the different BSE image grey levels to classify the minerals. As this mode does not acquire EDS spectra, it is the fastest analysis mode. However, there is no published information about the quality of results. The 'mapping analysis' is comparable to point counting in optical light microscopy. This full analysis method is accurate but slow. 'Spot centroid analysis' is a particle analysis routine and measures the centre of each segment in the BSE-segmented grains within particles. The composition of this analysis point is then assigned to the segment. Due to the low number of analysis points, the analysis is very fast, but susceptible to BSE image grey level related issues (see explanations in section TESCAN-TIMA-X). The 'feature scan' mode, an EDS spectral mapping routine also segments the particles, using BSE signal information, but uses an X-ray grid to measure the particles. The EDS spectra of each segment are summarised, and the average composition is assigned to the segment. This mode is placed between 'mapping analysis' and 'spot centroid analysis' in speed and accuracy. The 'line scan' analysis mode measures EDS spectra of points along horizontal lines which go through the centre of the particles. This is a fast analysis mode which provides valid bulk mineralogy results and gives an indication about textural information. The full quantitative EDS spectrum classification is performed already during the measurement. Thus, the analysis results are ready when the analysis is finished. The elemental composition of a mineral is quantified directly from its EDS spectrum [32,38].

The Advanced Mineral Identification and Characterization System AMICS was developed by Ying Gu, the original inventor of Mineral Liberation Analysis (MLA) software (Bruker Nano GmbH, Berlin, Germany), and purchased by Bruker in 2016. There is no direct bundle of AMICS to a hardware platform, as Bruker does not produce SEM instruments. AMICS analysis starts with a high-resolution BSE image acquisition. After several image processing steps, this is followed by EDS spectra acquisition. The mineral identification is performed online during analysis. This allows direct results extraction after analysis. AMICS software allows for calculating relative error for modal data. [39,40]

The INCAMineral software as a part of the INCAFeature particle analysis solution (Oxford Instruments plc, High Wycombe, UK) was launched in 2012. AZtecMineral was introduced in 2019 (Oxford Instruments) and is part of the AZtecFeature particle analysis system. Both products are not

SEM bundled but can be used with a wide range of SEM instruments. At the beginning of analysis, a BSE image is taken and processed. This is followed by EDS spectra acquisition. The mineral identification is made online during analysis based on measured chemical composition. The classification can be based on single element measurements or ratios of elements. The results of analysis can be explored in the software but can be exported to HSC Chemistry software too [30,41–43].

2.3. Uncertainties and Problems of Phase Identification by EDS Spectra

The SEM-based Automated Mineralogy usually claims that minerals can be identified by their EDS spectrum, as is exemplified in Figure 2 by the feldspar mineral albite. However, this cannot be fully correct, as minerals are characterised in the first place by their crystal lattice in XRD and only in the second place by their elemental composition as given by quantification through an EDS spectrum. In consequence, the identification of a mineral or a phase by its EDS spectrum remains incomplete, as it considers only the elemental composition. In addition, an identification by the chemical composition alone is severely hampered by the occurrence of numerous minerals which crystallise as solid solution. There are examples of minerals with identical chemical composition, but different crystal structure, as ilmenite and pseudorutile (reported below). A further challenge of mineral identification and discrimination are minerals having rather similar elemental composition. Magnetite (Fe_3O_4), for example, is composed of 72 wt% Fe and 28 wt% O, where hematite (Fe_2O_3) is composed of 70 wt% Fe and 30 wt% O. For both minerals, the EDS spectra look relatively similar and the slight differences in the Fe and O peaks cannot be resolved. In such cases, the BSE image grey level is used as an additional differentiation criterion. However, a specific BSE contrast and brightness calibration is needed for such a measurement.

Another aspect is that the detection range of EDS spectrometers does not cover the whole periodic system of elements. The first light elements, H, He, Li, and Be cannot be detected by EDS spectrometers, due to their design. This causes specific problems in mineral identification. Hydrogen and/or water are components of numerous minerals. As they cannot be detected, minerals such as gypsum ($CaSO_4 \cdot 2H_2O$) and anhydrite ($CaSO_4$) cannot be differentiated. In addition, albite cannot be distinguished from zeolites that contain H_2O (Figure 3a,b). Thus, a naming of the phase in the form of 'mineral A/mineral B' is required, in order to show that it can be either the one or the other. The elements lithium and beryllium occur in several minerals in significant amounts. As both are not detectable by EDS, additional data, such as X-ray diffraction (XRD) analysis or XRF bulk sample composition, are required to identify the Li or Be-bearing minerals. Fortunately, the majority of Li or Be-bearing minerals have often characteristic other elemental ratios outside of Li and Be, so that they can be identified and differentiated by these specific characteristics. For the lithium-bearing minerals petalite and spodumene, this has been shown [44], as they can be distinguished by their different Si/Al ratios (Figure 3c,d). The micas are a mineral family with a wide range of solid solution along different single and coupled element substitution vectors combined to the occasional occurrence of lithium (Figure 3e,f). Mineralogical expertise is required here for a clarification. There are numerous examples of complex composed minerals in nature, as for instance the REE-bearing minerals [33]. Such minerals have numerous peaks und subpeaks of EDS at different energy levels (Figure 3h). Additionally, the REE-bearing mineral groups are characterised by the occurrence of H_2O, CO_3 and F at low energy levels, whereas the peaks for the light REE show considerable interferences at intermediate energy levels (Figure 3h). Phases showing numerous peaks in their EDS spectrum can be expected in artificial products, as slags or ashes too (Figure 3g). Such phases often consist of glass or amorphous material which cannot be characterised and identified by XRD. In such cases, not a mineral naming but a generic labelling of the reference EDS spectra offers a pathway for the successful characterisation of these materials by SEM-AM [33,34].

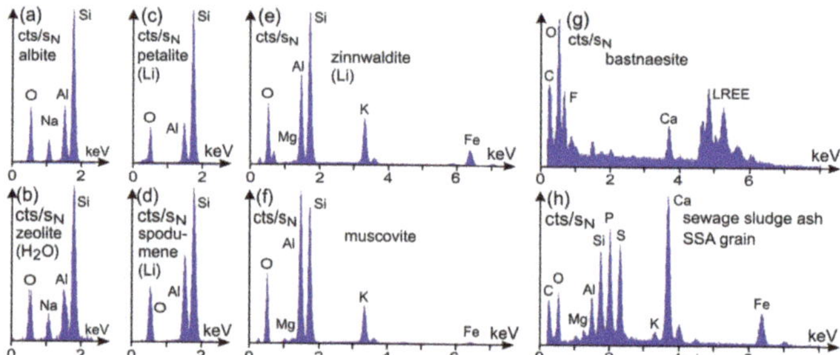

Figure 3. Challenges for classification of EDS spectra. (**a,b**) Two different phases have similar element contents. One of the phases contains H_2O which can not be detected by EDS. (**c,d**) The element Li cannot be detected by EDS. Different Al/Si ratios of the minerals and independent detection of Li by another method (XRF of the bulk sample) allow a distinction. (**e,f**) Example of the widely ramified mineral group of mica. Some members of the mica mineral group have the undetectable element Li. The Al/Si and Fe/Mg ratios vary due to substitutions. (**g,h**) EDS spectra with numerous peaks. In the REE-bearing mineral bastnaesite, many peaks are close together at the low energy range while overlapping peaks of LREE at higher energy ranges complicate element identification and quantification. Artificial materials like sewage sludge ash (SSA) bear many elements in unusual combinations which are rarely realised in nature.

3. Sample Preparation and Related Issues

An optimal sample preparation is a crucial step for any SEM-AM analytical success. There are several possible sample types that can be analysed with SEM-AM. It depends on the size of the SEM sample chamber and the configuration of the holder systems. For granular and particulate matter, preparations as grain mounts in round epoxy blocks with 25, 30 and 40 mm in diameter are mostly used. Petrographic glass-mounted thin or thick sections (standard size is 48 × 28 mm) are often used for preparation of compact and massive matter as rocks. It is also possible to produce thin grain mounts with 25 mm in diameter on glass. It is necessary that round blocks and samples on glass can be mounted parallel to the BSE detector and perpendicular to the electron beam.

For samples of granular, particulate and non-compact matter, as broken, grinded or hand-picked single grains, the grain mounts in epoxy blocks are the best form to prepare [45]. Because most of the AM software packages are not able to separate grains with the same grey scale in the BSE image, it is useful to stir graphite (quality fine and pure) as a distance material into the epoxy resin blocks [45,46]. Some granular sample materials show a wide range of densities among the phases. Thus, when the material is stirred with the graphite-saturated epoxy resin, there will occur phase separation with gradation of particles with high densities and with large grain sizes toward the bottom of the block. While analysing the polished bottom of such a block, the less dense and small grains will be missed. For these kinds of samples, the round block will be cut into vertical slices. These slices will be remounted as a vertical section [47,48]. With some EDS detectors, it is possible to also study material as coals or synthetic polymers. Because the BSE grey value of this organic matter is the same as for epoxy resin, one should use an alternative embedding material [49]. Carnauba wax can replace the epoxy resin [50], but it is a very soft substance and not easy to polish. For better stability, carnauba wax blocks of 25 mm can be double mounted in epoxy resin blocks of 30 mm diameter. Another possibility is to dope the epoxy resin with iodoform [49,51]. The epoxy resin then has a higher molecular number than the organic matter and can be therefore extracted as background. There are numerous sorts of epoxy resin available. In addition, the filler and hardener proportions can be varied. The main difficulty is to choose an epoxy resin which hardens within convenient time frames and temperature conditions, which does not evaporate under high vacuum conditions and which remains stable under the electron beam of 25 kV. Continuous application tests are the recommended way to solve the problem.

The preparation of thin and thick sections is quite simple if the sample material is massive, solid, compact and dry. If the sample is porous and or brittle, a previous impregnation with epoxy resin stabilizes the material before sawing. There are several authors, which describe the procedure of thin section production [52,53]. For the lapping of the sample material previous to the mounting on glass, mostly silicon carbide (600–1000 mesh) is used. For soft and brittle material, a SiC 1000 leads to better results with a minimum of substance loss, instead using the SiC 600 for the standard lapping procedure. Thin sections have the advantage that one can check the minerals or phases with an optical microscope, if there are minerals with a close chemical composition but different optical properties. In addition, glassy phases can be recognised under the microscope with polarised light by their optical isotropy. This information is helpful when creating the reference EDS spectra list.

All SEM-AM analyses of grain mounts in epoxy blocks and of thin and thick sections need a well-polished and plane surface. The polishing is always a work of craftsmanship because all materials need a specific treatment. Mostly the polishing procedure is performed with water. If there are minerals or materials which can be dissolved in water or react with it, they can be prepared with water free liquids like ethylene glycol [48]. This is recommended for many sorts of industrial ashes (e.g., sewage and power plant ashes), as these can contain anhydrite. To prevent the formation of a relief due to different grades of hardness among the particles, it is recommended to use polishing plates covered with hard textile cloths. Polishing plates with embedded diamonds reduce the relief. If there are ore minerals or soft metals or minerals, soft polishing cloths with long fibres will give convenient results. The polishing procedure is performed in several successive steps (3 to 5 steps, dependent on the material) with decreasing grain sizes, e.g., abrasive papers followed by grinding and polishing powders on textile cloth. It is not advised to use the traditional polishing plates with lead-bearing alloys. Although they produce relief-free surfaces due to the fixed polishing grains, a severe sample contamination by the lead cannot be excluded. Diamond powder with lubricant or diamond paste with a grain size of 1 μm on textile cloths is very effective for the last step of polishing. It is recommended to use a microscope with reflected light to control the results during the successive polishing steps.

A carbon coating of the polished samples for the dissipation of the impinging electrons is essential to obtain optimal BSE images. Otherwise, even a partial charging of the sample surface will severely hamper the analysis. The evaporation of a carbon-loaded thread under vacuum (10^{-4} Pa) leads to a carbon layer with a constant and reproducible thickness of several nanometres on the underlaying samples. The thickness of the carbon coating layer can be controlled by the carbon load of the thread, e.g., 27 g/m load of carbon produces a layer of ~6 nm thickness on a sample surface of 50 cm^2. Alternative carbon coating methods with carbon rods and electronic thickness control are less economic.

4. Case Studies

4.1. Applications in Petrology and Applied Sedimentology

Petrological investigations on igneous, metamorphic and sedimentary rocks are mainly based on polished thin sections. The thickness of a thin section is usually ~30 μm, which allows for optical microscopy under transmitted polarised light. Modern in situ analytical methods with a spatial resolution at the μm-scale as electron probe microanalysis (EPMA), laser ablation inductively coupled plasma mass spectrometry (LA-ICP-MS) and secondary ion mass spectrometry (SIMS) require a precise imaging and characterisation of the potential analytical target positions. Zircon, monazite and an increasing number of other minerals (e.g., rutile, sphene, garnet) are subject of in situ radiochronology. The suitable mineral grain sizes are <200 μm and often hardly to be detected and located by optical microscopy. Measurement routines of SEM-AM to find rare mineral grains, labelled as sparse phase search, initially designed for detection of gold and platinum group minerals, provide an efficient tool for the preparation of in situ dating techniques. In the case of EPMA-Th-U-Pb monazite dating, the potential target grains in meta-psammopelites were detected and located in thin sections by the sparse phase search routine using a BSE grey level trigger of bright grey (i.e., >100) at a gold-calibrated

0–254 grey scale [13,52]. A catalogue of all monazite, xenotime and zircon intermineral relationships is gained by the measurement (Figure 4a,b). This was used to select monazite grains for detailed investigation in BSE imaging under the SEM, and quantitative wavelength-dispersive (WDS) analysis with EPMA. The same routine can be also used for distinguishing different generations of zircon in a set of rhyolite samples [53].

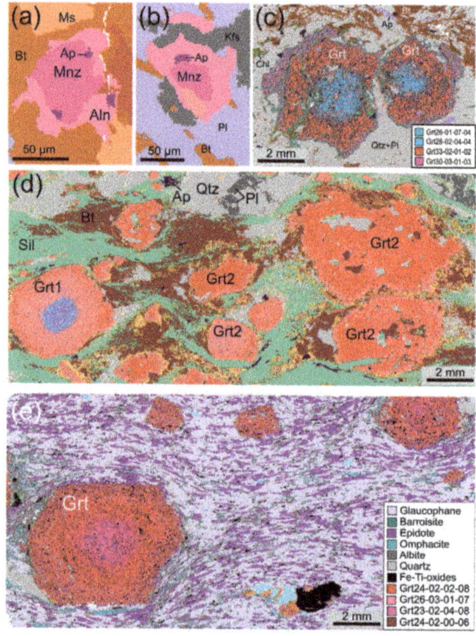

Figure 4. SEM-AM in metamorphic petrology applications. (**a,b**) Sparse phase search method as applied for the search for monazite in micaschists. Monazite (Mnz) for EPMA-Th-U-Pb dating is characterised by coronas of allanite (Aln) and apatite (Ap); matrix minerals are biotite (Bt), K-feldspar (Kfs) and muscovite (Ms). (**c**) EDS spectral map of garnet with chemical zonation in a micaschist, modified from [54]. The EDS spectra for garnet are generically labelled according to the contents in Fe-Mg-Mn-Ca of a normalised analysis at the same position where the reference EDS spectrum has been gained in a porphyroblast and the corresponding pixels are coloured (see text). Ap—apatite; Bt—biotite; Chl—chlorite; Pl—plagioclase; Qtz—quartz. (**d**) EDS spectral map of a cordierite-sillimanite-garnet gneiss (kinzigite) from Saxonian Granulite Massif. Garnet 1 (Grt1) and garnet 2 (Grt2) porphyroblast generations can be distinguished by the lacking of Ca-rich garnet 1 core with corresponding pixels of the EDS spectra labelled by blue colours. Sil—sillimanite. (**e**) EDS spectral map of a glaucophane eclogite from metamorphic Zone 1 in Ile de Groix. EDS spectra for classification of zoned garnet are labelled generically and the corresponding pixels are coloured.

The study of unzoned and zoned garnet in micaschists, gneisses and metabasites is of special interest for geothermobarometry [13,54] and age dating [55]. The WDS element mapping by EPMA is considered as the best method for the characterisation of zoned minerals. However, this method is time-consuming and thus hardly suitable to investigate a multitude of garnet grains in a 25 × 45 mm sized sample. Here, the EDS spectral mapping by SEM-AM provides a faster solution. The routine produces a narrow grid of ~1600 single EDS spectra per mm^2 in grains selected by a BSE grey scale trigger. For the classification of the sample EDS spectra, a list of reference EDS spectra was established by collecting spectra from defined parts of several zoned garnet porphyroblasts (core-mid-rim). It is indispensable to use sample garnets for this procedure, as the spectra from the available standard garnets and arbitrary other garnet-bearing rocks will not match due to the highly variable different element compositions in natural garnet solid solutions. Garnet reference spectra are further characterized by EDS single spot elemental analyses which revealed strong variations of Fe, Mg, Mn and Ca in the porphyroblasts. In a next step, the reference spectra were labelled in a generic way with the corresponding garnet Fe-Mg-Mn-Ca compositions. When the pixels which correspond to the labelled spectra are arranged in a colour scale, they visualise semi-quantitative garnet zonation maps (Figure 4c). The measurements were classified against the reference EDS spectra list with a high degree of probability of match [13,54]. The spectral mapping allowed for selecting garnets with a well-developed and complete concentrical chemical zonation out of dozens of porphyroblasts for quantitative WDS analysis by EPMA (Figure 4d). For the EDS spectral mapping of metabasites, it is recommended to adjust the BSE grey level trigger for an analysis of the complete sample excluding the glass and epoxy background at the margins (30–254, based on calibration at gold 254). The reason behind this is that the BSE grey levels of pyroxene, amphibole, biotite, chlorite, epidote and garnet in the metabasites are relatively similar (Figure 4e).

Coarse-grained rocks or materials with grain sizes above 5 mm are problematic to be analysed for their mineral mode and bulk rock chemical composition. For bulk rock chemical analysis, a corresponding large volume of rock has to be comminuted and then thoroughly homogenised. SEM-AM offers a novel way to relate the bulk material chemical data to a mineral or phase modal content by the analysis of grain mounts from the same pulverised sample. This has been used for the characterisation of lithologically zoned coarse-grained lithium-caesium-tantalum pegmatites [56].

The application of SEM automated mineralogy to sedimentary rocks is a fast evolving field. An important routine application is the investigation of heavy mineral separates in provenance studies or as a completion of bulk rock geochemical analyses [57]. Interesting applications of the EDS spectral mapping routine have been reported from the experimental petrology of sedimentary rocks. The mineralogical and physical property changes linked to geo-chemical alteration processes in chalk are of great interest for the oil and gas industry. Ultra-long-term tri-axial tests enclosing the flooding by $MgCl_2$-brines on 7 cm long cores of a reference material (outcrop chalk from Belgium) were performed under reservoir conditions [58]. It is shown that newly formed crystals of magnesite containing minor calcium impurities crystallised together with clay-minerals in the fine-grained calcite matrix. Dolomite or low- and high-Mg-calcite are not observed. Textures of larger micro-fossils are often preserved, but the mineralogy of their shells is altered. A sharp transition zone along the alteration front shows the highest porosity in the cores. This pattern resembles that the alterations are driven by phase dissolution and subsequent precipitation. Compositional variations of the injection fluid effectively control the amount of chemical reaction in chalk. This allowed for predicting changes in geo-mechanical parameters induced by mineral replacements [58,59].

4.2. Characterisation of Granular and Particulate Raw Materials

4.2.1. The Ilmenite-to-Leucoxene Alteration Process

The development of the SEM Automated Mineralogy analytical systems was mainly driven by the demands of mineral processing. This encloses the analysis of polished grain mounts with particle sizes mostly < 200 μm by the particle analysis routines and increasingly by the EDS spectral mapping measurements. The alteration of ilmenite ($FeTiO_3$) is a near-surface process which controls the economic value of placers. Degree and type of the ilmenite alteration also has an important influence on the beneficiation of Ti-bearing placer sediments by magnetic and gravity separation methods. It has been shown by ore microscopy and microprobe analyses that the alteration of ilmenite is characterised by a continuous loss of Fe and gain of OH that finally leads to leucoxene ($TiFe)_3O_6(OH)_6$. The process can be compared to an in situ leaching of ilmenite, as often porous phases with less density than ilmenite appear during intermediate stages. By involving powder X-ray diffraction (XRD), the alteration sequence has been established as ilmenite-leached ilmenite/pseudorutile-leached pseudorutile-leucoxene. The leucoxene was found to be a fine-grained polycrystalline aggregate of rutile [60]. By XRD, the leached ilmenite can be characterised as pseudorutile due to its changed lattice parameters. The XRD method cannot distinguish among large crystals as a single grain and aggregates of rutile microcrystals as leucoxene. The ilmenite alteration sequence was addressed in a novel study by SEM-AM methods by applying a list of generically labelled reference spectra which cover the Ti-Fe compositions of ilmenite s. str. (Ti 31, Fe 36 wt%), pseudorutile (Ti 36, Fe 28 wt%), leached pseudorutile (Ti 40, Fe 18 wt%) and leucoxene (Ti 60, Fe < 5 wt%). About 30,000 particles in polished grain mounts were analysed by a narrow grid (~10×10 µm) of single EDS spectra [61]. This measurement mode allows to resolve chemical heterogeneities within particles showing no or only slight variations of grey level values in their BSE image. Classification by the special Ti-Fe-spectra list provides the degree of ilmenite-to-leucoxene alteration for each particle. The ilmenite alteration features are visualised in detail by relating distinct colours to the pixel areas defined by EDS spectra related to different contents of Ti-Fe, reported as Fe in wt% in the legend (Figure 5a–e). Altered ilmenite particles are characterised by core-mantle structures with ilmenite in the cores and several successive coronas of pseudorutile followed by leached pseudorutile and then leucoxene Ti-Fe compositions toward the rims (Figure 5a,b). Although the leucoxene still contains small amounts of Fe (<5 wt%), it cannot be properly distinguished from rutile (TiO_2) by EDS and XRD spectra. As a consequence, particles with the "inverse" zonation, in detail visible by leucoxene/rutile cores mantled by pseudorutile or ilmenite-typical Ti-Fe compositions, then should be interpreted as preserved relics of initial magmatic or metamorphic origin (Figure 5c–e). Summarising the alteration in all Ti-Fe particles in combination with virtual particle size classing then provides indications for beneficiation concepts.

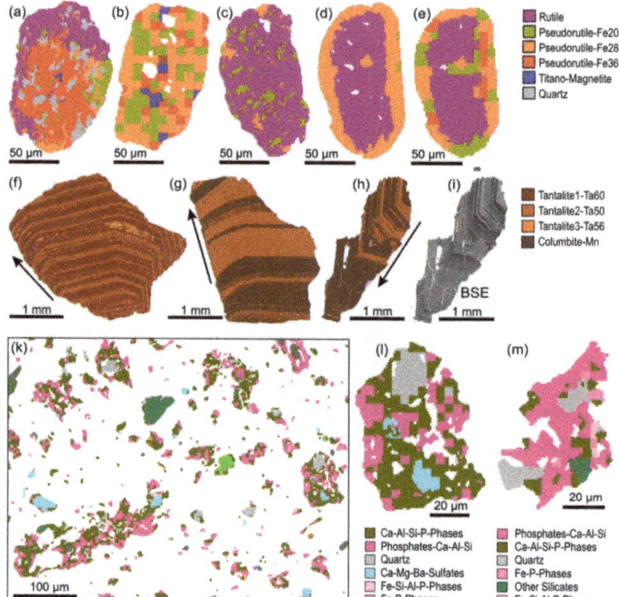

Figure 5. SEM-AM in particle analysis applications. (**a–e**) EDS spectral mapping and particle analysis (coarse pixels) for the characterisation of various grains in a sieved heavy mineral concentrate from a river sand. The pixels corresponding to generically labelled EDS spectra are coloured according to the ilmenite to leucoxene (rutile) natural weathering sequence. (**f–i**) EDS spectral map views and BSE image of zoned tantalite-columbite grains in an ore concentrate from Africa with an unknown provenance. Growth direction of oscillatory zonation is indicated by arrows. (**k**) Frame view of an EDS spectral map from a sewage sludge ash sample, see legends of (**l,m**). (**l,m**) Details of sewage sludge ash particles as seen in spectral map with legends combining mineralogical and generic spectra labelling. The legends below the particles are arranged according to decreasing pixel number representing the area %.

4.2.2. Nb-Ta Ore Fingerprinting

The Analytical Fingerprint (AFP) method is a scientific tool which can be used to check the documented origin and provenance of tin, tungsten, and tantalum (3T) ore mineral shipments. AFP is designed as an optional proof of origin within the framework of mineral certification. The process of AFP combines geochemical features of ore grains analysed by LA-ICP-MS with SEM automated mineralogy of ore concentrates in grain mounts [10,62–66]. Six grain mounts of Nb-Ta ore concentrate of unknown African origin were investigated using EDS analysis, SEM-AM as well as transmitted light and reflected light microscopy [67]. Costly further methods such as LA-ICP-MS and U-Pb dating of the mineral grains which are enclosed in routine AFP methodology by the Bundesanstalt für Geowissenschaften und Rohstoffe (BGR) were not applied. One aim of the investigation was to test the feasibility of SEM-AM alone to encircle the provenance of the samples despite the limited analytical methodology. Results were compared with published data from various authors. Tantalite, columbite-(Fe), columbite-(Mn) as well as microlith and tapiolite-(Fe) were determined as Nb-Ta minerals in the samples by EDS spectral mapping. The stoichiometrically overlaying Nb-Ta minerals can be sorted in the columbite quadrilateral with tapiolite ($FeTa_2O_6$), tantalite ($MnTa_2O_6$), columbite-(Fe) ($FeNb_2O_6$) and columbite-(Mn) ($MnNb_2O_6$) endmembers. Minor constituents and accessory minerals of the samples such as cassiterite, wolframite, garnet, iron oxides, tourmaline, staurolite, orthopyroxene, primary and secondary mica are indicators that the host rock of the samples may be a metasedimentary rock. The absence of feldspar, sulphides, secondary tourmaline, quartz, and the abundance of iron oxides are indications of possible greisenisation. The cumulative grain size distribution curves of the samples are indicators of artisanal mining, which is carried out in many coltan deposits in Africa. The mineral associations and a characteristic sequence of various oscillatory zonation trends of the columbite-tantalite minerals document a fractionation trend of Nb-Ta analogous to Fe-Mn described from Kibara deposits [65]. For visualising the zonations, a list of generically labelled reference EDS spectra which enclose variations of the Ta contents in tantalite has been established (Figure 5f–i). The oscillatory zonation trends provide evidence that the samples originate from the same deposit, despite their different modal compositions. Occurrence of tapiolite-(Fe) is a strongly limiting argument for the origin of the samples [9] on the African continent and signals Rwanda and the Democratic Republic of Congo. A further limitation based exclusively on tapiolite-(Fe) is not representative due to the limited availability of comparative data. The Manono-Kitotolo deposit (Democratic Republic of Congo) appears as a possible source for the samples, as both mineralogy of some of the pegmatite zones of the deposit match and an intense greisenisation is known. Due to the low number of investigated samples and the lacking complementary LA-ICP-MS data, this statement is not statistically representative. However, the case study documents the usefulness of the SEM-AM for materials fingerprinting when comparative data are available.

4.2.3. Sewage Sludge Ashes

Sewage sludge ashes (SSA) are an artificial product from the incineration of residual sludge from sewage treatment plants when previous disposal choices are no longer a legal option. The waste management sector started working on commodity recovery from materials previously regarded as disposable. In consequence, industrial ashes are now categorised as potential resources [68]. Due to its considerable contents, sewage sludge ash (SSA) turned into the focus of the recovery of phosphorous [22]. SSA is not only complex in composition, but also diverse, depending on provenance, incineration parameters and other factors. In order to optimise recovery techniques, it is necessary to characterise the primary ash particles in terms of their phosphorous and related element contents, their phase associations, particle and grain sizes, shapes and aggregations. In addition, after experiments with acid leaching, the residual particles need thorough investigation. The X-ray diffractometry (XRD) cannot measure all essential parameters, as the SSA are composed of fine-grained polyphase particles with many amorphous components. Therefore, the EDS spectral mapping method of SEM-AM has been applied as a novel method to powder samples of untreated, treated, digested, rinsed and dried SSA in polished grain mounts. The goal was to study phosphate phase associations and their behaviour towards chemical processes [22,34]. For a better resolution of the BSE image, for the analyses, the upper BSE grey level was calibrated at a maximum value of 254 with Cu instead of gold. Particles smaller than 10 μm were identified by a single measurement point. Fractions of the sample are given in area-% since the density of SSA material is difficult to assess. Keeping proportions representative, densities (even of identified crystalline minerals) were set to one, so that the mode in % of the sample refers to area-%. Proportions of the elemental contents in the reference spectra are in weight-%; thus, shares of phosphorus within the material are also in wt%. EDS spectra of SSA turned out to be complex (Figure 3h), with a multitude of peaks for Si, Al, Ca, Fe, S and P at various counts. Further elements found are Ba, Cl, Cr, Cu, K, Mg, Mn, Na and Ti. Due to small grain sizes and transitions between grains of different compositions, there are also some "mixed" spectra which display a combination of peaks from elements in adjacent grains. These conditions, in addition to amorphous, artificially produced material phases, render an interpretation of the spectra with mineral names as impossible. As a consequence, the list of reference EDS spectra is then composed by natural minerals as quartz, and generically labelled EDS spectra of particles according to their compositional ranges of elements (Figure 5k–m). A further refinement of the investigation is possible: Using the EDS elemental analysis for quantifying the Si, Al, Ca, Fe, S and P-contents, 53 spectra were collected from the samples, which were subsequently used as generically labelled reference spectra for the classification [34]. The classified sample spectra then were grouped under target-element aspects, phosphorus (P) in this case. Spectra groups summarising the spectra within a range of P-contents (in wt%) are then named descriptively according to the P contents, for example <1 %P, 5–10 %P, 10–15 %P, >15 %P. This step is called the target component grouping. Such a target-element focused classification has been also used for QEMSCAN for investigation of eudialyte acid digestion residue [69].

There is an interesting tendency of elemental complexity (number of elements detected) decreasing as P-content rises. With regard to incineration, the reorganisation of P-phases from those generated during P-precipitation in wastewater treatment to those present in ashes does currently not favour a clean recovery of calcium-phosphates. P-recovery always had to deal with Al, Fe and other major elements in the process [70]. In phosphate recycling, recovering a few other elements in addition to P is desirable as it reduces the need for product purification. Particles classified by three distinct EDS spectra are exclusively present in the untreated SSA material. In conclusion, these particles were completely resolved by acid digestion. The amount of such dissolvable particles varies and can be controlled by a pre-digestive thermochemical treatment. Thus, thermochemical treatment resulted in an improved P-recovery [22,34]. In that way, SEM-AM studies of SSA can help to assess different combustion regimes and their suitability towards P-recovery.

4.3. Slags and Ceramics

The characterisation of slags is a very challenging task. It starts by the sample preparation by mounting the often porous material in epoxy blocks and continues in choosing a suitable analytical method. A bulk chemical analysis appears as less problematic. The XRD-Rietveld methods may identify and quantify crystallised phases at modes of >1 wt%, dependent on the material, but mainly fail to characterise the often dominant amorphous or glassy phases. Slags are encountered in numerous domains. They are studied in an archaeological context [71] and for estimations of the environmental impacts and/or the raw material potential of slag heaps. The understanding and optimisation of pyro-metallurgical primary production and recycling processes is a recently growing field of slag characterisation. Waste of electrical and electronic equipment (WEEE) is one of the fastest growing waste streams globally. Therefore, recycling of the valuable metals of this stream plays a vital role in establishing a circular economy. The smelting process of WEEE leads to significant amounts of valuable metals and rare earth elements (REEs) trapped in the slag phase. The effective manipulation of this phase transfer process necessitates detailed understanding. Adequate process control is required to bring these metal contents into recoverable structures [72].

An alternative way for the phosphorus recycling strategy of sewage sludge is the transformation into a slag under high temperatures. Corresponding experiments were performed by the combustion of sewage sludge in ceramic crucibles. The ceramic crucibles with the slag were cut into plates, embedded in epoxy blocks, and then polished and studied by SEM-AM. The EDS spectral mapping is the recommended method, as the slag matrix and the embedded needle-like grains have relatively similar grey levels in BSE imaging (Figure 6a–c). The reference spectra have been generically labelled for the classification [34,73]. The matrix of the slag is dominated by an oxygen-bearing glassy phase with 11 wt% P, 9 wt% Ca and 7 wt% Fe (normalised to 100 wt%), where needle-shaped grains with 10 wt% P and 11 wt% Ca are randomly distributed (Figure 6b–d). Particles in the slag may contain maximal 18 wt% of P. Fe-rich particles are concentrated at the free surface of the slag (Figure 6d). This allows a quantified comparison of experimental slags generated under variable parameters like combustion temperatures, oxygen fugacity conditions and sewage pre-treatment processing.

Another emerging field where SEM-AM can be applied outside the traditional mineral processing domain is the study of ceramics, pottery and stoneware. As slags, the ceramics, pottery and stoneware are often investigated in an archaeological context in order to identify raw material sources and provenance [15,17,74]. Pottery from the Cycladic Bronze Age site of Akrotiri (Thera) was analysed by SEM-AM involving EDS spectral mapping measurements by QEMSCAN [15]. On the other hand, firing and calcination experiments give knowledge about the ancient production processes [75]. To obtain more precise estimates on the temperature regime of sequential ceramic transformations recorded in Sueki sherds from the Nakadake archaeological kiln site cluster (southern Japan), a series of firing experiments along well-controlled temperature-time paths and reducing conditions has been performed [17]. The firing was carried out in sloping single-chamber tunnel kilns at reducing conditions and temperatures exceeding 1000 °C. Set in a vitrified matrix (glass, mullite, minor spinel and cristobalite), quartz and feldspar constitute ubiquitous clast components. The varying degree of vitrification (15–60 wt% glass phase) and reversely correlated modal proportions of quartz and feldspar clasts likely reflect superposed effects of varying exposure time to peak-firing temperature and temperature gradients in the kilns [17]. Comparison of the feldspar melting phenomena in experiments with the archaeological samples suggests that firing temperatures did not exceed ca. 1150 °C [17]. Apart from temperature, the soaking time may influence the reaction process during firing. Experiments with an increased soaking time of 48 h have been performed on clay samples from the Nakadake archaeological kiln site. The illitic clay closely matches the bulk chemistry and particulate mineralogy of low-fired sherds recovered from the site [76]. Firing experiments on dried clay plates (25 × 45 × 5 mm) were performed in 50 °C steps over the temperature range from 700 °C to 1250 °C, using a N_2-flushed tube furnace. After a soaking time of 48 h, the furnace was switched off and the ceramic slabs allowed to cool-down slowly [77]. Petrographic thin sections were produced from the firing experiments and investigated by the EDS spectral mapping mode with SEM-AM. As the material is composed by natural mineral grains as quartz and feldspars, and also by glass phases induced by the melting of feldspars and the clay minerals, two sets of reference EDS spectra were combined for classification (Figure 7). One set of spectra includes the minerals of the raw material. The other set was established by generic labelling of EDS spectra from glassy areas. The latter was completed in a stepwise manner by several classification steps. After a first classification, some larger domains of "unknown" remained in the sample. Then, further spectra were gained from these "unknown" domains and added to the reference spectra set for a second round of classification. By repetition of the steps, a high degree of areal coverage of classified sample domains has been achieved. By means of SEM-AM, a quantitative comparison of the experiments at different temperatures to the archaeological material is possible. The SEM-AM results allow for comparing the experimental results not only for the temperature, but also for other parameters like the soaking time [77].

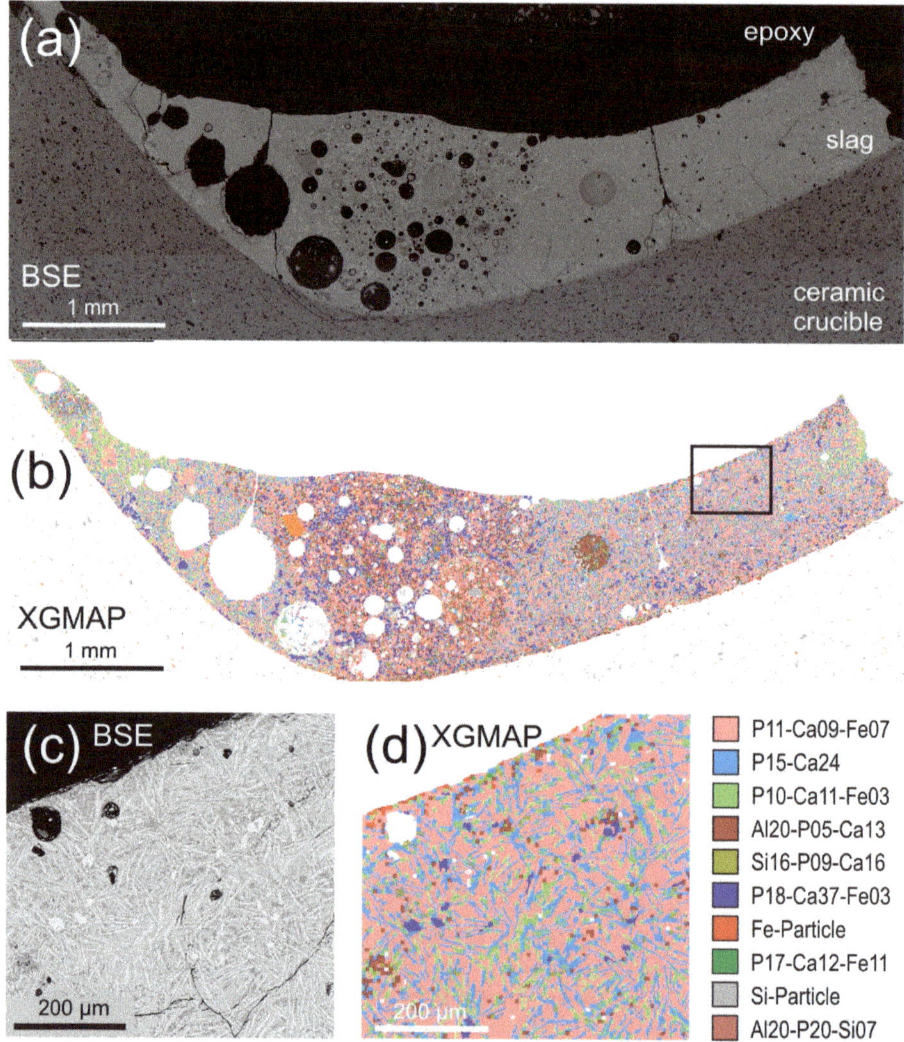

Figure 6. SEM-AM applied for the characterisation of a slag produced by melting of sewage sludge in a ceramic crucible. (**a**) Slag in a backscattered electron image (BSE). (**b**) Corresponding EDS spectral map of the slag. (**c**) BSE image of square marked in (**b**). (**d**) Details of slag in spectral map from (**b**). Needles and grains of P-Ca phases appear with random spatial distribution in a glassy phosphorous-calcium matrix. Fe-particles are enriched at the surface of the slag.

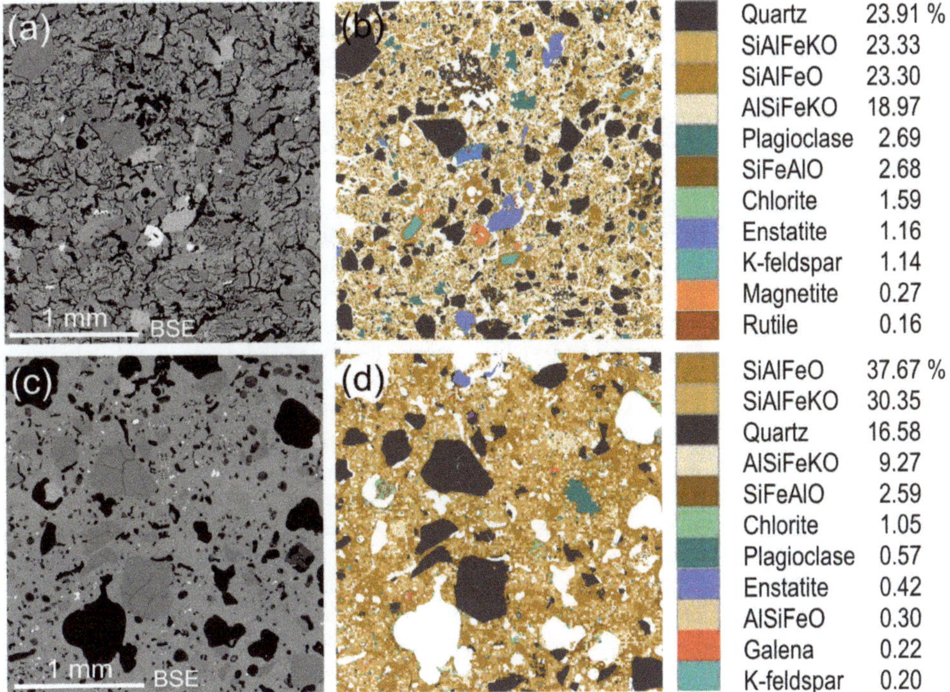

Figure 7. SEM-AM applied for the characterisation of firing experiment products for reproducing the historical stoneware production, modified from [17,77]. (**a,b**) BSE image and related EDS spectral map of a measurement frame of the experimental product at 950 °C. (**c,d**) BSE image and related spectral map of experimental product at 1250 °C. Angular quartz fragments, plagioclase, K-feldspar, chlorite and enstatite grains are embedded in a matrix composed of diverse glassy phases. Matrix components are characterised by generically labelled EDS spectra and correspondingly coloured pixels. Area percent of components and matrix refer to the complete sample. Note the differences in modal proportions of matrix phases in area %.

5. Conclusions

SEM-AM is a combined analytical tool of upgrading an SEM. It was initially developed and adopted for the fast and effective characterisation of metal-bearing ores and their processing products. A main driving force of SEM-AM development was to control and improve the effectivity of mineral processing.

The SEM-AM is organised in several successive steps (1–3) which can be controlled and adopted by the users. A measurement begins by the collection of a BSE image and its processing by various image analysis software routines (step 1). In step 2, the electron beam is focussed to induce EDS spectra at selected points according to the adjustments of the BSE image analysis. Step 3 is the classification of the gained sample EDS spectra against a list of approved reference EDS spectra. Different algorithms of the EDS spectra classification are realised by the various SEM-AM software providers.

Combinations of steps 1 and 2 lead to the principal SEM-AM measurement routines (1) point counting modal analysis; (2) particle analysis; (3) sparse phase search, and (4) EDS spectral mapping. These can be applied for the characterisation of natural and artificial electron-beam stable particulate, granular and solid massive materials. The routine (1) follows the traditional point counting method for estimation of the modal contents of different phases in a sample and provides no further information on particle properties and geometries. The particle analysis (2) measurement routines are designed for the analyses of grain mounts with 10^4–10^6 particles for addressing the special tasks in mineral processing and mining. This measurement routines allows the quantification of modal mineralogy, calculated assay, sizes and geometries of particles and grains as well as association and liberation of particles with different chemical compositions within a bulk sample and without previous phase separations. This is not possible by conventional sedimentological particle size and size distribution analysis methods. The sparse phase search measurement routines (3) were developed to find rare phases like small grains of gold and platinum group minerals which is hard and time-consuming in a manual search. BSE triggers allow for focusing the SEM-AM sparse phase search on other interesting phases. The EDS spectral mapping methods appear as the most versatile routines to resolve intra-particle details, chemical zonation and phase relationships in combination with the full particle sizes and geometries.

The application of SEM-AM requires a high-quality preparation of the samples. Grain mounts with epoxy raisins in blocks and on glass, epoxy-embedded hard material fragments and epoxy-glued thin (30 µm) and thick (100 µm) sections on glass are the mostly used compounds. The detection and selection of a suitable epoxy type and mixture which does not evaporate under high vacuum conditions and remains stable under the electron beam after hardening is one of the principal challenges in sample preparation. The other difficulty is the development of sample-specific multi-step polishing procedures to produce relief-free planar surfaces in samples containing phases of very different hardness.

As demonstrated by several case studies in this contribution, the EDS spectral mapping measurement methods appear to have the most promising potential for novel applications apart from metal-bearing ores and mineral processing. This especially concerns the SEM-AM applications to artificial materials such as ashes, slags and ceramics. It has to be clarified that the complete identification of minerals and their denotation through the classification of their EDS spectra remains incomplete. For this, a full and reliable mineral identification requires additional XRD data. It also appears that XRD data identifies phases which appear in microcrystalline aggregates and not in larger grains. Many of the materials which potentially could be characterised by SEM-AM consist of microcrystalline, amorphous and glassy phases. In such cases, the generic labelling of reference EDS spectra and their subsequent target component grouping allow an interesting utilisation of the SEM-AM for novel studies not feasible by other analytical methods.

Author Contributions: Conceptualisation, B.S., D.S.; methodology, D.S., S.G.; investigation, B.S., D.S., S.G.; resources, B.S.; data curation, B.S., DS, S.G.; writing—original draft preparation, B.S., D.S.; writing—review and editing, B.S., D.S., S.G.; visualisation, B.S., D.S.; project administration, B.S., D.S., S.G. All authors have read and agreed to the published version of the manuscript.

Funding: This research was funded by the Helmholtz Institute Freiberg for Resource Technology, the Deutsche Forschungsgemeinschaft (DFG) Grant SCHU676/20 and through contract work for numerous enterprises. The Open Access Funding and APC was funded by the Publication Fund of the TU Bergakademie Freiberg.

Acknowledgments: The authors acknowledge the great expertise of Andreas Bartzsch, Roland Würkert and Michael Stoll in preparation of numerous standard and special geomaterials at Helmholtz Institute Freiberg for Resource Technology. The constructive comments of three anonymous reviewers considerably contributed to improvement of the manuscript.

Conflicts of Interest: The authors declare no conflict of interest.

References

1. Baum, W.; Lotter, N.O.; Whittaker, P.J. Process mineralogy—A new generation for ore characterization and plant optimization. In Proceedings of the 2004 SME Annual Meeting, Denver, CO, USA, 23–25 February 2004; Society for Mining, Metallurgy and Exploration: Englewood, CO, USA, 2004; pp. 73–77.
2. Lastra, R. Seven practical application cases of liberation analysis. *Int. J. Miner. Process.* **2007**, *84*, 337–347. [CrossRef]
3. Hoal, K.O.; Stammer, J.G.; Appleby, S.K.; Botha, J.; Ross, J.K.; Botha, P.W. Research in quantitative mineralogy: Examples from diverse applications. *Miner. Eng.* **2009**, *22*, 402–408. [CrossRef]
4. Ford, F.D.; Wercholaz, C.R.; Lee, A. Predicting process outcomes for Sudbury platinum-group minerals using grade-recovery modeling from mineral liberation analyzer (MLA) data. *Can. Mineral.* **2011**, *49*, 1627–1642. [CrossRef]

5. Grammatikopoulos, T.; Mercer, W.; Gunning, C.; Prout, S. *Quantitative Characterization of the REE Minerals by QEMSCAN from the Nechalacho Heavy Rare Earth Deposit, Thor Lake Project, NWT, Canada*; SGS Minerals Services: Lakefield, ON, Canada, 2011; p. 11.

6. Rollinson, G.K.; Andersen, J.C.O.; Stickland, R.J.; Boni, M.; Fairhurst, R. Characterisation of non-sulphide zinc deposits using QEMSCAN®. *Miner. Eng.* **2011**, *24*, 778–787. [CrossRef]

7. MacDonald, M.; Adair, B.; Bradshaw, D.; Dunn, M.; Latti, D. Learnings from five years of on-site MLA at Kennecott Utah Copper Corporation. In Proceedings of the 10th International Congress for Applied Mineralogy (ICAM 2011), Trondheim, Norway, 1–5 August 2011; Broekmans, M.A.T.M., Ed.; Springer: Berlin/Heidelberg, Germany, 2012; pp. 419–426. [CrossRef]

8. Anderson, K.F.E.; Wall, F.; Rollinson, G.K.; Moon, C.J. Quantitative mineralogical and chemical assessment of the Nkout iron ore deposit, Southern Cameroon. *Ore Geol. Rev.* **2014**, *62*, 25–39. [CrossRef]

9. Melcher, F.; Graupner, T.; Henjes-Kunst, F.; Oberthür, T.; Sitnikova, M.; Gäbler, E.; Gerdes, A.; Brätz, H.; Davis, D.; Dewaele, S. Analytical Fingerprint of Columbite-Tantalite (Coltan) Mineralisation in Pegmatites—Focus on Africa. In Proceedings of the Ninth International Congress for Applied Mineralogy (ICAM 2008), Brisbane, Australia, 8–10 September 2008; Australasian Institute of Mining and Metallurgy (AusIMM): Carlton, VIC, Australia, 2008; pp. 615–624.

10. Gäbler, H.-E.; Melcher, F.; Graupner, T.; Bahr, A.; Sitnikova, M.A.; Henjes-Kunst, F.; Oberthür, T.; Brätz, H.; Gerdes, A. Speeding Up the Analytical Workflow for Coltan Fingerprinting by an Integrated Mineral Liberation Analysis/LA-ICP-MS Approach. *Geostand. Geoanal. Res.* **2011**, *35*, 431–448. [CrossRef]

11. Lamberg, P.; Rosenkranz, J.; Wanhainen, C.; Lund, C.; Minz, F.; Mwanga, A.; Parian, M. Building a Geometallurgical Model in Iron Ores using a Mineralogical Approach with Liberation Data. In Proceedings of the Second AusIMM International Geometallurgy Conference (GeoMet 2013), Brisbane, Australia, 30 September—2 October 2013; Australasian Institute of Mining and Metallurgy (AusIMM): Carlton, VIC, Australia, 2013; pp. 317–324.

12. Lund, C.; Lamberg, P.; Lindberg, T. Practical way to quantify minerals from chemical assays at Malmberget iron ore operations—An important tool for the geometallurgical program. *Miner. Eng.* **2013**, *49*, 7–16. [CrossRef]

13. Schulz, B. Polymetamorphism in garnet micaschists of the Saualpe Eclogite Unit (Eastern Alps, Austria), resolved by automated SEM methods and EMP-Th-U-Pb monazite dating. *J. Metamorph. Geol.* **2017**, *35*, 141–163. [CrossRef]

14. Hrstka, T.; Gottlieb, P.; Skála, R.; Breiter, K.; David Motl, D. Automated mineralogy and petrology—Applications of TESCAN Integrated Mineral Analyzer (TIMA). *J. Geosci.* **2018**, *63*, 47–63. [CrossRef]

15. Knappett, C.; Pirrie, D.; Power, M.R.; Nikolakopoulou, I.; Hilditch, J.; Rollinson, G.K. Mineralogical analysis and provenancing of ancient ceramics using automated SEM-EDS analysis (QEMSCAN®): A pilot study on LB I pottery from Akrotiri, Thera. *J. Archaeol. Sci.* **2011**, *38*, 219–232. [CrossRef]

16. Šegvić, B.; Ugarković, M.; Süssenberger, A.; Mählmann, R.F.; Moscariello, A. Compositional properties and provenance of Hellenistic pottery from the necropolis of Issa with evidences on the cross-Adriatic and the Mediterranean-scale trade. *Mediterr. Archaeol. Archaeom.* **2016**, *16*, 23–52. [CrossRef]

17. Raith, M.M.; Hoffbauer, R.; Spiering, B.; Shinoto, M.; Nakamura, N. Melting behaviour of feldspar clasts in high-fired Sue ware. *J. Eur. Mineral.* **2016**, *28*, 385–407. [CrossRef]

18. Bevins, R.E.; Pirrie, D.; Ixer, R.A.; O'Brien, H.; Pearson, M.P.; Power, M.R.; Shail, R.K. Constraining the provenance of the Stonehenge 'Altar Stone': Evidence from automated mineralogy and U–Pb zircon age dating. *J. Archaeol. Sci.* **2020**, *120*, 105188. [CrossRef]

19. Pirrie, D.; Butcher, A.R.; Power, M.R.; Gottlieb, P.; Miller, G.L. Rapid quantitative mineral and phase analysis using automated scanning electron microscopy (QemSCAN); potential applications in forensic geoscience. *Geol. Soc. Spec. Publ.* **2004**, *232*, 123–136. [CrossRef]

20. Pirrie, D. Forensic geology in serious crime investigation. *Geol. Today* **2009**, *25*, 188–192. [CrossRef]

21. Pirrie, D.; Crean, D.E.; Pidduck, A.J.; Nicholls, T.M.; Awbery, R.P.; Shail, R.K. Automated mineralogical profiling of soils as an indicator of local bedrock lithology: A tool for predictive forensic geolocation. *Geol. Soc. Spec. Pub.* **2019**, *492*. [CrossRef]

22. Greb, V.G.; Guhl, A.; Weigand, H.; Schulz, B.; Bertau, M. Understanding phosphorous phases in sewage sludge ashes: A wet-process investigation coupled with automated mineralogy analysis. *Miner. Eng.* **2016**, *99*, 30–39. [CrossRef]

23. Hoang, D.H.; Kupka, N.; Peuker, U.A.; Rudolph, M. Flotation study of fine grained carbonaceous sedimentary apatite ore—Challenges in process mineralogy and impact of hydrodynamics. *Miner. Eng.* **2018**, *121*, 196–204. [CrossRef]

24. Sandmann, D.; Jäckel, H.G.; Gutzmer, J. Clues to Greater Recycling Efficiency—Characterization of a Crushed Mobile Phone by Mineral Liberation Analysis (MLA). *Mater. Sci. Forum* **2019**, *959*, 134–141. [CrossRef]

25. Gottlieb, P.; Wilkie, G.; Sutherland, D.; Ho-Tun, E.; Suthers, S.; Perera, K.; Jenkins, B.; Spencer, S.; Butcher, A.; Rayner, J. Using quantitative electron microscopy for process mineralogy applications. *JOM* **2000**, *52*, 24–25. [CrossRef]

26. Petruk, W. *Applied Mineralogy in the Mining Industry*, 1st ed.; Elsevier Science: Amsterdam, Netherlands, 2000; p. 288.

27. Gu, Y. Automated scanning electron microscope based mineral liberation analysis. An introduction to JKMRC/FEI Mineral Liberation Analyser. *J. Miner. Mater. Charact. Eng.* **2003**, *2*, 33–41. [CrossRef]

28. Fandrich, R.; Gu, Y.; Burrows, D.; Moeller, K. Modern SEM-based mineral liberation analysis. *Int. J. Miner. Process.* **2007**, *84*, 310–320. [CrossRef]

29. Pirrie, D.; Rollinson, G.K. Unlocking the applications of automated mineral analysis. *Geol. Today* **2011**, *27*, 226–235. [CrossRef]

30. Liipo, J.; Lang, C.; Burgess, S.; Otterström, H.; Person, H.; Lamberg, P. Automated mineral liberation analysis using INCAMineral. In Proceedings of the Process Mineralogy '12, Cape Town, South Africa, 7–9 November 2012; Minerals Engineering International (MEI): Falmouth, UK, 2012; p. 7.

31. Sylvester, P. Use of the Mineral Liberation Analyzer (MLA) for Mineralogical Studies of Sediments and Sedimentary Rocks. In *Quantitative Mineralogy and Microanalysis of Sediments and Sedimentary Rocks*; Sylvester, P., Ed.; Mineralogical Association of Canada (MAC): St. John's, NL, Canada, 2012; Volume 42, pp. 1–16.

32. Graham, S.D.; Brough, C.; Cropp, A. An Introduction to ZEISS Mineralogic Mining and the correlation of light microscopy with automated mineralogy: A case study using BMS and PGM analysis of samples from a PGE-bearing chromitite prospect. In Proceedings of the Precious Metals '15, Falmouth, UK, 11 May 2015; Minerals Engineering International (MEI): Falmouth, UK, 2015; p. 11.

33. Schulz, B.; Merker, G.; Gutzmer, J. Automated SEM Mineral Liberation Analysis (MLA) with Generically Labelled EDX Spectra in the Mineral Processing of Rare Earth Element Ores. *Minerals* **2019**, *9*, 527. [CrossRef]

34. Guhl, A.C.; Greb, V.-G.; Schulz, B.; Bertau, M. An improved evaluation strategy for ash analysis using scanning electron microscope automated mineralogy. *Minerals* **2020**, *10*, 484. [CrossRef]

35. Haberlah, D.; Owen, M.; Botha, P.W.S.K.; Gottlieb, P. SEM-EDS based protocol for subsurface drilling mineral identification and petrological classification. In Proceedings of the 10th International Congress for Applied Mineralogy (ICAM 2011), Trondheim, Norway, 1–5 August 2011; Broekmans, M.A.T.M., Ed.; Springer: Berlin/Heidelberg, Germany, 2012; pp. 265–273. [CrossRef]

36. FEI Company. *QEMSCAN® 650F Automated, Quantitative Petrographic Analyzer*; Brochure DS0038 12-2011; FEI Company: Hillsboro, OR, USA, 2011; p. 4.

37. TESCAN. *TIMA−X—TESCAN Integrated Mineral Analyser*; Brochure 2017.04.10; TESCAN: Brno, Czech Republic, 2017; 16p.

38. Zeiss—Microscopes for Automated Mineral Analysis. Available online: https://web.archive.org/web/20200724134840/https://www.zeiss.com/microscopy/int/products/scanning-electron-microscopes/mineralogic-systems.html (accessed on 24 July 2020).

39. Bruker. *AMICS Software—Advanced Mineral Identification and Characterization System*; Brochure DOC-H82-EXS018, Rev.1; Bruker: Berlin, Germany, 2017; p. 2.

40. Bruker—AMICS—Advanced Mineral Identification and Characterization System. Available online: https://web.archive.org/web/20200810115608/https://www.bruker.com/de/products/x-ray-diffraction-and-elemental-analysis/eds-wds-ebsd-sem-micro-xrf-and-sem-micro-ct/quantax-eds-for-sem/amics-software.html (accessed on 10 August 2020).

41. Oxford Instruments INCAMineral. Available online: https://web.archive.org/web/20200810110842/ (accessed on 10 August 2020).

42. Oxford Instruments—AztecMineral: Dedicated Mineralogy on Multi-Purpose SEM. Available online: https://web.archive.org/web/20200810111028/ (accessed on 10 August 2020).

43. Oxford Instruments. *INCAFeature. High Performance Feature Detection, Analysis and Classification*; Brochure OINA/075/E/0412; Oxford Instruments: High Wycombe, UK, 2012; 4p.

44. Sandmann, D.; Bachmann, K.; Gutzmer, J. From ore to metal—Advanced Materials Characterisation by Automated Mineralogy. *World Min. Surf. Undergr.* **2019**, *71*, 283–291.

45. Jackson, B.R.; Reid, A.F.; Wittemberg, J.C. Rapid production of high quality polished sections for automated image analysis of minerals. *Proc. Australas. Inst. Min. Metall.* **1984**, *289*, 93–97.

46. Røisi, I.; Aasly, K. The effect of graphite filler in sample preparation for automated mineralogy—A preliminary study. *Mineralproduksjon* **2018**, *8*, A1–A23.

47. Kwitko-Ribeiro, R. New Sample Preparation Developments to Minimize Mineral Segregation in Process Mineralogy. In Proceedings of the 10th International Congress for Applied Mineralogy (ICAM 2011), Trondheim, Norway, 1–5 August 2011; Broekmans, M.A.T.M., Ed.; Springer: Berlin/Heidelberg, Germany, 2012; pp. 411–417. [CrossRef]

48. Bartzsch, A.; Gilbricht, S.; Bachmann, K.; Heinig, T. A Method of Preparing a Sample Preparation, Sample Preparation, and Method of Assaying a Sample Material. Patent DE102017128355, 17 January 2019.

49. Rahfeld, A.; Gutzmer, J. MLA-Based Detection of Organic Matter with Iodized Epoxy Resin—An Alternative to Carnauba. *J. Miner. Mater. Charact. Eng.* **2017**, *5*, 198–208. [CrossRef]

50. O'Brien, G.; Gu, Y.; Adair, B.J.I.; Firth, B. The use of optical reflected light and SEM imaging systems to provide quantitative coal characterisation. *Miner. Eng.* **2011**, *24*, 1299–1304. [CrossRef]

51. Gomez, C.O.; Strickler, D.W.; Austin, L.G. An iodized mounting medium for coal particles. *J. Electron Microsc. Tech.* **1984**, *1*, 285–287. [CrossRef]

52. Green, O.R. Thin Section and Slide Preparation Techniques of Macro- and Microfossil Specimens and Residues. In *A Manual of Practical Laboratory and Field Techniques in Palaeobiology*; Springer: Dordrecht, The Netherlands, 2001. [CrossRef]

53. Grundmann, G.; Scholz, H. Preparation Methods in Mineralogy and Geology: The Preparation of Thin Sections, Polished Sections, Acetate Foil Prints, Preparation for Elutriation Analysis and Staining Tests for the Optical and Electron Microscopy. 2015. Available online: https://www.researchgate.net/publication/275948069_Preparation_methods_in_Mineralogy_and_Geology_The_preparation_of_thin_sections_polished_sections_acetate_foil_prints_preparation_for_elutriation_analysis_and_staining_tests_for_the_optical_and_electro (accessed on 10 November 2020). [CrossRef]

54. Schulz, B.; Krause, J.; Zimmermann, R. Electron microprobe petrochronology of monazite-bearing garnet micaschists in the Oetztal-Stubai Complex (Alpeiner Valley, Stubai). *Swiss J. Geosci.* **2019**, *112*, 597–617. [CrossRef]

55. Burisch, M.; Gerdes, A.; Meinert, L.D.; Albert, R.; Seifert, T.; Gutzmer, J. The essence of time—Fertile skarn formation in the Variscan Orogenic Belt. *Earth Planet. Sci. Lett.* **2019**, *519*, 165–170. [CrossRef]

56. Dittrich, T.; Seifert, T.; Schulz, B.; Hagemann, S.; Gerdes, A.; Pfänder, J. *Archean Rare-Metal Pegmatites in Zimbabwe and Western Australia*; Geology and Metallogeny of Pollucite Mineralisations; Springer Briefs in Worlds Mineral Deposits; Springer: Cham, Switzerland, 2019; p. 125. [CrossRef]

57. von Eynatten, H.; Tolosana-Delgado, R.; Karius, V.; Bachmann, K.; Caracciolo, L. Sediment generation in humid Mediterranean setting: Grain-size and source-rock control on sediment geochemistry and mineralogy (Sila Massif, Calabria). *Sediment. Geol.* **2016**, *336*, 68–80. [CrossRef]

58. Minde, M.W.; Zimmermann, U.; Madland, M.V.; Korsnes, R.I.; Schulz, B.; Gilbricht, S. Mineral replacement in long-term flooded porous carbonate rocks. *Geochim. Cosmochim. Acta* **2020**, *268*, 485–508. [CrossRef]

59. Andersen, P.Ø.; Wang, W.; Madland, M.V.; Zimmermann, U.; Korsnes, R.I.; Bertolino, S.R.A.; Minde, M.W.; Schulz, B.; Gilbricht, S. Comparative study of five outcrop chalks flooded at reservoir conditions: Chemo-mechanical behaviour and profiles of compositional alteration. *Transp. Porous Media* **2018**, *121*, 135–181. [CrossRef]

60. Mücke, A.; Bhadra Chaudhuri, J.N. The continuous alteration of ilmenite through pseudorutile to leucoxene. *Ore Geol. Rev.* **1991**, *6*, 25–44. [CrossRef]

61. Schulz, B.; Haser, S. The ilmenite-pseudorutile-leucoxene alteration sequence in placer sediments in the view of automated SEM mineral liberation analysis. In *Abstracts of GeoBerlin 2015*; Joint Meeting of DGGV and DMG; GFZ German Research Centre for Geosciences: Potsdam, Germany, 2015; p. 336. [CrossRef]

62. BGR—Mineral Certification at the BGR. Available online: https://www.bgr.bund.de/EN/Themen/Min_rohstoffe/CTC/Home/CTC_node_en.html (accessed on 8 September 2020).

63. Gäbler, H.-E.; Schink, W.; Goldmann, S.; Bahr, A.; Gawronski, T. Analytical Fingerprint of Wolframite Ore Concentrates. *J. Forensic Sci.* **2017**, *62*, 881–888. [CrossRef]

64. Melcher, F.; Graupner, T.; Sitnikova, M.A.; Oberthür, T.; Gäbler, H.-E.; Bahr, A.; Henjes-Kunst, F. *Herkunftsnachweis von Columbit-Tantalit-Erzen: Status-quo-Bericht*; BGR Open File Report no. 11082/09; Bundesanstalt für Geowissenschaften und Rohstoffe (BGR): Hannover, Germany, 2009; p. 154.

65. Melcher, F.; Graupner, T.; Sitnikova, M.; Henjes-Kunst, F.; Oberthür, T.; Gäbler, H.-E.; Bahr, A.; Gerdes, A.; Brätz, H.; Rantitsch, G. Ein Herkunftsnachweis für Niob-Tantalerze am Beispiel afrikanischer Selten-Element-Pegmatite. *Mitt. Österr. Miner. Ges.* **2009**, *155*, 231–268.

66. Schütte, P.; Melcher, F.; Gäbler, H.-E.; Sitnikova, M.; Hublitz, M.; Goldmann, S.; Schink, W.; Gawronski, T.; Ndikumana, A.; Nziza, L. The Analytical Fingerprint (AFP) Method and Application, Process Manual Version 1.4, Bundesanstalt für Geowissenschaften und Rohstoffe (BGR). 2018. Available online: https://www.bgr.bund.de/EN/Themen/Min_rohstoffe/CTC/Downloads/AFP_Manual.pdf?__blob=publicationFile&v=7 (accessed on 26 September 2020).

67. Weber, A. Charakterisierung afrikanischer Niob-Tantal-Erze mit REM-MLA und Eingrenzung der unbekannten Herkunftsorte. Master Thesis, TU Bergakademie, Freiberg, Germany, 2019; p. 107.

68. Werther, J.; Ogada, T. Sewage sludge combustion. *Prog. Energy Combust. Sci.* **1999**, *25*, 55–116. [CrossRef]

69. Ma, Y.; Stopic, S.; Gronen, L.; Friedrich, B. Recovery of Zr, Hf, Nb from eudialyte residue by sulfuric acid dry digestion and water leaching with H_2O_2 as a promoter. *Hydrometall.* **2018**, *181*, 206–214. [CrossRef]

70. Fang, L.; Li, J.-S.; Donatello, S.; Cheeseman, C.R.; Wang, Q.; Poon, C.S.; Tsang, D.C.W. Recovery of phosphorus from incinerated sewage sludge ash by combined two-step extraction and selective precipitation. *Chem. Eng. J.* **2018**, *348*, 74–83. [CrossRef]

71. Pietranik, A.; Kierczak, J.; Tyszka, R.; Schulz, B. Understanding heterogeneity of a slag derived Technosol: The role of automated SEM-EDS analyses. *Minerals* **2018**, *8*, 513. [CrossRef]

72. Buchmann, M.; Borowski, N.; Leißner, T.; Heinig, T.; Reuter, M.A.; Friedrich, B.; Peuker, U.A. Evaluation of Recyclability of a WEEE Slag by Means of Integrative X-Ray Computer Tomography and SEM-Based Image Analysis. *Minerals* **2020**, *10*, 309. [CrossRef]

73. Guhl, A.C.; Brett, B.; Schulz, B.; Bertau, M. Particle responses of stabilised fly ash to chemical treatment for resource extraction: An automated mineralogy investigation. *Miner. Eng.* **2020**, *145*. [CrossRef]

74. Heimann, R.B.; Maggetti, M. *Ancient and Historical Ceramics: Materials, Technology, Art, and Culinary Traditions*; Schweizerbart Science Publishers: Stuttgart, Germany, 2014; p. 550.

75. Rasmussen, K.L.; de la Fuente, G.A.; Bond, A.D.; Mathiesen, K.K.; Vera, S.D. Pottery firing temperatures: A new method for determining the firing temperature of ceramics and burnt clay. *J. Archaeol. Sci.* **2012**, *39*, 1705–1716. [CrossRef]

76. Raith, M.M.; Euler, H.; Spiering, B.; Hoffbauer, R.; Nakamura, N.; Shinoto, M. Firing conditions in Anagama kilns: Constraints from the archaeometric study of kiln wall samples from the Nakadake-Sanroku archaeological site. In Proceedings of the Abstracts of GeoBremen 2017. Joint Meeting of DGGV and DMG, Bremen, Germany, 24–29 September 2017; p. A-178.

77. Raith, M.M.; Euler, H.; Spiering, B.; Hoffbauer, R. Ceramic transformations during firing of ancient Japanese stone ware (Sueki): Insights from firing experiments. In Proceedings of the Abstracts of GeoMünster 2019. Joint Meeting of DGGV and DMG, Münster, Germany, 22–25 September 2019; p. 286.

Publisher's Note: MDPI stays neutral with regard to jurisdictional claims in published maps and institutional affiliations.

MDPI

St. Alban-Anlage 66

4052 Basel

Switzerland

Tel. +41 61 683 77 34

Fax +41 61 302 89 18

www.mdpi.com

Minerals Editorial Office

E-mail: minerals@mdpi.com

www.mdpi.com/journal/minerals